THE

BETHE-PEIERLS CORRESPONDENCE

THE BETHE-PEIERLS CORRESPONDENCE

Sabine Lee
University of Birmingham

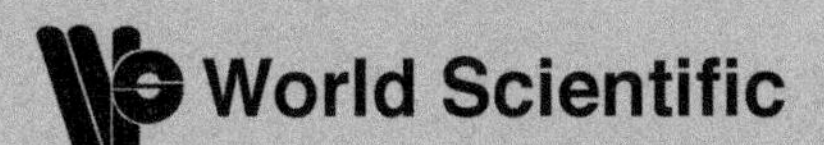

World Scientific
NEW JERSEY • LONDON • SINGAPORE • BEIJING • SHANGHAI • HONG KONG • TAIPEI • CHENNAI

Published by

World Scientific Publishing Co. Pte. Ltd.
5 Toh Tuck Link, Singapore 596224
USA office: 27 Warren Street, Suite 401-402, Hackensack, NJ 07601
UK office: 57 Shelton Street, Covent Garden, London WC2H 9HE

British Library Cataloguing-in-Publication Data
A catalogue record for this book is available from the British Library.

THE BETHE–PEIERLS CORRESPONDENCE

ISBN-13 978-981-277-135-3
ISBN-10 981-277-135-2

Printed in Singapore.

To Gerald E. Brown
with gratitude

Contents

Acknowledgements

In the summer of 1992, as a final-year doctoral student in history, I was encouraged by a college friend to accompany him to the 1992 Dirac Memorial Lecture at Cambridge University. I had little idea that this afternoon excursion into theoretical physics would have a significant impact on my own work as a historian. The lecture on 'Broken Symmetries' was given by Sir Rudolf Peierls. I was immediately fascinated by the presentation and the person who delivered the lecture, and I decided to find out more about this scientist who, well into his eighties, still pursued active research and succeeded in bringing to life his subject for scientists and non-scientists alike.

Almost three years later, in the spring of 1995, a few months before his death, I had the opportunity to meet Rudolf Peierls. Among the many details I learnt about him and his work in our two-hour conversation, two names stood out as keys that would further unlock his biography. First, his long-time friend whom he had first met when they were both students at Munich in the 1920s, and who had since become Professor at Cornell, Hans Bethe. And second, Gerry Brown, who had been a trusted colleague at Birmingham in the 1950s and later moved on to become professor at NORDITA, Princeton and eventually Stony Brook, New York.

As I was to discover soon, Rudolf Peierls had left an extensive collection of written material, of papers, notes, and — perhaps most importantly — correspondence. Among the thousands of letters in Peierls' collection was a sizeable file of letters exchanged with Hans Bethe. This correspondence went back to their student days in Munich in the 1920s and had continued throughout their distinguished careers into retirement. The documents are a fascinating source that throw light not only on the lives of these two remarkable individuals, but also on the historical circumstances in which they lived and worked and on the academic and political environments in which they operated. Those letters form the core of this document collection.

In October 2003, Gerry Brown, then Distinguished Professor of Physics at Stony Brook, attended the opening of the Rudolf Peierls Centre for Theoretical Physics at Oxford. I finally had the opportunity to meet the person of whom Rudolf Peierls had remarked that he was best placed to tell me about himself and his friend Hans Bethe. Our initial meeting, a morning of reminiscing and discussing the lives and work of both Peierls and Bethe, was followed by many more such encounters and was the beginning of a fruitful collaboration. Gerry Brown, having not only been a lodger in the Peierls household and a close collaborator within Peierls' theoretical physics department at Birmingham in the 1950s, but having also been a long-term collaborator of Hans Bethe in the last twenty years of the latter's life, was indeed best placed to explain the lives and work of both. Without Gerry Brown's encouragement, his insights and his continuing support, this book would not have been written.

Research for this publication has been supported by grants of the Royal Society and the British Academy. The School of Historical Studies at the University of Birmingham provided me with a generous flexible research arrangement that made possible the smooth progress of my work. The translation of the German language material into English was funded by a Royal Society grant, and the translations were carried out by WriteScienceRight, Las Vegas, Nevada.

The papers of Rudolf Peierls and Hans Bethe are held at the Western Manuscripts Department, Bodleian Library Oxford, and the Division of Rare and Manuscript Collections, Cornell University Library, respectively. I am grateful to Colin Harris and his colleagues in the Western Manuscripts Department (Bodleian Library) and Elaine Engst, David Corson and their colleagues at the Division of Rare and Manuscript Collections (Cornell University Library) for all their support and advice, which greatly facilitated the completion of the project.

I am very grateful to the families of Rudolf Peierls and Hans Bethe for their support of this edition. In particular I am indebted to Mrs Joanne Hookway (née Peierls) and to Mrs Rose Bethe for helpful discussions. The letters exchanged by Rudolf Peierls and Hans Bethe are reproduced by kind permission of the Peierls family and of Mrs Rose Bethe.

Birmingham, 2007 Sabine Lee

Two Full and Good Lives

In August 1927 while enjoying his customary summer holiday with his mother in the Alps, Hans Bethe sent a postcard to Rudolf Peierls, a fellow student at Arnold Sommerfeld's institute[1] of theoretical physics in Munich.[2] Little did the two students know that their meeting in Munich in 1927 would be the beginning of a deep friendship and a productive work relationship; little, also, did they know that when Hans Bethe, shortly before his friend's death in 1995 would reminisce about their lives[3] as those of two of the giants of 20th century physics he would look back on more than 68 years of uninterrupted contact most clearly visible in the hundreds of pages of letters exchanged between the two.

Bethe and Peierls were born within a year of each other in 1906 and 1907 respectively. They came from similar social backgrounds. Hans Bethe's mother was Jewish, the daughter of a professor of medicine; she converted to Christianity even before meeting Hans's father Albrecht, a successful physiologist, who himself was a Protestant. Religion did not play an important role in the Bethe household; culture and learning, on the other hand, did.[4] Peierls was born into an assimilated Jewish family. His parents had converted to Christianity in 1912, but religion — Jewish or Christian — played little role in the Peierls household, either. Peierls' mother died, when Rudolf was fourteen years old, and his father soon remarried. His second wife was Aryan, and the Peierls's

[1]Arnold Sommerfeld (1868–1951), received his Ph.D. from Königsberg in 1891. After positions at Göttingen, Clausthal-Zellerfeld and Aachen, he became Professor of Physics and Director of the Theoretical Physics Institute at the University of Munich in 1906 where he stayed until his retirement (and the completion of the appointment of his successor) in 1939.

[2]Letter [1].

[3]Letter [214].

[4]See e.g. letters [6–7].

moved as easily in Jewish as in non-Jewish circles. And again, in these environments, religion was of little consequence; in contrast to culture and learning which were highly valued.

When Bethe and Peierls came to Munich to join Arnold Sommerfeld's Mecca of theoretical physics in 1927, they had both studied at their 'home universities' for some time: Bethe, who arrived in Munich just ahead of Peierls, had spent about two years at Frankfurt University; Peierls, a year younger and seemingly much junior to Bethe at the time, had just completed a year's study at Berlin. The correspondence between Bethe and Peierls, much of which has survived and is kept in the large collections of their respective private papers passed on to the libraries of Cornell University (Bethe) and Oxford University (Peierls) is an invaluable source. The letters shed light on the lives and the work of the two scientists, whose achievements span from significant scientific discoveries to enthusiastic and committed university teaching, from the development of the Atom Bomb to the fight against the nuclear arms race, from the establishment of institutes of theoretical physics in their adopted home countries, the US and the UK, to the development of well-functioning scientific exchange networks which linked physics institutes in Britain and the US during early post-war decades. The letters enlighten us about the lives and the plight of young Jewish scientists who were forced to leave their native countries as a result of National Socialist racial policies; they demonstrate the strength of scientific collaboration during one of the most exciting periods of scientific discovery, the years of the development of quantum mechanics; and they illustrate how these early bonds between members of an exceptional generation of theoretical physicists survived the war-years and paved the way for further scientific progress in the second half of the twentieth century.

This collection presents the vast majority of letters exchanged between Hans Bethe and Rudolf Peierls. They are arranged chronologically in four chapters, which represent four phases in their academic development and their working and personal relationship. Chapter 1 contains letters written between 1927 and 1933. During this period, both young men completed their doctorates, Bethe in 1928 under Arnold Sommerfeld's supervision and Peierls in 1928 under one of Sommerfeld's star students, Werner Heisenberg[5] in Leipzig. Both then engaged in research and teaching at various European centres of theoretical physics. Hans Bethe went to Frankfurt (1928–29), Stuttgart (1929) and Munich (1929–1930), from where he visited Cambridge (1930–31) and Rome (1931) on a Rockefeller Fellowship, before taking on a position as acting professor in Tübingen (1932). In April 1933, with the enactment of the racial laws in Germany which prohibited Jews from holding government positions, Bethe lost his job, and although Arnold Sommerfeld provided him with a fellowship in Munich for the summer of 1933, Bethe gladly accepted an invitation from William Lawrence Bragg to come to Manchester for a year.

Rudolf Peierls followed in some of his friend's footsteps. After completion of his doctorate in 1929, he initially became Wolfgang Pauli's[6] assistant in Zurich. Like Bethe, he travelled to Rome (1932) and Cambridge (1932–33) on a Rockefeller Fellowship. He chose not to return from England to a position in Hamburg which had been offered to him for 1933, as the political situation in Germany had become so precarious for Jews that staying abroad was considerably safer an option despite the uncertainty of job prospects. During the academic year 1933/34 both Bethe and Peierls held temporary positions in Manchester. Rudolf Peierls, in 1931, had married the Russian physicist Eugenia Kannegiser, whom he had met at a conference in Odessa the previous summer, and Bethe gladly accepted the couple's invitation to share a house with them during their Manchester year. This explains the gap in their correspondence, similar to the gap which would occur in the mid 1940s, when both were engaged in war work, most notably at Los Alamos.

[5]Werner Heisenberg (1901–1976), studied at Göttingen and Munich where he obtained his Ph.D. in 1923. After research at Göttingen and Copenhagen he became Professor of Theoretical Physics at Leipzig in 1927. In 1941 he became Director of the Max Planck Institute of Physics in Berlin. After the war he held positions at Göttingen and Munich.

[6]Wolfgang Pauli (1900–1958), received his doctorate from Munich in 1921. After research at Göttingen and Copenhagen, he taught at Hamburg for five years. In 1928 he accepted is professorship at the ETH Zurich. Having spent the war years in the United States, Pauli returned to Zurich in 1946 where he remained for the rest of his career.

Chapter 2 covers the years 1935 to 1940. After the completion of his year at Manchester, Bethe initially accepted a fellowship at Bristol to work with Nevill Mott, before taking up an assistant professorship at Cornell in February 1935 — the place where he would remain, except for shorter intervals, for the remainder of his long and illustrious career. Peierls, in the spring of 1935 moved back to Cambridge where he was offered a position at the Mond Laboratory. This laboratory for magnetism and low temperature physics had been built for Peter Kapitza, who in 1934 during one of his home visits to Russia was detained and prevented from returning to the UK. The money earmarked for Kapitza's unclaimed salary was then used to fund two additional fellowships, one of which was taken up by Peierls. After two stimulating years in Cambridge, Peierls secured his first permanent position, a professorship in Mathematical Physics at Birmingham University, the place he would stay at, except for short intervals, for over quarter of a century. The Bethe-Peierls correspondence of the second half of the 1930s is the most intense part of their exchange. Spurred by the highly productive year they had spent together in Manchester, Bethe and Peierls communicated about ongoing research, but also increasingly also about the difficult political situation in Europe and the prospects of other refugee scientists who were still looking for employment in the UK and the US. But the letters also contain more personal and family concerns. By 1935, the two friends had moved from the more formal 'Sie' to the informal 'Du' in their addressing each other, and the year spent together in Manchester had, without doubt, intensified their friendship. Peierls by then had two children, Gaby born in August 1933 and Ronnie in 1935. And Hans, in 1938, had married Rose Bethe, the daughter of his former Stuttgart mentor, Paul Ewald.[7] The depth of their friendship is best demonstrated by the offer of Rose and Hans Bethe, after the outbreak of the war in Europe and with the danger of a German

[7]Paul Peter Ewald (1888–1985), studied physics, chemistry and mathematics at Cambridge, Göttingen and Munich where he obtained his doctorate in 1912. After positions at Göttingen, Copenhagen and Munich, he took of a professorship at Stuttgart. In 1937 he emigrated, intially to Cambridge and later to Belfast. In 949 he became Professor of Physics at the Polytechnic Institute of Brooklyn.

invasion of England looming, to put up Gaby and Ronnie in their home in Ithaca.[8]

The third chapter contains letters written in the first two post-war decades between 1946 and 1963, the period when both physicists were at their most effective in building world-class institutes of theoretical physics in their respective adopted homes: Hans Bethe at Cornell and Rudolf Peierls at Birmingham. The correspondence illustrates the return to peacetime research, the endeavours to build on pre-war and wartime scientific achievements, the spirit of international collaboration cultivated among many of the scientists of the Manhattan project (if not by the political leadership), the disappointment at the ongoing east-west divide which made the revival of contacts with Soviet colleagues all but impossible, and the activities of both men in the area of arms control, nuclear disarmament and the initiatives for world peace such as in the Pugwash Movement.

The final chapter contains letters written between 1963 and 1995. During this final period, the frequency of letters was much reduced; and the nature of the exchange was much altered with personal and political rather than scientific concerns dominating the letters, especially during the 1980s and 1990s. The change in content and frequency of the correspondence was partly due to the fact that there was less official university business, especially after the retirement, in 1974, of Peierls from Oxford University, where he had spent the last eleven years of his career, and, in 1975, of Bethe from the Cornell faculty. Furthermore, when Peierls retired from Oxford, he took on a part-time position at the University of Washington, Seattle, where he spent six months in every twelve for the subsequent three years. His regular extended visits to the U.S. allowed Peierls to re-establish more regular direct contacts with his American friends and colleagues, not least with the Bethes.

The four chapters are prefaced by brief introductory remarks explaining the background to some of the developments mentioned in the correspondence. Although the collection is remarkable for both its duration and for its coverage of almost the entire period of Bethe's and

[8]In the event, the children went to Canada as part of a university evacuation scheme. See letters [59–62].

Peierls' acquaintance, important phases when the scientists were working in close proximity, are not reflected in their letters and therefore not visible in this collection. For a fuller picture of Bethe and Peierls, their friendship and working relationship, please refer to the biographical works by Lee, Brown, Bernstein and Schweber.[9]

What had begun with a light-hearted postcard in the summer of 1927 ended with a letter of reminiscences looking back on almost seven decades of friendship of collaboration which Hans Bethe sent to Rudolf Peierls whose life, as Bethe was well aware, was nearing its end in early September 1995. The spirit of their relationship was so aptly captured in Bethe's closing sentence, a sentence which, without the shadow of a doubt, Peierls would have subscribed to, had the situation been reversed: "You had a full and good life, and I thank you for letting me participate in it."

Editorial Comments

1. The letters are listed chronologically, and, unless stated otherwise, reproduced in toto.

2. The transcription is limited to the text content, not to layout or graphic details.

3. The early correspondence is in German, and letters are reproduced in the original followed by an English translation. Editorial and other notes are in English and added only to the English texts and translations.

---

[9]Jeremy Bernstein, *Hans Bethe, Prophet of Energy*, Basic Books, New York, 1980 (hereinafter cited as Bernstein, *Hans Bethe, Prophet of Energy*); S.S. Schweber, *In the Shadow of the Bomb. Oppenheimer, Bethe and the Moral Responsibility of the Scientist*, Princeton University Press, Princeton, 2000; G.E. Brown and Chang-Hwan Lee (eds.), *Hans Bethe and His Physics*, World Scientific, Singapore, 2006 (hereinafter cited as Brown and Lee: *Hans Bethe and His Physics*); S. Lee (ed.), *Sir Rudolf Peierls. Selected Private and Scientific Correspondence*, 2 volumes, World Scientific, Singapore, 2007 (hereinafter cited as Lee: *Selected Correspondence* and volume number).

4. Typos and misplaced letter sequences are corrected.[10]

5. The letters reproduced in this collection are held in the Bodleian Library[11] and Cornell Universitry Library.[12] Often Bethe and Peierls kept carbon copies of their typed letters, and therefore both original letter and carbon copy survive. Unless otherwise stated, the letters reproduced here are the originals. Letters from Bethe to Peierls are taken from the Peierls Papers, letters from Peierls to Bethe are taken from the Bethe Papers.

6. Handwritten additions, where included, are identified as such.

7. Editorial additions are made in square brackets.

8. Text which was deleted in the manuscript is not transcribed.

9. The reproduced material makes no distinction between single and multiple underlining.

10. Gaps arising from punching or filing are filled and only noted in square brackets, if there are doubts. Abbreviations are only completed where necessary.

11. Illegible words are noted as [??], with [?] for one word, [??] for two words and [???] for three or more words.

12. Footnotes in the text are marked with asterisk and added at the end of the letter. Editorial notes appear at the bottom of the page.

13. Repeated words at the page break are transcribed as one word.

---

[10]This applies in particular to correspondence in the late 1980s and 1990s. As a result of failing eyesight in later life, many of Rudolf Peierls' word-processed letters written in those years contain a large number of typographical errors which have been corrected.

[11]Sir Rudolf Peierls, Papers and Correspondence, 1898–1996; CSAS catalogue no 52/6/77 and NCUSACS supplementary catalogue 57/6/96.

[12]Division of Rare and Manuscript Collection, Collection no. 14-22-976.

1. The Brave Men Who Fearlessly Climb the Harsh Mountains

The first card sent from Bethe to Peierls from his vacations in Bad Gastein,[13] where he had spent the summer shortly after his parents' divorce, is a first indication of one of the common interests of the two young men: a love of the mountains. Neither of them being athletic as such, the two nevertheless enjoyed mountaineering in the summer and skiing in the winter, a passion which they shared with many of their generation of young physicists. And indeed, many of the following letters describe the various alpine adventures, and frequent references are found to the 'brave men who fearlessly climb the harsh mountains', an odd phrase that had its origin in a remark of a young boy who saw Bethe, Peierls, and their two friends, Werner Sachs and Hans Thorner, climb some mountains in 1927.[14]

But clearly, the single most important interest was physics, and already in 1928/29, when Bethe and Peierls were still working towards their doctorates, their letters were full of comments on physics and their work. After the Christmas vacations 1927, which Bethe had spent with Peierls and Thorner, he was suffering the after-effects of severe frostbite and was forced to convalesce at home for several weeks, during which he was working on his doctoral thesis, a discussion of the diffraction of electrons by crystals.[15] He joined his colleagues at Munich in February 1928,[16] and soon afterwards, Peierls moved on to Leipzig where he joined Werner Heisenberg while Arnold Sommerfeld was embarking of an extended world lecture tour. The correspondence in the summer of

[13]Letter [1].
[14]See, e.g., letters [5,11–12,24].
[15]Letters [2,8].
[16]See letter [7].

that year reflects the special situation of Bethe revising for, passing and later extensively celebrating the 'summa cum laude' completion of this doctoral examinations, and exchanging gossip about the Munich crowd, while at the same time enquiring about the impressions that Peierls had gained in Leipzig. Rudolf Peierls then spent the summer of 1928 in England, primarily to perfect his English, but also to gain experiences among the England-based colleagues, among others Dirac[17] and R.H. Fowler.[18]

The following years brought more travel for both. Peierls returned to Leipzig to tackle his first more serious research task, an examination of the galvano-magnetic effects, a study of the anomalous or positive Hall effect.[19] While working with Heisenberg was not the only attraction of the Leipzig seminar; during Peierls' stay at Heisenberg's institute, Georg, Wentzel, Friedrich Hund,[20] Peter Debye[21] and Guido Beck[22] were also working at Leipzig. Still, Peierls decided to move to Pauli in Zurich, when Heisenberg accepted an invitation to lecture abroad in the spring of 1929. While at Zurich he completed a transla-

[17]Paul Adrien Maurice Dirac (1902–1984), studied electrical engineering and mathematics in Bristol and received his Ph.D. from Cambridge in 1926; he served as Lucasian Professor of Mathematics at Cambridge between 1932 and 1969. Later he moved to Florida State University, Tallahassee where he remained until his death.

[18]Ralph Fowler (1889–1944), studied mathematics at Trinity College Cambridge; after the First World War, under the influence of Rutherford, (whose daughter Eileen he was to marry in 1921) he turned his attention to physics. In 1932 he took up the post of the newly created Plummer Chair of Theoretical Physics.

[19]R. Peierls, 'Zur Theorie der galvanomagnetischen Effekte', *Z. Phys.* **53**, 255–66 (1929).

[20]Prof. Friedrich Hund (1896–1997) had studied mathematics, physics and geography at Marburg and Göttingen, taught at Göttingen before becoming Professor of Physics at Rostock (1927) and Leipzig (1929).

[21]Peter J.W. Debye (1884–1966), studied mathematics and physics at Aachen and obtained his Ph.D. in physics from Munich in 1908. After positions at Zurich, Utrecht, Göttingen and Leipzig, he became Director of the Physics Section of the Kaiser Wilhelm Institute in Berlin. In 1939, he moved to Cornell where he remained for the rest of his career.

[22]Guido Beck (1903–1988), studied at Vienna where he received his doctorate in 1925. After positions in Leizig, Prague, Odessa and France, he emigrated to Latin America in 1943. He worked at the National Astronomic Observatory in Cordoba and at the Centro Brasileiro de Pesquisas Físicas in Rio de Janeiro.

tion from French into German of de Broglie's *Introduction to the Study of Wave Mechanics*, a book that first appeared in Peierls' translation published by a Leipzig company[23] and only appeared in French after de Broglie had received the 1929 Nobel Prize for Physics. At the same time Peierls approached with enthusiasm and remarkably quick success the research project which would earn him, just a few months later in July 1929, his Leipzig doctorate: a study of thermal conductivity in crystals with its recognition of the so-called 'Umklapp-processes' at low temperatures.[24] Following the completion of his own doctorate, Bethe had taken up his first fellowships in Frankfurt,[25] Stuttgart,[26] and from October 1929 onwards back in Munich with Sommerfeld, and appointment which promised him plenty of research time and an early habilitation.[27]

From 1929 onwards, the correspondence between Bethe and Peierls is more strongly focussed on physics again. Both Bethe and Peierls followed each other's work on ionisation potentials, thermal conductivity, or work on the electron affinity of hydrogen to name but a few topics discussed between them. They met during the Christmas vacations when they overlapped in Arosa for a few days.[28] In the summer of 1930, Rudolf Peierls followed an invitation to attend the 7th All Union Conference in Odessa. The previous autumn, the young Soviet theoretician, Lev Landau, had visited Pauli at the ETH on a government scholarship, and Peierls had been thoroughly impressed with Landau's very mature understanding of physics, and Peierls was intrigued by the Russian scientific community. He was keen to increase his contacts with the very active community of young (and older) Russian physicists. However, even more lasting than the scientific contacts made and enhanced at Odessa was his acquaintance with the only girl in the circle among Lev Landau, Genia Kannegiser. Following their first encounter at Odessa

[23]De Broglie, L. *Einführung in die Wellenmechanik*, translated by R. Peierls, Leipzig: Akademische Verlagsgesellschaft, Leipzig, 1929.

[24]R. Peierls, 'Zur kinetischen Theorie der Wärmeleitung in Kristallen', *Ann. Phys.* V **3**, 1055–1101 (1929).

[25]Letter [14].

[26]Letters [15–16].

[27]Letter [16].

[28]Letters [20–21].

and the subsequent holiday — with some of their scientist friends — in the Soviet Union, Genia and Rudi engaged in an extensive and intensive correspondence.[29] The following spring Rudi returned to Russia and married Genia.[30]

Meanwhile, Hans Bethe had taken up a Rockefeller Fellowship half of which he spent in Cambridge working primarily with Fowler[31], and the second half of which he spent in Rome, working with Enrico Fermi.[32] The latter had profound impact on Bethe, who enthusiastically reported to Peierls: 'Der Clou von Rom ist aber unbedingt Fermi, der meiner Meinung nach in der physikalischen Welt noch lange nicht genug geschätzt wird. ... sein Urteil über theoretische wie experimentelle (!) Literatur ist eigentlich unfehlbar.'[33]

Christmas 1931 would be the last one, which both Bethe and Peierls spent in the south of Germany and Switzerland respectively, and the two duly met up with friends in the Alps for a skiing trip.[34] Here Peierls had the opportunity not just to 'talk physics' but also to gain more impressions of Bethe's experiences in Cambridge and Rome, and in fact he followed in Bethe's footsteps choosing the same two locations for his Rockefeller Fellowship year, as he later claimed only improving on the order of the visits, enjoying the autumn and winter 1932/33 in Rome before moving on to the Cambridge spring and summer (1933).[35]

In 1933, the correspondence is taken up primarily by communications about the future employment of Bethe, Peierls and some of their physicist friends who had to leave Germany after the advent of power

---

[29]See Lee, *Selected Correspondence*, 1, chapter 2.

[30]Letter [25].

[31]Nachlass Arnold Sommerfeld. Deutsches Museum München, Archiv HS 1977-28/A, 19.

[32]Enrico Fermi (1901–1954), obtained his Ph.D. from Pisa. After research at Göttingen, Leiden and Florence, he took up a professorship in Rome in 1926. In 1938 he emigrated to the US, working at Columbia University and the University of Chicago, where, after contributing to the Manhattan Project as a consultant, he returned and became professor in 1946.

[33]Letter [25].

[34]See letters [26–27].

[35]R.E. Peierls, *Bird of Passage*, Princeton University Press: 1985, (hereinafter cited as Peierls, *Bird of Passage*) p. 82.

of the National Socialists. Peierls who had the good fortune of being in Cambridge in January 1933, chose not to return to a post which he had been offered at Hamburg. Instead he tried to apply for positions in the UK following the end of his Rockefeller Fellowship. Bethe, who had completed his habilitation under Sommerfeld in Munich, had taken up a professorship at Tübingen in 1932. After the enactment of the racial laws which prohibited the government employment of Jews, Bethe, whose mother was Jewish and who was therefore affected by the regulations, lost his position. Arnold Sommerfeld, who worked tirelessly to secure positions for his many outstanding Jewish and non-Jewish students and colleagues,[36] offered Bethe a fellowship at Munich, but Bethe, as his friend Peierls, managed to secure temporary appointments at Manchester, working with William Bragg.[37] Hans Bethe often referred to the year at Manchester as his most productive period, and Peierls, too, came to see this as his most prolific period. It was certainly a phase of fruitful collaboration between Bethe and Peierls, when both of them turned to nuclear physics and published several joint papers in this field.[38] Of those collaborative efforts, one was recalled often by the two scientists and many of their colleagues, and is even referred to in the last letters exchanged sixty years later.[39] Chadwick[40] had just discovered the neutron and he was experimenting with deuterium, which

---

[36]Michael Eckert and K. Märker (eds.), *Arnold Sommerfeld. Wissenschaftlicher Briefwechsel*, vol. 2, GNT-Verlag, Diepholz, 2004 (hereinafter cited as Eckert and Märker, *Sommerfeld Briefwechsel*), pp. 353–377.

[37]Letters [29–32].

[38]R.E. Peierls and H.A. Bethe, 'The "Neutrino"', *Nature* **133**, 532–33 (1934); H.A. Bethe and R.E. Peierls, 'The Neutrino', *Nature* **133**, 689–690 (1934); H.A. Bethe and R.E. Peierls, 'Photoelectric Disintegration of the Diplon', *Proc. Int. Conf. Phys. London, 1934; vol. 1 Nucl. Phys.*, 93–94; H.A. Bethe and R.E. Peierls, 'Quantum Theory of the Diplon', *Proc. Roy. Soc. Lond.* A**148**, 146–56 (1935); H.A. Bethe and R.E. Peierls, 'The scattering of neutrons by protons', *Proc. Roy. Soc. Lond.* A**149**, 176–83 (1935).

[39]See letter [214].

[40]James Chadwick (1891–1974), studied at Manchester, Cambridge and Berlin. After further research at Cambridge he became Professor of Physics at Liverpool in 1935. He worked as the Head of the British Mission attached to the Manhattan Project 1943–1946 and later had an influential role in the UK Atomic Energy Authority.

he bombarded with gamma rays in the expectation that the nucleus would be broken up. On one of their visits to Cambridge, Chadwick challenged Bethe and Peierls that they would not be able to develop a theory of this deuteron photo-effect, which, as legend has it, the two young men took up immediately, i.e. on their way home on the train to Manchester, and before the train had arrived at its destination, they had explained the phenomenon theoretically.[41]

The Manchester period, successful as it was for Bethe and Peierls, was predestined to be a temporary relief, as the appointments were only short-term with no hope of a significant extension. Hans Bethe moved on to Bristol and soon after across the Atlantic, where he had been offered a position as acting professor at Cornell. Here he would remain — with the exception of short intervals of teaching and researching in a visiting capacity elsewhere — until the end of his life. Peierls returned to Cambridge for another temporary appointment in the autumn of 1935, but it would not be long until he, too, would move to his first permanent position, a chair at the University of Birmingham in 1937.

[41] See Brown and Lee, *Hans Bethe and His Physics*, p. 9.

[1] Hans Bethe to Rudolf Peierls

Bad Gastein, 27.8.1927
(postcard)

Schöne Gegend kann man sehen
(Bitte Karte nur zu drehen!)
Doch darin ein Häuserblock
So von fünf bis dreizehn Stock,
Drinnen Menschen über achtzig
(Was nicht hindert, dass man macht sich
Zur Promnade äusserst fein)
Dieses nennt der Mensch Gastein.
Wünsche Ihnen viel Vergnügen
In Carezzas Sonn zu liegen.

Bad Gastein, 27.8.1927
(postcard)

You can see a nice view
(If you just turn the card!)
But therein a block of houses
Of about five to thirteen floors,
Inside people over eighty
(Which does not keep them from dressing up
And going to the promenade).
This is what you call Gastein.[42]
Wishing you a lot of fun,
Lying in Carezza's sun.

[42]Hans Bethe spent the holidays with his mother in Bad Gastein. This little verse indicates both his wit and his dislike for Gastein, which he regards as an architectural disaster, made worse by the 'typical tourists' described in his poem. See also letter [2].

[2] Hans Bethe to Rudolf Peierls

Seite 1055

Frankfurt/M, 19.9.1927

Lieber Peierls!
Ich fühle mich entsprechend schuldbewußt, und dabei rege ich mich normalerweise sogar noch auf über mangelhafte Schreibtätigkeit bei anderen Leuten, und behaupte, ich schriebe gern Briefe. Wofür ich als Beweis nur die Zahlen oben in der Ecke anführen kann (gerechnet weder von der Erschaffung der Welt noch von Christi Geburt an, sondern von meinem Auszug nach München am 26 IV 26). Aber angesichts des vorliegenden Tatbestandes nehme ich alles zurück, was ich in den letzten 40 Jahren gegessen und getrunken habe und behaupte das krasse Gegenteil. (Was Sie anstelle einer Entschuldigung nehmen sollten, denn die sind doch nie wahr). Um historisch anzufangen, belastete der Schöpfersche Balken mein Gemüt noch längere Zeit in Baden-Baden, bzw. lag mit im Magen (wo er wohl in der Magensäure schwamm), wo er aber wenig Platz fortzunehmen schien, da sich für das vorzügliche Essen der Selighofs in Baden noch genügend Platz fand. Als ich dort meine Warendepots genügend aufgefüllt hatte, liess ich mich vier Tage in Frankfurt sehen, und fuhr dann mit meinem Vater über München, wo ich Frau Höchstetter mit meinem Besuch beglückte, nach Gastein, dem baulich scheusslichsten Ort, den ich kenne, und der um so scheusslicher wirkt, weil seine zehnstöckigen Häuser in einer fabelhaften Umgebung liegen. Dementsprechend habe ich dann auch eine Autotour mit Herrn und Frau Prof. Sachs gemacht, zum Glocknerhaus bei strahlendster Sonne und entsprechender Stimmung, und zum Ausgleich mit meinem Vater eine Hochtour bei Nebel und Erkältungsentwicklung. (war aber natürlich doch sehr schön, und vor allem sehr übend, weil auf dem Grat ein Windchen wehte, das zwar ohne Schnee und Kälte, aber sonst von ähnlicher Güte war wie das an der Zugspitze, und infolgesdessen die Standfestigkeit übte). Als drittes Ereignis leistete ich mir noch eine Tour allein auf einen Aussichtsberg, der aber durch einen Steingrat ohne gebahnten Weg + Neuschnee interessant gemacht wurde.

Sonst bestand der Gasteiner Aufenthalt meist aus Jausen mit Familienbeigabe, die weniger lohnend als auf Dauer langweilig waren, und aus anderen Jausen mit Herrn und Frau Sachs, die die "bessere Sorte repräsentierten". Zum Schluss machte ich es Ihnen nach, in dem ich als der "bessere junge Mann" über die Hauptkette der Alpen, in diesem Fall durch den Tauerntunnel, für zwei Tage an die Kärntener Seen fuhr. Zuhause hatte ich dann die angenehme Aufgabe, 460 Renditen für die Wirtschaftskurve auszurechnen, und dann meine diversen trauernden Witwen zu besuchen, letzteres in Ermangelung augenblicklich hier vorhandener männlicher Wesen, die mit mir befreundet sind. Allmählich will ich sogar jetzt anfangen, etwas zu arbeiten. (es wird Zeit!) und ich habe vorläufig zwei Arbeiten studiert, die über das Thema meiner Doktorarbeit gehen, die sich aber beide als relativ ungefährlich erwiesen. Womit sich aber nicht erreichen liess, dass meine Arbeit bisher erschienen ist. Was damit passiert, sage ich Ihnen nicht, aber es ist ja wohl anzunehmen, dass die Geschichte bald mal rauskommen wird. Ich werde jedefalls etwa am 10. X. nach München fahren, weil experimentell erwiesenermassen das Arbeiten dort immer noch wesentlich besser geht als hier. Dann werde ich Berlin die Ehre geben (wann sagt die Redaktion Ihres Blattes mir vielleicht). Worauf vermutlich der Semesterbeginnkater wegen Zuweniggearbeitethaben folgen wird. Das wäre also das Programm. Ich glaube zwar noch nicht, dass dieser Brief eine normale Länge hat, aber es ist nachmittags 17h und da kann man noch nicht richtig schreiben, das geht erst ab 21. In diesem Sinne wünsche ich Ihnen dass Sie sich ebenso gut erholt und ebensowenig gearbeitet haben, aber nicht ebenso lang mit der Antwort warten.
Herzlichst Ihr

Hans Bethe

page 1055[43]

Frankfurt/M., 19.9.1927

Dear Peierls!
I feel quite guilty; normally it is I who gets upset with people who do not write often enough, as I claim to love writing letters. As proof, I can only point to the numbers in the upper corner (which are counted neither from the creation of the world nor from the birth of Jesus Christ, but from my exodus to Munich on April 26th 1926). But in view of the given facts, I take back everything that I have eaten and drunk over the last 40 years and now claim just the opposite. (Which you might take as an excuse, since these claims are never true.) To begin historically, the Creative Beam weighed heavily on my mind for a long time in Baden-Baden, or rather it lay in my stomach (where it swam in my stomach acid, I guess). It seemed to take up only a little space, since there was still enough room for the delightful meal that I had at the Selighof in Baden.[44] After having replenished my stocks, I showed up in Frankfurt for four days and then travelled to Gastein via Munich with my father, where I pleased Mrs. Höchstetter with a visit. Gastein has the worst architecture I have ever seen, which looks even more horrible given that its ten-story houses are located in fabulous surroundings. I therefore drove to the Glocknerhaus with Prof. and Mrs. Sachs.[45] The weather was beautiful, and so was our mood. As a kind of compensation, I also went on an alpine tour with my father. It was quite foggy and I was on my way to getting a cold. (But it was beautiful anyway and good training for my balance, as the ridge was similar to the Zugspitze: windy

[43]Hans Bethe was in the habit of numbering his letters and pages thereof. This first letter is p. 1055 of letter 217. Peierls later used a numbering system in his correspondence with his future wife, Genia, when every letter contained a number with date and the same information about previous letter. See Lee, *Selected Correspondence*, 1, chapter 2.

[44]Bethe's parents had divorced earlier that year; his mother lived in Baden-Baden, his father in Frankfurt.

[45]Parents of their friend Werner Sachs, one of a group of Munich students who often spent their vacations hiking in the Alps. See Chris Adami, 'Three weeks with Hans Bethe', in Brown and Lee, *Hans Bethe and His Physics*, pp. 45–110; here pp. 74–75, also photo on p. 24.

without being snowy or chilly.) My third trip, which I took on my own, was to a mountain with a nice view that was made more interesting by a stony ridge with fresh snow and no beaten track. Aside from these trips, my stay in Gastein basically consisted of some family luncheons, which became boring after a while and were less interesting, and some other luncheons with Mr. and Mrs. Sachs which were of a better sort. In the end I followed your example; after passing the main mountain range of the Alps, in this case through the Tauerntunnel, I went to the Carithian Lakes for two days. At home I had the nice challenge of calculating 460 yields for the economy curve. I then had to visit all my dolorous widows, since there are no manly creatures among my friends at the moment. I can now even think about starting to work (it is time!). Up to now I have studied two papers related to the subject of my dissertation, which both turned out to be not too dangerous.[46] But this has not helped in publishing my work. I will not tell you what is happening with that, but you can assume that the story might come out soon. I will go to Munich on October 10th, since it has been experimentally confirmed that it is easier to work there than here. Then I will honour Berlin with a visit. (The editorial staff of your journal will probably tell me at what time.) This might be followed by the early-semester-hangover as a result of having worked too little. Well, these are my plans. I still don't think that this letter is of normal length, but it is 5 o'clock in the afternoon. At this time one cannot write properly; this is only possible after 9 o'clock in the evening. In this spirit I hope you have recovered as well as I have and have not worked any more than I have either, but that you don't wait so long to answer.
Sincerely Yours,

Hans Bethe

[46]Bethe was one year Peierls' senior; after two years of studying physics at Frankfurt, he had joined the 'Sommerfeld school' in Munich in the spring of 1926. By the summer of 1927, Sommerfeld had suggested to Bethe to develop a theory of diffraction of electrons by crystals. See N. David Mermin and Neil W. Ashcroft, 'Hans Bethe's Contribution to Solid State Physics', in Brown and Lee, *Hans Bethe and His Physics*, pp. 189–200, here p. 191.

[3] Hans Bethe to Rudolf Peierls

Frankfurt/M, 1.10.1927
Inliegend meine Arbeit!
212/1063

Lieber Peierls!
Vielen Dank für das Fachschaftsprogramm, das mich zur Umstossung meines Programms veranlasste, indem und dadurch dass ich meinen Plan, in München im Oktober zu arbeiten, aufgab, und statt dessen etwa am 16. nach Berlin kommen will, und gleich nach der Tagung nach München fahre. Ihre nette Einladung, bei Ihnen zu wohnen, kann ich vorläufig noch nicht annehmen, da Werner Sachsens Verwandte noch "ältere Rechte" auf mich haben. Seien Sie mir also bitte nicht böse, wenn ich Ihnen zunächst einen Korb gebe; unser Zusammensein braucht aber dadurch in keener Weise gestört zu werden, da ich sowieso ja wohl den ganzen Tag in der Stadt sein werde.

Sie werde nicht darum herumkommen, mir eingehend die "scheene Jejend zu erklären". Hier ist momentan Physiologenkongress, bei dem ich väterlicherseits beteiligt bin. Immerhin hält sich diese Beteiligung, von einem Vortragsvormittag abgesehen, an dem mein Vater selbst sprach und als Attraktion einen Film vorführte, in den relativ engen Grenzen der Teilnahme an allgemeinen Abspeisungen und nachfolgenden Tanzvergnügen. Im übrigen war ich im Wesentlichen mit Werner Sachs und Eltern zusammen, obwohl ich zunächst erklärte, mit einem schnurrbärtigen Menschen, der aussieht wie ein unsympathischer Filmheld, nicht befreundet sein zu können.

Gearbeitet habe ich natürlich wieder nichts. Nichtsdestotrotz auf Wiedersehen in dem trauten Städtchen Berlin.
Herzlichst Ihr

Hans Bethe

Frankfurt/M, 1.10.1927

Find enclosed my work!
212/1063

Dear Peierls!
Thank you for the program of the student society, which made me change my plans since I have given up the idea of working in Munich in October and will come to Berlin around the 16th instead. I will go straight to Munich after the conference. I can not accept your nice invitation to stay at your place at the moment, because I already promised Werner Sachs' relatives that I would stay with them. Don't be upset if I have to reject your offer; we can meet as often as we want, as I will be staying in the city for the whole day anyway.

You will not get around showing me the "nice area"[47] in detail. The conference of physiologists is also taking place at the moment, which I am involved in because of my father.[48] But at least, except for one morning, when there were some lectures and my father spoke and showed a film as an attraction, my attendance at this conference has been restricted to going to the common meals and dances in the evening. Aside from this, I have basically spent time with Werner Sachs and his parents even though I once declared that I could not be the friend of an moustached man who looked like an dislikeable film star.[49]

Of course, once again I haven't done any work. Anyway, see you again in the nice little city of Berlin.
Sincerely yours,

Hans Bethe

---

[47]Written in Berlin slang in German.

[48]Bethe's father was professor of physiology at Frankfurt University. Hans was close to his father (see S.S. Schweber, 'Writing the Biography of a Living Scientist: Hans Bethe', Special Collections, Oregon State University, OR 97331-4501. In fact, Hans Bethe's first published paper was a joint paper with his father and one of the latter's colleagues: A. Bethe, H.A. Bethe, and Y. Terada, 'Versuche zur Theorie der Dialyse', *Zeitschrift f. physikalische Chemie* **112**, 250-69 (1924)) and he occasionally joined him for conferences.

[49]Reference to Werner Sachs' new hairstyle and moustache.

[4] Hans Bethe to Rudolf Peierls

[ohne Ort], 16.12.1927
(postcard)

Lieber Peierls!
Ich verlasse heute anlässlich des bestandenen Examens von Werner Sachs meine Pflicht und München und betätige mich skilaufenderweise. Ich erscheine spätestens um 21 $\frac{3}{4}$ im Hotel Union, und wollte Sie bitten, mir meine Insignien als Weihnachtsmann entweder dorthin oder in meine Wohnung zu besorgen. Gebrauchsanweisung können Sie mir ja in den reichlichen Pausen erteilen.
Mit herzlichem Dank, Ski-Heil und Gruss
Ihr

Hans Bethe

[location unspecified], 16.12.1927
(postcard)

Dear Peierls!
Since Werner Sachs passed his exam today, I'm shirking my duty and Munich to go skiing. I will show up at the Union Hotel at 9:45 pm at the latest, and I wanted to ask you whether you could bring me my insignia as Santa Claus either there or at my apartment. You may give me instructions for its use in the breaks.

Thank you very much, good luck with the skiing and take care.
Yours,

Hans Bethe

[5] Hans Bethe to Rudolf Peierls

Baden-Baden, 24.12.1927
(postcard)

Lieber Peierls, lieber Thorner!
Sie werden jetzt irgendwo im wilden Gerlostal als wackere Männer auf die rauhen Berge steigen, aber da ich "Ihnen jetzt nicht sage", wo Sie

sind, schreibe ich Ihnen nach München. Sie haben sich so rührend in München um mich gekümmert, dass ich Ihnen trotz des ungewissen Schicksals dieser Karte gleich dafür danken muß. Ich bin also glücklich hier angekommen, von Oos nach Baden im Rollstuhl per Gepäckwagen und dann teils per Auto, teils per Schultertransport. Jetzt bin ich in der Pflege meiner Mutter, die trotz des Schrecks, den sie zuerst bekam, froh ist, mich hier zu haben, und eines Arztes, der alle 2 Tage kommt. Der linke Fuss heilt schon und dem rechten geht es etwas besser. Also jetzt nochmals herzlichen Dank für Alles; fröhliche Tage und Ski-Heil!
Herzlichst Ihr

Hans Bethe

Baden-Baden, 24.12.1927
(postcard)

Dear Peierls, dear Thorner![50]
You brave fellows are climbing rough mountains somewhere in the wild Gerlos valley,[51] but as I "won't tell you" where you are I'm sending this letter to Munich. In Munich you took care of me in such a nice way that I had to thank you immediately, even though the fate of this postcard is quite uncertain.[52] I am happy to have arrived here. I came via Oos to Baden in a wheelchair, by baggage car, and then by car, and I was even carried on shoulders. Now my mother is taking care of me. She is happy that I'm here, even if she was shocked at the beginning. There is also a doctor who comes every two days. My left foot is healing already, and the right one is getting better too. So thanks again for everything; enjoy your time and take care!
Sincerely yours,

Hans Bethe

---

[50] Hans Thorner, medical student and friend of Peierls' living in Heidelberg.

[51] This phrase became an ongoing joke among Bethe, Peierls and their friends. On one of their excursions, a young boy had admiringly remarked to his mother: "These are real men who fearlessly climb the harsh mountains."

[52] After a weekend skiing trip, Bethe suffered from frostbite causing an infection in his foot. This was to trouble him well into the New Year, and he did not return to Munich until mid-February 1928.

[6] Hans Bethe to Rudolf Peierls

248/1295

Baden-Baden, 12.1.1928

Lieber Peierls!
Von 0 bis ∞ laufen meine Füsse leider noch nicht, sondern bestenfalls hüpfe ich abends auf dem linken einmal um den Tisch herum. Da ich aber sonst noch keine Wege machen kann, befinde ich mich doch wenigstens auf dem Wege der Besserung. So stellten wenigstens alle Autoritäten fest, nämlich der Doktor, der mir alle zwei Tage an den Stellen, wo sich Eiter bildet, die Haut abzieht, eine Prozedur, die zwar gut tut, aber nicht sehr, deren Dauer sich aber allmählich von einer Stunde auf $\frac{1}{4}$ St. reduziert hat. Daneben habe ich neuerdings jeden Tag das Vergnügen eines Kamillenfussbades von der sanften Hand meiner Mutter (ganz reizend), und sonst liegt mein Füsschen Nr. 45 ruhig im Bett und heilt. Schmerzen habe ich seit etwa fünf Tagen ganz abgeschafft, nachdem sie vorher auch nur noch selten aufgetreten waren. Meine Mutter stellt fest, ich ware "über den Berg", was mir komisch vorkommt, da ich nicht gehen kann. Aber der Abstieg ins Tal, um bei dem Bild zu bleiben, dauert wahrscheinlich noch einige Zeit, und ich taxiere, dass ich vom Januar in München wenig mehr sehen werde.

Ich tue inzwischen ganz, als ob ich hier zuhause wäre, bin sehr glücklich, dass ich damals mit Ihrer Hilfe hierher gereist bin und nicht als verlorene Nummer im Münchener Krankenhaus liegend mit fettem Schweinefleisch und harten Knödel gefüttert werde sondern statt dessen durch Wildenten, Fasanen und Karpfen Ihnen den Mund wässerig und mir die Wiedergewöhnung an Alt-Wien schwer mache.

Ich bin, solange meine Mutter nicht irgendwelche Tätigkeiten wie Bankkorrespondenz und Kamillenfussbäder von mir verlangt, sehr fleissig und habe in meiner Arbeit eine langwierige Rechnerei vor der mir immer graute, mit Todesverachtung und Erfolg durchgeführt. Jetzt kommt aber bald die Stelle, wo ich wahrscheinlich ohne Literatur nicht mehr weiterkomme. Vielleicht fange ich dann an zusammenzuschreiben, was ich bisher gemacht habe, was auch als nützliche Beschäftigung gilt.

Gelangweilt habe ich mich jedenfalls nicht, oder wenigstens nicht mehr als ich es in München bei dieser Rechnerei getan hätte.

Daneben habe ich so allerhand gelesen, ein Quartal gemischte Post, die sämtlich unbeantwortet in meiner Schublade liegt, Schickeles "Erbe am Rhein", dann die "Vlämischen Legenden" von de Coster das "Zweite Gesicht" von Löns und den "Jud Süss" von Feuchtwanger. Allerhand schöne Sachen aus den verschiedensten Kategorien, das fabelhafteste davon ist aber nach meinem Geschmack unbedingt der "Jud Süss". Darin liegt eine kolossale Kraft und Grösse, alle Menschen stehen leibhaftig vor einem, und es ist nicht eine billige Charakterisierung von guten und bösen Menschen, mit denen man Sym- und Antipathie haben soll, sondern man bekommt alle Fehler dieses Jud Süss vorgeführt, und bewundert ihn trotzdem. Daneben eine ausgezeichnete Schilderung der Juden der damaligen Zeit, und der Zeit überhaupt.

Abendliche Gesänge von Schlagern, deren Texte ich aus 10-Pfg-Büchern entnehme, während meine Mutter über die Verderbnis der Zeit jammert, vervollständigen mein Programm. So schöne Sachen wie Ihre Skireise mit wackerem auf die rauhen Berge steigen habe ich nicht aufzuweisen, und Skireisen ist für diesen Winter sowieso Essig. Naja — so pleite, so pleite. Das Geld für meine Skireise im März kriegt statt dessen der Arzt.

Bitte grüssen Sie Thorner herzlich von mir, ich schreib ihm nächstens auch. Schicken Sie mir mal wieder "gemischte Post" und schreiben Sie, wie Seyffahrt den Doktor gemacht hat!

Herzlichst Ihr

Hans Bethe

248/1295

Baden-Baden, 12.1.1928

Dear Peierls!

From 0 to ∞ my feet unfortunately do not yet run; at most I hop around the table on my left foot once in the evening. While I cannot go on my way, I am on the way to getting better. At least, that's what all the authorities (namely, the doctor) told me. The doctor removes skin in places where pus has formed every two days, a procedure which feels good but not very much so. The time it takes has slowly decreased from one hour to 15 minutes. Besides this, I enjoy a daily footbath of chamomile which my mother prepares for me (very nice). Little foot no. 45 is quietly lying in bed and getting better. I have totally abolished pain for the past five days, but it wasn't too painful even before this. My mother says I'm "over the hump"[53], which seems odd to me as I can't walk. But to use the same metaphor, climbing down from the mountain will presumably take a while and I guess that I won't see much of Munich in January.

In the meantime I act as if I were at home here. I am very happy that I came here with your help at the time, and that I am not lost in a hospital in Munich being fed with fat pork and hard dumplings. Instead I can whet you appetite with wild duck, pheasant, and carp which make it hard for me to get used to Old Vienna again.

As long as my mother does not ask me to correspond with the bank or have chamomile footbaths, I am working hard and have successfully finished a long calculation for my thesis which I had anticipated with horror in defiance of death and with success.[54] But I will soon be at a point where I cannot go on without literature. Probably I will then start writing down what I have done so far, which is also held to be quite useful. I haven't been bored at any time, or at least not more I would have been if I had done this calculation in Munich.

Besides that I have been reading a lot: three months' worth of letters, which are all lying unanswered in my drawer; and Schickele's "Erbe am Rhein" (Heritage on the Rhine)[55] followed by "Flemish legends" by de

[53]The German expression used literally means 'over the mountain', hence Bethe's comment on the metaphor.

[54]Refers to his doctoral thesis. In this he explained why electrons in certain energy intervals were observed to be reflected. He used Paul Ewald's work on the reflection of x-rays by crystals, and using certain properties of the wave function, he established that for some incident directions and energy intervals, there did not exist any wave functions that corresponded to an electron in the crystal. His work was published later as H. Bethe, 'Theorie der Beugung von Elektronen an Kristallen', *Ann. Phys.* **87** (1928), 55–128. In this letter he refers to the complex calculation underlying this result.

[55]Rene Schickele (1883–1940), writer from Alsace Lorraine. His novel *Erbe am Rhein* (1925) was the first book of a trilogy about a German-French Alsacian family.

Coster,[56] "Das Zweite Gesicht" (Second Face) by Löns[57] and "Jud Süss" by Feuchtwanger.[58] All these are beautiful things of different kinds, but to my taste the most fabulous is definitely "Jud Süss". It is of enormous power and splendour. All the people appear vividly before you, and it is not a primitive characterisation of good and evil people towards whom you are supposed to feel sympathy or antipathy. All the failings of this Jud Süss are shown to you, but you admire him anyway. Furthermore, it is a great description of the Jews of that time and of the period in general.

The evening hits, whose lyrics I can read in 10-penny books while my mother is complaining about our wicked times, complete my program. I cannot come up with such beautiful things as your ski tours and trips to rough mountains, and ski tours wouldn't be good for me this winter anyway. Well — so broke, so broke. The money for my ski tour in March goes to the doctor instead.

Please send my warm greetings to Thorner, I promise to write to him soon. Send me "mixed letters" again some time, and tell me how Seyffahrt got on with his doctorate![59]

Sincerely yours,

Hans Bethe

---

[56]Charles de Coster (1827–1879), Belgian author, *Flämische Legenden* was first published in 1861.

[57]Hermann Löns (1866–1914), journalist and writer is best known for his famous folksongs, and his novel *Der Werwolf. Das Zweite Gesicht*, a love story, was written in 1911, shortly after he separated from his wife.

[58]Lion Feuchtwanger (1884–1958), German-Jewish novelist who openly criticised National Socialism and was forced to escape Nazi-occupied Germany in 1940. *Jus Süß* was his first popular novel.

[59]H. Seyfahrt, one of Sommerfeld's graduate students, had been working on x-ray analysis of crystals.

[7] Hans Bethe to Rudolf Peierls

Baden-Baden, 28.1.[19]28

Lieber Peierls!
Ich liege immer noch kontinuierlich auf der Matratze, während Sie darüber vortragen. Zur Abwechslung, und zur Vervollständigung des Lazaretteindrucks hat sich meine Mutter jetzt auch ins Bett gelegt, mit einer Angina, aber nicht sehr. Dafür mache ich schüchterne Versuche, auch mit dem rechten Bein aufzutreten, verfüge aber unter dem kleinen Zehlein (das ich jedoch nicht falte), noch über eine offene Stelle. Da ich bisher die vom Doktor genannten Heilungszeitem immer versechsfachen musste, könnte ich eigentlich nicht vor dem 1. IV. erwarten, nach München zu kommen. Aber ich löse Kreuzworträtsel mit Tinte, d.h. ich bin Optimist und schrieb demgemäss an Unsöld, ich käme about 13 II, es wäre mir aber sympathischer, erst am 20. auf dem Podium erscheinen zu müssen. On verra.

Da ich keine Tiegel quantitativ kaputtschlagen kann (Ihre diesbezüglichen Ausführungen haben mich übrigens sehr an mein eigenes quantitatives Arbeiten erinnert) begnüge ich mich mit dem Erwerb mehr oder weniger tiefgründiger mathematischer Kenntnisse aus Bieberbachs Differentialgleichungen. Wobei ich wieder meine mangelnde Begabung für die reine Mathematik sehe, in dem ich meine, der Mond geht unter und der Beweis wäre zu Ende, wenn der strenge und kluge Mathematiker noch drei grosse Haken findet. Ich könnte, glaube ich nie solche Beweise selbst bis ins Letzte herauskriegen, d.h. so, dass Bochner es glauben würde. Ganz so schlimm wie Ihre principia mathematica sind aber Differentialgleichungen nicht; die Probe, die Sie mir davon gaben, scheint mir auf eine sinnvolle Mischung des "grossen Lalula" von Morgenstern und von "Fisches Nachtgesang vv-v" zu deuten nach Belieben garniert mit Pfeilen ↓ abwärts (Konfektionsabteilung, Trikotagen, Erfrischungsraum).

Meine Hauptbeschäftigung ist jetzt, solange ich nicht gerade darüber nachdenke, dass ich es noch beliebig lange im Bett aushalten würde, und meine eigene Geduld lobe, zu lesen, und ich finde diese Beschäftigung a) sehr angenehm, b) sehr zeitraubend, c) sehr notwendig von wegen

Spezialwissenschaft und Allgemeinbildung, d) sehr unangenehm, weil alle guten Gedanken schon gedacht worden sind. Als letztes habe ich in vier Tagen die "Kristin Lavranstochter" besiegt, und bin sehr begeistert davon. Das ist ein Buch, in dem so viele Freuden und Qualen der Menschen geschildert sind, so wie wir sie selbst erleben und um uns sehen, wie ich kaum ein zweites kenne. Das sind wirklich tiefe Menschen, die da vor einen hingestellt werden, und ich verstehe nicht recht, wie z.B. die "Forsyte Saga", die ein stellenweise gutgeschriebener Roman von uninteressanten Menschen ist (ich kenne übrigens nur den ersten Band), damit an Qualität verglichen wird. (Wobei ich eben die ekelhafte Angewohnheit mancher Menschen nachahmte, von jedem anderen die Kenntnis aller "Neuerscheinungen" vorauszusetzen.)

Dazwischen bin ich mal ganz klassisch geworden, und habe gefunden, dass der alte Shakespeare auch was gekonnt hat. Trotzdem muss ich (zu meiner Schande?) gestehen, dass mir die modernen Sachen doch sehr viel näher liegen, und die Probleme von Shakespeare oft ganz fernstehen. Schiller kann ich ja überhaupt nicht mehr lessen.

Ihrem und Unsölds gemeinsamen Schluss, dass die Welt nicht existieren kann, weil sonst die Quantenmechanik gelten müsste, muss ich zustimmen. Aber das dürfen wir nicht weitersagen, denn sonst entzieht man uns die Gelder, und was soll dann aus all den Physikern werden? Aber vielleicht finden Sie auf diese Feststellung hin eine Anstellung bei Lenard, und werden mit dem Uräther verwandt.

Sehen Sie Thorner eigentlich mal? Oder vertieft er sich so sehr in Examensvorbereitungen? Und kann man Ski laufen? D.h. bei Arosa kann mans, wie mir Erwin Strauss Karte mitteilte. Und ich kanns diesen Winter nicht mehr. Wann ich komme, sage ich Ihnen also noch nicht, und bin für Post stets gern Nehmer.

Herzlichst Ihr

Hans Bethe

Baden-Baden, 28.1.[19]28

Dear Peierls!

I am still lying on the mattress while you are holding your lectures. For a change and to complete our sick bay atmosphere, my mother is now also in bed with angina but it is not too dramatic. I am now timidly trying to stand on my right foot as well, but under the little toe (which does not bend yet) there is still an open wound. As so far I have had to multiply the doctor's healing time by six, I cannot expect to come to Munich before the first of April. But I do crosswords with ink, which means that I am an optimist and therefore wrote to Unsöld[60] that I would come around February the 13th. It would be nicer for me, if I did not have to be on stage before the 20th, however. *On verra.*

As I cannot quantitatively smash any crucibles (your comments on that reminded me very much of my own quantitative work, by the way) I restrict myself to acquiring a more or less profound mathematical knowledge of Bieberbach's differential equations.[61] In doing so I was forced to realise once more that I am not very gifted in pure mathematics, because I see the moon go down and think the proof is given, while a real, smart mathematician still manages to find three bigger snags. I think I could never really bring such proofs to completion, which is to say in a way that Bochner[62] would accept. But these differential equations are not as bad as your *principia mathematica*; the example you have given to me seems to be a reasonable mixture of Morgenstern's "Big Lalula"[63] and "Fische's night song vv-v",[64] sometimes crowned, just as you want, with downward arrows ↓ (clothes section, knitting wares, refection room).

When I am not thinking about staying in bed as long as I want and praising my own patience, I mainly spend my time reading, and I think this is a) very pleasant, b) takes up a lot of time, c) is highly

---

[60]Albrecht Unsöld (1905–95), studied physics under Sommerfeld, completed his Ph.D. at the age of 21 in Munich, where he became Sommerfeld's assistant. In 1932, aged 27, he became professor of theoretical physics at Kiel. He later made his most important contributions to the theory of stellar atmospheres.

[61]L. Bieberbach, *Theorie der gewöhnlichen Differentialgleichungen*, Springer, Berlin, 1923.

[62]Salomon Bochner (1899–1982). At the time lecturer of mathematics in Munich working mostly in harmonic analysis. In 1933 he emigrated to Princeton where he stayed until his retirement.

[63]Christian Morgenstern (1871–1914), author and poet who was best known for his nonsense poems. 'Das große Lalula' (1905) was one of these comical poems consisting entirely of non-existent words.

[64]'Fisches Nachtgesang' was a parody on Goethe's 'Ein Gleiches', consisting only of two symbols (- and u).

necessary for both specialized science and general education, and d) is very annoying, as all good thoughts have already been thought. I finally read "Kristin Lavransdsatter"[65] over the last four days and I am very excited about it. It is a singular book, which describes so much of the human joy and pain that we experience on our own and see all around us. These are really deep characters that are shown to us, and I do not understand how (for example) the "Forsyte Saga",[66] which is partly a well-written novel about uninteresting people (I know only the first part, by the way), can be compared with it. (I just adopted a disgusting habit of some people, namely to expect everybody to know all "new publications".)

In between I have become classical, and found out that old Shakespeare wasn't so bad either. Nevertheless, I have to confess (what a shame?) that I like modern things more and that Shakespeare's problems seem quite distant from me. Schiller I can't read at all anymore.

I have to agree with your and Unsöld's conclusion that the world must not exist, since then quantum mechanics would have to be right. But we shouldn't say this, because then our grants would be cancelled, and what would happen to all the physicists? But Lenard[67] will probably offer you a job because of this conclusion, and you will become a relative of the primary ether.

Do you happen to see Thorner from time to time? Or is he too busy preparing for his exams? And do you go skiing? In Arosa you can, it seems, as Erwin Strauss[68] wrote on a postcard to me. But this winter I can't. When I will come I can't say, but I am always a grateful recipient of your letters.

Sincerely yours,

Hans Bethe

---

[65] *Kristin Lavransdatter*, a trilogy set in 14th century Norway by Sigrid Undset, for which she received the Nobel Prize for Literature in 1928.

[66] John Galsworthy's trilogy, *Forsyte Saga*, chronicling the lives of three generations of a monied, middle-class English family at the turn of the century, would be awarded the Nobel Prize in 1932.

[67] Philip Lenard (1862–1947), Nobel laureate and professor of physics at Heidelberg University until his retirement in 1931; well-known and outspoken supporter of national socialism.

[68] Fellow student and friend of Bethe's and Peierls' from Munich.

[8] Hans Bethe to Rudolf Peierls

München, 4.5.[19]28

Lieber Peierls,
Ich bin mir bewusst, dass ich die Verpflichtung hätte, mich zu Tode zu schämen. Das tu ich aber nicht, da ich das Leben momentan recht angenehm finde. Andererseits liegt aber in dieser Annehmlichkeit des Lebens der tiefere Grund für das zu Tode schämen, sodass es sich um so was ähnliches handelt wie bei den immer lügenden Kretern. Ich habe es eben etwas zu gut gehabt, und infolgedessen meine Erfahrung insofern bereichert, als man nicht nur nicht zur Arbeit kommt, wenn man zu Hause ist, sondern nicht mal zum Schreiben — ausser wenn man sich längere Zeit ins Bett legt. Wann soll man aber auch schreiben, wenn man die ganze Zeit in Heidelberg und Frankfurt von einem Harem grösserer Dimension umgeben ist, den man als einziger Mann zu unterhalten verpflichtet und berechtigt ist, sodass man nur die Sorge hat, genügend rasch von einer Verabredung zur nächsten zu kommen. Gemein ist eben nur, dass ich mir Sie als Objekt zweckes Gewinnung der Erkenntnis von der Unmöglichkeit des Schreibens zu Hause verwendet habe. Was ich hierdurch untertänigst zu entschuldigen bitte.

Ich habe mich sogar beinahe als Organisator bewährt: Nachdem ich die Theorie aufgestellt hatte, dass man immer nur drei Tage bei denselben Leuten bleiben soll, um den vollen Genuss davon zu haben, und abreisen muss, bevor man sich langweilig wird, tat ich also: drei Tage Schwarzwaldtour mit Werner Sachs + Schwester u. Vetter, unter Produzierung beliebig grossen Stusses und amtlicher Prüfung meiner Füsse auf ihre Haltbarkeit. (Ohne wesentliche Beschwerden konnte ich schon wieder fünf Stunden gehen, trotzdem muss ich aber immer noch vorsichtig sein und darf vor allem keine Hochtouren machen, die in Schnee führen.) Dann zu Ostern bei meiner Mutter in Baden-Baden mit abwechselnder Bestaunung der Eleganz in der Lichtenthaler Allee und der ganz fabelhaften Auslagen von Edelsteinen, phantastischen Kleidern, moderner Keramik und gefärbten Gesichtern und zu anderen Zeiten völliger Zurückgezogenheit in einem einsamen grünen Seitental mit Ostblüte, der Villa meiner Mutter und grossen Spaziergängen.

Nach Ostern erst Heidelberg mit Werner Sachs + Schwester und einer gemeinsamen Freundin, Spaziergänge, Kino, Theater mit gleichbleibend meschugger Stimmung. Dann drei Tage Frankfurt, Kinobesuch mit 1–4 Mädchen und ebensolche Abendessen. Ich bin dabei auf den Gedanken gekommen, dass es eigentlich eine sehr weise Verteilung ist, im Semester ungestört arbeiten zu können, und in den Ferien zu den trauernden Hinterbliebenen zu reisen und sich da zu amüsieren, ohne die Arbeit anzusehen. Jetzt habe ich aber das Prinzip durchbrochen, da eine Frankfurter Freundin zum Studieren hierher gekommen ist. Ich bin noch nicht ganz meiner Meinung, ob das angenehm ist oder nur störend, neige aber, bis der Katzenjammer von wegen zu wenig Arbeit kommt, zu ersterer Ansicht.

Gearbeitet habe ich selbstverständlich nichts. Eben habe ich versucht, die Arbeit von Dirac im Februar- und Märzheft der Proc. Roy. Soc. zu lesen. Nach mehreren Tagen habe ich ahnungsweise erfasst, worum sich's handelt, aber verstanden habe ich keine Spur. Schon ganz rein rechnungsmässig verstehe ich höchstens 25% der Matrizenoperationen, die er macht, und warum das alles geschieht, davon verstehe ich Bahnhof, und schliesse daraus, dass die Arbeit sehr gut ist. Aber ich habe keine Zeit, mich in Matrizenmechanik und Dirac zu vertiefen, sondern muss jetzt meine Arbeit endgültig abschliessen und dann — lernen. Blödsinn! Ich schicke Ihnen übrigens auch meine gesammelten wissenschaftswissenschaftlichen Werke.

Hoffentlich gefällt Ihnen Leipzig gut, und Sie werden bald ein berühmter Mann $\geq$ Heisenberg!
Herzlichst Ihr

Hans Bethe

Munich, 4.5.[19]28

Dear Peierls,
I know that I should feel totally ashamed. But I do not, because life is quite acceptable these days. But on the other hand, these conveniences of life are the deeper reason for feeling so ashamed, so that it is something similar to the ever-lying Cretes. I have had it a little too easy, so I have experienced the phenomenon that one just doesn't find the time to work at home; but what's more you don't even find the time to

write unless you have to lie in bed for a prolongued period of time. But how can you write anything, if you are always surrounded by a biggish harem in Heidelberg and Frankfurt, which can and must be entertained only by you, as the only man, so that all you can think about is how to get from one appointment to another in time? I know that it is not fair to have chosen you as my object for demonstrating the impossibility of writing at home. That's why I ask you submissively to excuse me.

I have almost become a good manager. Following my theory that you should always stay with the same people for only three days to have an optimal outcome, and that you have to leave before getting bored of each other, I did this: a three-day trip to the Black Forest with Werner Sachs + sister and cousin, with the production of much nonsense and an official proof of the soundness of my feet. (I could walk for five hours without serious problems but I still have to be careful, especially with alpine tours that lead to the snow.) At Easter I was in Baden-Baden with my mother, gazing at the elegance of Lichtenthaler Avenue and its fabulous presentations of jewelery, fantastic clothes, modern ceramics, and tanned faces. At other times I was totally isolated in a lonesome valley with flowers on the eastern slopes, or at my mother's villa with the long walks. After Easter I was in Heidelberg with Werner Sachs + sister and a common friend, with walks, cinema, and theatre in a consistently strange atmosphere. I then spent three days in Frankfurt, going to the cinema with between 1 and 4 girls and having similar dinners. I thought that it would be a good idea to work throughout the semester without a break, and to visit the dolorous bereaved during vacations to enjoy the time and not care about work. But now I have given up this principle, since a friend from Frankfurt has come here to study. I am not sure yet whether this is nice or just annoying, but I prefer the first until the hangover comes, as I will have worked too little.

I haven't worked at all, of course. I have just tried to read the work of Dirac, which was published in the February or March edition of the Journal of the Proc. Roy. Soc.[69] After a few days I have begun to have

[69]P.A.M. Dirac, 'The Quantum Theory of the Electron', *Proc. Roy. Soc.*, **A117** (1928), 610; and P.A.M. Dirac, 'The Quantum Theory of the Electron. Pt. II', *Proc. Roy. Soc.*, **A118** (1928), 351.

an inkling of what it deals with, but I haven't understood a single word. Regarding the calculations, I understand at most 25% of the matrix operations. Why they are done I don't understand at all, which I guess proves that the paper is very good. However, I don't have the time to really dig into matrix operations and Dirac, right now I have to finish my thesis and then — learn. Humbug! I am sending you my collected scientific works, by the way.

I hope you like Leipzig and will soon be a famous man $\geq$ Heisenberg![70]

Sincerely yours,

Hans Bethe

[9] Hans Bethe to Rudolf Peierls

302/1571
Zug Baden-Frankfurt, 5.6.[19]28

Lieber Peierls!

Sie sehen, ich reise mal wieder spazieren, und meinen, ich hätte es gut. Aber ich bin darüber noch nicht ganz meiner Meinung. Denn es ist eigentlich sehr traurig, warum ich jetzt nach Frankfurt fahre: Die Mutter einer Freundin von mir ist gestorben, und ich will sehen, ob ich nicht ein bisschen helfen kann. Das ist da wohl nicht so ganz einfach. Aber manchmal tut einem wohl Aussprache und Gesellschaft in solchen Momenten sehr wohl. Und manchmal braucht man auch jemand, der einen ein wenig aufmuntert, etwas zu tun, um den Grübeleien zu entgehen, die ja die grösste Gefahr für einen Menschen nach einem Schmerz sind. Sie scheint übrigens ziemlich ruhig zu sein.

Vorher war ich in Baden-Baden bei meiner Mutter, um mich zu erholen. Dafür ist Baden sehr geeignet, denn der Wald ist ein angenehmer Aufenthalt, und man meint immer, wenn man dort ist, man

[70] Arnold Sommerfeld was travelling the world and had advised Peierls to continue his studies with Werner Heisenberg at Leipzig.

wäre bei sich selbst zur Sommerfrische. Ausserdem bietet die Küche bei meiner Mutter eine rein angenehme Abwechslung gegenüber dem unter den verschiedenen Namen auftretenden Münchener Kalb- und Rindsbraten. Immerhin – ich bin nicht ohne Entschuldigung in Baden gewesen sondern habe mich vorher – offenbar um einen Grund zu haben nach Baden zu fahren, zehn Tage lang ins Bett gelegt, mit Fieber beliebiger Qualität (ich habs bis 40,3 gebracht!) und einer selbst duch Kapazitäten des Schwabinger Krankenhauses nicht näher erklärten Krankheit. Also sagen wir Grippe. Aparterweise habe ich infolgedessen im Schwabinger Krankenhaus gewählt, aber die Niederlage der Demokraten nicht verhindern können. Dafür habe ich bei der bayrischen Landtagswahl "gesiegt", indem ich in weiser Voraussicht der Tatsache, dass die Demokratische Partei Bayerns keine hervorragende Bedeutung im Landtag Bayerns haben wird, sozialdemokratisch gewählt habe. Wenn meine Tante in Stettin das wüsste, würde sie mich enterben, da sie aber sowieso nichts hat, macht das nichts. Im übrigen bin ich mit dem deutschen Volk relativ zufrieden, ausser mit der Zunahme der Kommunisten. Nach Münchener Plakatsäulen zu urteilen, müsste übrigens der Reichstag mindestens 200 Nationalsozialisten enthalten: Hitler liess nicht weniger als 10 Plakate los, die ein guter Karrikaturist hätte illustrieren sollen. Welch schönes Bild könnte man z.B. zu den Worten machen: "Der internationale Börsenjude triumphiert".

Im Institut ist der Personalbestand beinahe unverändert, nur die Amerikaner sind bis auf unseren Freund four times u (WW) Sleator entschwunden. Der ist aber nicht magerer geworden, obwohl er gerade eine Arbeit publiziert, über den Halleffekt und noch einen anderen Effekt mit wohlklingendem Namen. Ich selbst schwelge darin, dass mein Name in Arbeiten zitiert wird, und habe ausserdem nach den Satz "jeder sein eigener Amanullah" meine Doktorarbeit abgeschrieben (in Baden-Baden). Über die Länge bin ich erschrocken. 70 Seiten, und Sommerfeld wird gewiss dasselbe tun. Gelernt habe ich noch immer Null, aber habe jetzt den festen Vorsatz und wenig Zeit, um ihn auch auszuführen.

Letzten Samstag war ich mit Thorner zusammen, der, umgeben von Frl. Ster, die dick aber nett und witzig war (das sind Mädchen selten), von einer Schwarzwaldtour mit 3h-Aufstehen und derlei Stuss nach

Baden-Baden kam zum Neurologentag. Er schien von Krehl recht befriedigt und hatte sich offenbar mit dem Thema seiner Arbeit schon etwas befreundet.

Es freut mich, dass Ihnen das Arbeiten in Leipzig so gut gefällt. Heisenberg muss rührend sein, wenn er um 11 abends noch Unterhaltungen über Arbeiten führt. Wie ist denn Wentzel? Und Sie sind sehr experimentell interessiert mit Arbeiten über Hg. Ich glaube, Sie sind seit langem der erste Physiker, der gleichzeitig experimentiert und theoretisert. Mit meinen Geistesprodukten sind Sie hoffentlich glücklich geworden!

In diesem Sinne herzlichst Ihr

Hans Bethe

302/1571

On the Baden-Frankfurt train, 5.6.[19]28

Dear Peierls!

As you see, I am on a nice trip again and you might think that I am having a good time. But I am not yet sure what to think about this, as my reason for going to Frankfurt now is quite sad: the mother of a friend of mine has died, and I want to try to help a little bit. This will not be very easy, but sometimes it is helpful just to talk to people and not be alone in these moments. And sometimes you need somebody who will encourage you to do something and stop thinking, which is the biggest danger for a person that feels sad. She seems to be quite calm, by the way.

Before this I visited my mother in Baden-Baden to gather my strength. Baden is good for that, as the forest is a nice place to be, and you always think when you are there that you are visiting yourself for a summer holiday. And my mother's cooking is a nice change from Munich's various kinds of roast veal and roast beef labelled under different names. At least I haven't been to Baden without an excuse, because before — apparently to have a reason to go to Baden — I lay in bed for ten days with all kinds of fever (I reached 40.3!). Not even the specialists at Schwabing Hospital could explain what the disease was. So let us call it the flu. Curiously, as a result, I voted from this hospital in Schwabing, but I couldn't fight off the defeat of the democrats regardless.[71] Instead I "won" the Bavarian election, since I voted for the social democrats knowing that the democratic party of Bavaria wouldn't play an important role in the Bavarian parliament.[72] If my aunt in Stettin knew this she would disinherit me, but as she has nothing anyway, it doesn't matter. Besides that I am quite pleased with the German people, except for the rise of the communists. According to advertising pillars in Munich, parliament should have at least 200 national socialists: Hitler put out no less than 10 posters, which a good cartoonist should have illustrated. What a nice picture could have been made to go with the words: "The international stock market Jew is triumphing", for example.

The staff at the institute has hardly changed; only the Americans have disappeared, except for our friend quadruple u (WW) Sleator.[73] But he isn't any skinnier, although he has just published a paper about the Hall effect and some other effect with a melodious name.[74] I enjoy being quoted in other papers, and following the motto "everybody is his own Amanullah" I have now finished my dissertation (in Baden-Baden). I am slightly shocked by the length. 70 pages, and Sommerfeld will also be shocked. I have still learned nothing, but now I will really try to learn something and have little time to do so.[75]

---

[71]The early letters between Bethe and Peierls contain primarily comments on physics, cultural issues or personal remarks. This is one of the few examples of expression of political opinion. These only resurfaced in the mid-1930s when both had settled outside Germany, Bethe in America and Peierls in England. The German General Elections in May 1928 resulted in a strengthening of the Social Democrats, but they failed to gain a majority of seats.

[72]In the Bavarian state elections, which coincided with the General Elections, the Social Democrats increased their share of the vote significantly.

[73]W.W. Sleator, American student of Arnold Sommerfeld's until 1928.

[74]In Munich already, Rudolf Peierls had been interested in physical interpretations of quantum mechanics. In February 1928 he had given a paper entitled: 'Neuere Arbeiten zur physikalischen Deutung der Quantenmechanik (prinzipielle Unschärfe der Beobachtung)'. At Leipzig the quantum theory of magnetism was being developed and Peierls published his important paper on the Halleffekt. R. Peierls, 'Zur Theorie der galvanomagnetischen Effekte', *Z. Phys.* **53**, 255–66 (1929).

[75]Hans Bethe was preparing for his doctoral examination which was due to take place on 24th July 1928.

Last Saturday I met Thorner, who came (after a trip to the Black Forest, with getting up at 3 and other nonsense like that) to Baden-Baden for a neurology conference. He was together with Miss Ster, who is a little fat but nice and funny (which girls are usually not). He seemed to be quite satisfied with Krehl,[76] and has obviously become more familiar with the issues of his thesis.

I am glad that you like your work in Leipzig so much. Heisenberg has to be very nice, if he talks about work at 11 in the evening. How is Wentzel?[77] And I see you are interested in experiments, with your work on Hg.[78] I presume that you are the first physicist in a while who is doing experiments and theory at the same time. I hope that you like my brainchild as well!
Sincerely yours,

Hans Bethe

[10] Hans Bethe to Rudolf Peierls

München, 12.7.[19]28
(postcard)

Lieber Peierls,
Vielen herzlichen Dank für Briefe, Geburtstagsgratulationen und sonstige gute Wünsche. Sie werden aber Verständnis für Doktorvorbe-

---

[76]Hans Thorner was interested in general medicine as well as psychosomatic problems. At Heidelberg he was working with Ludwig von Krehl (1861–1937), director of the Heidelberg Medical School and founder of the so-called 'Heidelberg School' which stood for a more anthropological approach to medicine.

[77]Georg Wentzel (1898–1978), studied at the Universities of Freiburg and Greifswald before moving to Sommerfeld's institute in Munich where he took his doctorate in 1921 and his habilitation a year later; in 1926 he became professor of mathematical physics at the University of Leipzig, in 1928 he moved to the University of Zurich as successor to Erwin Schrödinger. After the war, Wentzel became professor at the University of Chicago where he taught between 1948 and 1970.

[78]While at Leipzig, Peierls took part in the advanced practical laboratory, a rather unusual step for a theorist. One of his tasks in the lab was the measurement of the fine structure of the mercury lines by using a simple spectrometer and a parallel-plate interferometer. See Peierls, *Bird of Passage*, p. 38.

reitungen haben und entschuldigen dass ich Ihnen erst nach dem grossen Ergeignis etwas "Besseres" schreibe. Ich bin sehr fleissig, und finde es sogar ganz nett, mal was zu lernen, weil man so sichtbar vor sich hat, ob man vorwärtsgekommen ist. Übrigens schreiben Sie ruhig mal an Sommerfeld, er wird sich gewiss freuen. Es wirkt nicht aufdringlich, mal was von sich hören zu lassen. Am 24. ist mein grosser Tag.
Bis dahin besten Gruss Ihr

Hans Bethe

Munich, 12.7.[19]28
(postcard)

Dear Peierls,
Thank you for your letters, for the congratulations on my birthday, and all your other good wishes. You will understand that I am preparing for my viva and excuse my writing something "better" only after the big event. I am hard-working, and I even like learning something, as you can notice precisely whether or not you have advanced. You might send a letter to Sommerfeld, by the way, who will certainly be happy about that. It's not too presumptuous to keep in touch every now and again. The 24th is my big day.[79] Till then, I wish you the best.
Yours,

Hans Bethe

[11] Hans Bethe to Rudolf Peierls

München, 25.7.[19]28
(postcard)

Lieber Peierls!
Gestern was der grosse Tag, den ich summa cum laude überstanden habe. Brück hat "magna" gemacht; er hatte Pech, in dem er Sonntag

[79] Refers to oral examination as part of the doctorate.

vorher sich ausgerechnet mit Fieber ins Bett legte. Wien war schwer, ich wusste gerade eben alles Wesenliche ausser einer Frage. Trotzdem bekam ich meine Eins bei ihm. Perron überging alle Fragen, die ich nicht sofort beantwortete, geflissentlich. Sehr anständig! Fajans war an sich am einfachsten und ging am besten. Sommerfeld pflegte eine halbe Stunde Unterhaltung mit mir. Was machen Sie in den Ferien? Ich gehe mit Werner Sachs und einigen anderen etwa zehn Tage in die "rauhen Berge". Adresse am besten B. Baden, Gunzenbachstr. 27.
Herzlichst Ihr

Hans

Munich, 25.7.[19]28
(postcard)

Dear Peierls!
Yesterday was my big day, and I finished *summa cum laude.* Brück[80] got a "magna"; he was unlucky, as he had fever and unfortunately had to stay in bed last Sunday. Wien[81] was difficult, but I knew everything important except for one question. I got a first, anyway. Perron[82] intentionally ignored all the questions I couldn't answer. Very fair! Fajans[83] was the easiest and went best. Sommerfeld just talked to me for half an hour. What are your plans for the vacations? I am planning to go to the "rough mountains" with Werner Sachs and some others for ten days. Send the post to B. Baden, Gunzenbachstr. 27.
Sincerely yours,

Hans

[80]Hermann Alexander Brück studied at Kiel, Bonn and at Munich where Arnold Sommerfeld supervised his doctoral studies on the wave mechanics of crystals. He later turned to astronomy, moved to Cambridge and later to Dublin before becoming Astronomer Royal of Scotland in 1957.

[81]Wilhelm Wien (1864–1928), since 1920 professor of physics at Munich.

[82]Oskar Perron (1880–1975), at the time professor of mathematics at Munich.

[83]Kasimir Fajans (1887–1975), Polish-born chemist who worked in Germany between 1911 and 1935. In the 1930s he was director of the Munich Institute of Physical Chemistry.

[12] Hans Bethe to Rudolf Peierls

[ohne Ort], 12.8.[19]28
(postcard)

Lieber Peierls!
Sie jagen in England den neuesten Errungenschaften der Quantenmechanik nach und analysieren vermutlich Kreiselektronen und sowas. Wir beschäftigen uns mit grösseren Gegenständen wie Bergen, Felsen, Gletschern und Seen, und markieren "wackere Männer". Ich hab's auch verdient. Nebenbei vielen Dank für Ihre wohlgezielten Glückwünsche und viel Erfolg!
Ihr

Hans Bethe. Baden-Baden, Gunzenbachstr. 27

Gruss Werner Sachs, Ilse Sachs, Franz Jacobsohn

[location unspecified], 12.8.[19]28
(postcard)

Dear Peierls!
You are chasing the latest achievements of quantum mechanics in England, and presumably analysing orbital electrons or something like that.[84] We are dealing with bigger objects like mountains, rocks, glaciers, and lakes, and try to be "the brave guys". But I really deserve it. Thank you very much for your hearty congratulations and good luck!
Yours,

Hans Bethe. Baden-Baden, Gunzenbachstr. 27

Greetings from

Werner Sachs, Ilse Sachs, Franz Jacobsohn[85]

---

[84]Peierls spent the summer of 1928 in England where he met, among others, Fowler, Dirac, and Mott.
[85]Franz Jacobsohn, cousin and good friend of Rudolf Peierls.

[13] Hans Bethe to Rudolf Peierls

320/1649

Baden-Baden, 8.9.[19]28

Lieber Peierls,
Eigentlich sind ja deutsche Briefe unerzieherisch für Sie, wenn Sie in England sind, aber ich habe Sie à conto Doktor lange genug warten lassen — also sei's drum! Es ist augenblicklich auch zufällig mal eine Zeit, in der ich einige Zeit habe; im übrigen ist mir nämlich die Vergänglichkeit des menschlichen Daseins immer klarer zum Bewusstsein gekommen, und insbesondere die Vergänglichkeit von Ferientagen. Immerhin habe ich das erhebende Bewusstsein, mich insgesamt um $2{,}7 \cdot 10^{13}$ Å.E. über den Meeresspiegel erhoben und während dieses Sommers eine Steigleistung von 450 000 kg m (bei 75 kg Lebendgewicht) vollbracht zu haben. Dabei meine geographische Kenntnis von Tirol um das Sebraintal, unter Ötztal, Unterengadin, Patznauntal, Montavon, Sasaplana, bereichernd, wozu die zugehörigen Übergänge, blauer Himmel, weisse Gletscher (die zwei von uns zum ersten Mal betreten haben) und volle Hütten dazuzudenken sind, und ebenso eine kleine Sonne, die zu meiner Färbung angestellt war und ihre Sache gut gemacht hat. Bis sie nachher in Braunwals im Kanton Glarus offenbar genug hatte — vielleicht war sie auch von der Wetterhexe, bei der wir zwei Tage Notquartier hatten und die so unverständliche Schweizer Worte in ihren Bart brummte, dass sie alles bedeuten konnten, verhext worden — und sich zeitweise verschleierte.

Zwischendurch konnte man sich aber auch den ganzen Tag auf die Hotelterasse setzen und Studien über Reflexion nicht von Röntgenstrahlen an Kristallen sondern von Sonnenstrahlen an Tödischnee machen, bis in der Reisekasse eben nur noch gerade der Billetpreis war. Dann musste es sein (Dies z.B. war eine Gelegenheit zum Konstatieren der menschlichen Vergänglichkeit) und ich liess zu Hause zum fünften Mal meinen neugeborenen Doktor feiern, der immerhin schon so entwickelt war für sein Alter, dass ihm Sekt und Reh schmeckt und er sogar eine Rede seines Vaters (aber nicht Doktorvaters) zu verstehen imstande ist. Seine Geburt hat aber — ausser fünf Wochen

Vorbereitung, in welcher Frist man, meinen soliden Experimenten zufolge, die gesamte Physik, Mathematik und Chemie zu lernen imstande ist, besonders wenn der hohe Chef so nett ist, einem zu sagen "Lernen Sie nur ja nichts für mich, lernen Sie für Wien und den Mathematiker" — auch genügend gekostet: 215 M Prüfungsgebühren, 25 M Trinkgeld für Schieners und — damit, dachte ich, wars getan, aber es kommt erst: 170 M für die Separata der Arbeit, die auf die Kleinigkeit von 75 Druckseiten angeschwollen ist, aber dafür auch das allerhöchste Gefallen von Sommerfeld und mir selbst erregt. Bei dieser Aufstellung fehlen dann noch die beiden besonders teuren und für Münchener Verhältnisse sogar guten Mittagessen, die ich mir in weiser Erkenntnis der Dinge vor dem Ereignis zuführte, nachdem ich raffinierterweise drei Tage vorher das letzte Buch zugemacht hatte, und ausserdem die obenerwähnte Sommerreise mit ca. 300 M. Summa 720 M = 9,60 M pro Druckseite. Dafür kann man wohl beinahe drei Monate Engländer kennenlernen und — für mich beruhigenderweise — nichts arbeiten, was sicher viel vernünftiger ist als den bischen Doktor zu machen, wo man nachher nur einen ziemlich langweiligen Titel hat, nach dem weder ein Hahn noch ein gewaltiger Gebieter über Stellen kräht. Denn ich habe statt einer Stelle nur einen sehr netten Brief von Ewald, der sich sehr freut, dass ich nach Stuttgart komme, obwohl ich ihm das garnicht geschrieben habe. Ich gehe aber trotzdem wohl hin, und glaube, dass ich sehr viel von ihm habe, besonders wo er persönliches Interesse für mich hat. Einstweilen will ich mich auf dem Naturforscherkongress in Hamburg überflüssig fühlen, da ich dort einen Vortrag zu spät angemeldet habe und dazu nicht mehr zugelassen wurde. Also kann ich meinen Grundsatz "Jeder sein eigener Vortragender nicht durchführen". Nachher will ich in Frankfurt die Mathematik in die Physiologie einführen, und mich bei meinem Vater ebenso überflüssig machen wie vorher in Hamburg und nachher in Stuttgart. Trotz all dem Überfluss "geht es mir aber nach wie vor noch gut genug" um hier mit meiner Mutter in den Hügeln spazierenzugehen und die grossartige Entdeckung zu machen, dass Sonnenuntergänge hier in ganz besonders hervorragender Qualität angefertigt werden. In diesem Sinne herzliche Grüsse

Ihr

Hans Bethe

320/1649

Baden-Baden, 8.9.[19]28

Dear Peierls,
Since you are in England, German letters are actually not very educational for you, but you had to wait long enough because of my doctorate — who cares! This is a time when I happen to have some time; besides I more and more understand the fugacity of human life, especially of vacations. At least I am enjoying the impressive realisation of being at $2.7 \cdot 10^{13}$ Å.E. above sea level, and that I have achieved 450 000 kg m (at 75 kg weight). At the same time I have improved my geographical knowledge of the Sebrain Valley, the Ötz Valley, the Lower Engadin, the Patznaun Valley, Montavon, and Sasaplana. You should also think of paths that are part of the scenery, blue skies, white glaciers (which two of us climbed for the first time) and crowded huts, and of a little sunlight which was turned on because of my tan and it did a good job. In Braunwals, the sun had finally had enough and covered itself up — but it was probably bewitched by a weather witch we had to stay with for two days, who grumbled in such an incomprehensible Swiss dialect that she could have meant anything.

For a change you could sit on the terrace of the hotel all day long and study not the effect of X-rays on crystals, but the effect of sun rays on the Tödi snow. We stayed there until we only had enough money left for a ticket back home. Then it had to end (which was an opportunity to realise that human life is fleeting), and I celebrated my newborn doctorate for the fifth time at home. It has developed so well compared to its age that it liked deer and sparkling wine, and was even able to understand its father's speech (but not that of its doctoral father[86]). The birth of my dissertation was expensive enough! In addition to five weeks of preparation (which according to my experiments is long enough to learn all of physics, mathematics and chemistry, especially if your big boss is nice enough to tell you "Please learn nothing for me, only for Wien and the mathematician!"), it cost 250 M in examination

[86]This is a pun on the German term for Ph.D. supervisor which is 'doctoral father'.

fees, 25 M in tips for Schieners, and (I thought that this was it, but there was more to come) 170 M for the reprints of my thesis, which is 75 pages long and highly appreciated by Sommerfeld and myself. In this calculation the expensive, and by Munich standards even nice, meals have been omitted. I enjoyed them just before the big event after having ingeniously closed the last book three days earlier. In addition to that, the summer holidays I mentioned cost another 300 M. Altogether this adds up to 720 M = 9.60 M per page. That would be enough money to spend three months in England without doing any work — which is good for my conscience — and which would definitely have been more sensible than my little doctoral degree. This is only a rather boring degree that nobody cares about, not even the mighty rulers of acceptable jobs. Instead of a job I received a nice letter from Ewald, who is pleased that I will come to Stuttgart even though I have not told him that I will.[87] I will work with him anyway, and I think that he will be good for me, especially considering that he is personally interested in me. In the meantime I will feel superfluous at the conference of natural scientists in Hamburg, as I applied for a presentation too late and was not accepted. So I can't stick to my principle: "Everybody is his own lecturer". After that I will introduce mathematics into physiology, and I will be as unnecessary to my father as I was earlier in Hamburg and will later be in Stuttgart. "I still feel good enough", however, to walk through the hills with my mother and make the important discovery that the sunsets produced here have an especially brilliant quality.
Sincerely yours,

Hans Bethe

[87]Hans Bethe had accepted a position in Frankfurt at Madelung's institute, but Paul Ewald was interested in him joining the University of Stuttgart.

[14] Hans Bethe to Rudolf Peierls

F[rankfurt], 2.12.[19]28

Lieber Peierls,
Ich bin also jetzt schon so alt und dumm geworden, dass ich für den Staatsdienst geeignet erscheine. Immerhin ist die Art Staatsdienst, die ich hier bei Madelung zu leisten habe, recht harmlos, und meine Befürchtungen, mit dem Doktor wäre die bequemste Zeit des Lebens vorbei, war unrichtig. Im Grunde genommen verlangt Madelung nichts weiter von mir, als dass ich jeden Morgen etwa zwei Stunden mit ihm schmuse, teils über vernünftige Dinge wie Quantenmechanik, teils indem ich einige philosophisch-physikalische Aperçus anhöre, die mir meistens sehr in ihrer Geistesfülle imponieren, aber doch nicht ernstzunehmen sind. M[adelung] ist an sich ein ziemlicher Skeptiker im Gegensatz zu Sommerfeld, und es macht ihm mehr Vergnügen als die ganze Physik, wenn er aus einpaar sowieso schon vagen Hypothesen ein Gebäude von kühnen Aphorismen aufbaut, das er dann sozusagen von unten her ansehen, bestaunen und sich darüber moquieren kann. Dazwischen macht er verblüffend einfache Erfindungen für physikalische Apparate, Elektrometer, Viskosimeter, Bolometer usw., die er mit Vorliebe so erklärt, dass man erst ganz am Schluss merkt, worum es sich im Grunde handelt, und dann natürlich umso erstaunter ist über die Einfachheit des Prinzips. Den grössten Teil des Tages bin ich allein im Institut, was mir nach dem Münchener Betrieb etwas komisch vorkommt, und kann, wenn ich will, einen Kaffee oder einen dancing tea veranstalten. Das Alleinsein hat sich vorläufig übrigens nicht bewährt, ich arbeite dabei weniger als wenn ich mit mehreren zusammen in einem Zimmer sitze, gehe im Zimmer spazieren und denke mitunter garnicht an Physik bzw. Fourierintegrale, mit denen ich mich momentan beschäftige. Es kommt aber recht wenig bisher heraus. Ich möchte für praktische und ev[entuell] auch theoretische Zwecke eine vereinfachte Methode zur Fourieranalyse eines willkürlich gegebenen Kurvenzugs herausbekommen, und ausserdem das Fourierspektrum gewisser willkürlicher Punktanordnungen berechnen, Aber, wie gesagt, es wird noch nichts, und ich bewundere die Leute, die jeden Monat ihr Arbeits-Ei legen.

Daneben muss ich das physikalische Anfängerpraktikum leiten, und staune, trotz minimalster Erwartungen – über die Unkenntnis der einfachsten Dinge. Als neulich ein Mädchen nicht einmal die Hebelgesetze kannte, und zwei lineare Gleichungen mit zwei Unbekannten nicht auflösen konnte, nützte bloss noch Anschreiben. Als ich mitten dabei war, kamen aus dem Nebenzimmer so etwa 10 Leute hereingestürzt, als ob der Teufel hinter ihnen her wäre: Zwei Mädchen hatten Wasser im zugekorkten Gefäss gekocht, und dem Wasser wurde Mazedonien zu klein. Ausserdem stellte man bei der Bestimmung des spez[ifischen] Gewichts nach dem Archimedischen Prinzip regelmässig das Wassergefäss, in das der Körper eintaucht, mit auf die Waage. Der Chef von der Experimentalphysik ist übrigens nicht klug.

Frankfurt hat immer noch den Vorzug, dass ich hier nette Bekannte habe. Für das Arbeiten ist das allerdings ein Nachteil, selbst wenn nach den drei Jahren, die ich weg war, bloss noch drei Leute übriggeblieben sind – die aber umso intensiver. Dafür ist aber auch zur Abwechslung ganz schön, mal bis $\frac{1}{2}$ 5h morgens zu tanzen, wie ich das gestern tat, besonders wenn die Gesellschaft trotz viel Frankfurt-W so nett und amüsierlich ist wie diese es war. Man ist zwar am nächsten Tag ziemlich unfähig, aber schön wars doch. Auch habe ich gefunden, dass Theater mit Freundin schöner ist als allein, und bin derartig snobistisch geworden, dass ich in jede Premiere gehe – Dreigroschenoper, Krankheit der Jugend, Lederköpfe von Georg Kaiser etc. Manche Sachen sind sehr reizend gewesen, davon z.B., "Es liegt in der Luft", was bei Ihnen in Berlin schon im Sommer in der Luft lag. Die Dreigroschenoper hat mir in ihrer Originalität auch sehr imponiert; sie ist etwas so ganz anderes als das Gewohnte und wurde sogar hier in Frankfurt einigermassen annehmbar gespielt. Die "Krankheit der Jugend" (Bruckner) ist schon ein schwieriger Fall, besonders weil das Stück, das wirklich von recht krankhaft veranlagten jungen Menschen handelt, von der Frankfurter Kritik meiner Ansicht nach ganz falsch aufgefasst wird als symptomatisch für die "Jugend von heute". Nicht etwa im anklagenden Sinn der Leute, die den verlorenen Krieg, die böse Inflation und die heutige Jugend als die Grundübel unserer augenblicklichen Lage sehen, und die ich neulich auf einer Gesellschaft in grosser Schar beobachten konnte, wie sie erst die Probleme der heutigen Jugendverderbnis erörterten und sich ach so einig waren, um sich nachher über den Streitpunkt "Auto oder Fussgänger?" gründlichst zu verkrachen. Sondern die Kritiker sind hier auf die Idee gekommen, dass gerade die Besten, Begabtesten und Nachdenklichsten jungen Leute in der Auseinandersetzung mit den Problemen des Lebens besonders mit der Sexualität zu Grundegehen und beim Selbstmord enden. Da kann ich nicht mit. Ebensowenig wie bei Kaisers Lederköpfen, die mir physischen Ekel erregt haben und das will allerhand heissen. Ausserdem war die Aufführung schlecht wie hier meistens.

Die Geschichte vom Hilbert und dem trostlosen jungen Mann, die Sie mir schrieben, hat auch mich einen Tag lang erheitert. Er muss überhaupt ein Original sein: Eines Abends hatte er Gesellschaft, und sollte sich noch eine andere Krawatte anziehen. Nach einer halben Stunde, als alle Gäste da waren, war er noch nicht zurück. Seine Frau suchte ihn und fand ihn – im Bett. Der Reflex des Krawatteausziehens hatte in ihm die Vorstellung "ins Bett gehen" erzeugt. Und das tue ich jetzt auch.
Herzlichst Ihr

Hans Bethe

F[rankfurt], 2.12.[19]28

Dear Peierls,
I seem to have become so old and silly that I now work as a civil servant. At least the kind of job I have to do here for Madelung is quite easy,[88] and I needn't have been afraid that the most convenient time of my life would come to an end with my doctoral exam. Madelung basically expects nothing from me but to chat with him for two hours every morning. Sometimes we discuss reasonable things like quantum mechanics, and sometimes I have to listen to some of his philosophical-physical specialities. Most of these are very impressive and full of wit, but can't be taken seriously. In contrast to Sommerfeld, M[adelung] is quite a skeptic, and instead of discussing physics he has more fun constructing an

[88]Erwin Madelung (1881–1972) had succeeded Max Born as professor of theoretical physics at Frankfurt University in 1921.

edifice of audacious aphorisms based on some vague hypothesis which he can then observe from the outside, gaze at, and complain about. In between he makes surprisingly simple discoveries regarding physical apparatus such as electrometers, viscosimeters, bolometers, etc. He likes to explain all these things in such a way that you only understand at the end what it is all about, which makes the simplicity of the principle seem even more amazing.[89] For most of the day I am on my own in the institute, which coming from Munich feels a bit strange to me as there it was the other way round. If I want, I can have a coffee or a tea party. Being on my own is not good for me, by the way. I work less than I would if I shared the office with others. I walk around the room, and sometimes don't even think about physics or the Fourier integrals that I'm working on at the moment.

Up to now I haven't been very successful with them. For practical (and possibly theoretical) applications I am trying to find an easier method to perform the Fourier analysis of an arbitrary series of curves, and I also want to calculate the Fourier spectrum of an arbitrary arrangement of points. But as I said, I haven't succeeded yet and I wonder about those people that come up with something every single month.

I also have to teach experimental classes in physics for the first-year students, and even though I expect almost nothing I am surprised that they don't even know the simplest things. Recently a girl didn't even know the law of levers, and as she couldn't solve a pair of linear equations with 2 unknowns I had to write down the answer on the board. Just as I was doing that, about 10 people came in from the next room looking mad. Two girls had boiled water in a corked vessel, and for

[89]The remarks demonstrate that Bethe was not very impressed with Madelung. In an interview he later summarised his time at Frankfurt: 'Madelung didn't suggest anything. He suggested that I should paint in India ink some wiggles on paper which one could then use as a grating and Fourier analyse light going through it. It couldn't have worked. It didn't work. And the other thing Madelung was interested in was the different interpretations of Dirac's equation. In fact, he wrote a couple of papers about it and he talked to me about it. It didn't seem to me then, and it does not seem to me now that it added anything to Dirac's equation. So I was entirely on my own.' Interview by Lillian Hoddeson with Hans Bethe, 29.4.1981, Niels Bohr Library, AIP.

the water Macedonia had become too small. When measuring specific weights according to the principle of Archimedes, they regularly put the vessel on a set of scales along with the water into which the object was submerged. The head of the experimental physics department is not very smart, by the way.[90]

Frankfurt still has the advantage that I know nice people here. For my work this is a disadvantage; even though after the three years I was away only three of those people are left, but with them I socialise even more intensively. It is nice to dance until 4:30 in the morning for a change, as I did yesterday, especially when the company is as nice and amusing as it was this time (despite its being from Frankfurt). On the following day you are quite unable to do anything, but it was still nice. I also found out that it is nicer to go to the theatre with a girl-friend than on your own. I am such a snob now that I go to every single premiere — Dreigroschenoper, Krankheit der Jugend, Lederköpfe by Georg Kaiser, etc.[91] Some of them were very nice, for example "*Es liegt in der Luft*[92]", which was in the air in Berlin already this summer. The Threepenny Opera impressed me a lot in its originality; it is something very different from what you already know, and the performance was acceptable even for Frankfurt. "Krankheit der Jugend" (Bruckner) is a difficult case. It is really about young people with a sick disposition, but it was totally misunderstood by the Frankfurt critics as being symptomatic of "the youth of our days". It is not accusing, in the manner of those people who regard the lost war, vicious inflation, and youth as the evils of our time. At a social event I recently had the opportunity to watch several of these people and see how they first discussed the wickedness of our youth, agreeing totally about this, then fell out over the "car or pedestrian?" issue. The critics' interpretation was that the best, most

[90]In 1925 Karl-Wilhelm Meissner succeeded Walter Gerlach as professor of experimental physics in Frankfurt, before becoming professor of astronomy in 1928.

[91]Refers to the Kurt Weill's *Three Pennies Opera* which had its premiere in 1928, and two recent plays, Ferdinand Bruckner, *Krankheit der Jugend* (1926) and Georg Kaiser, *Die Lederköpfe* (1928).

[92]Cabaret by Marcellus Schiffer and Mischa Spolanski captured the spirit of Berlin in the 1920s with its satiric approach to the portrayal of Berlin Society during the Golden Twenties. It was first performed in Berlin in 1928.

gifted, and most thoughtful young people would perish in the struggles of life and sexuality, and finally commit suicide. I don't see that. It was the same with Kaiser's *Lederköpfe*, which caused me physical disgust which is really saying something. Besides that, the performances here were as bad as usual.

The story of Hilbert[93] and the miserable young man that you sent to me cheered me up for a day. He seems to be very peculiar: one evening he was entertaining, and had to put on a new tie. Half an hour later, by which time all the guests had arrived, he still had not come back. His wife went looking for him and found him — in bed. The action of taking off his tie had triggered the thought of "going to bed". And that's what I am doing now.

Sincerely yours,

Hans Bethe

[15] Hans Bethe to Rudolf Peierls

Frankfurt, 17.2.[19]29

Lieber Peierls!

In Göttingen war ich nicht, wie Sie wohl festgestellt haben. Ich wäre wahrscheinlich auch nicht hin, wenn ich nicht erkältet gewesen wäre und in erbittertem Kampf mit der Grippe läge, die ich zu vermeiden versuche. Denn wenn dieser Kampf nicht wäre, hätte ich wahrscheinlich wacker die rauhen Berge des Taunus mit Skiern bestiegen, was sich trotz verbreiteter gegenteiliger Ansichten mangels besserer Tätigkeitsfelder bewährt hat. Der Schwarzwald ist allerdings schöner; jedenfalls schien es mir vom 1.–5. Januar dieses Jahres auf dem Feldbergerhof also, obwohl Petrus mir fünf Sonnentage schuldig blieb. Die liefert er jetzt bei -22 deg, also quasi in entwerteter Valuta, nach (ob wohl eine Aufwertungsklage da Erfolg hätte?).

[93]David Hilbert (1862–1943). At the time professor of mathematics at the University of Göttingen where he stayed until his death in 1943. He had a reputation of being the stereotypical absent-minded professor, and many anecdotes to that effect were being recounted by his students and younger colleagues.

Trotzdem war Göttingen sicher sehr interessant, d[as] h[eißt] die Sitzungen, Göttingen selbst wohl weniger. Viele Sachen schlugen sogar in mein Spezialgebiet, wie Elektronenbewegung, Austrittsarbeit, und so. Aber Göttingen ist weit, und die Anziehungskraft von Kongressen ist direkt proportional den Reizen der Stadt und umgekehrt dem Quadrat der Entfernung.

Meine Produktivität ist inzwischen auf einem Minimum angelangt. Die Fourieranalyse unvollkommen periodischer Funktionen habe ich abgeschlossen, ohne etwas herausbekommen zu haben, was des Publizierens wert wäre. Jetzt versinke ich in Fermischer Statistik, und komme nicht über die unangenehme Feststellung hinaus, dass andere Leute gute Arbeiten schreiben, z[um] B[eispiel] Bloch, und ich nichts herauskriege. Vor allem möchte ich mein mittleres Potential in Ordnung bringen, das Herr Aalpern mir nicht glauben will. Dass es bei hoher Ordnungszahl nicht stimmt, liegt einfach daran, dass die Paulingschen Abschirmungszahlen falsch sind. Also beschäftige ich mich erst einmal mit Abschirmungszahlen, für die ich die Fermische Statistik verwenden will. Dann schwebt es mir vor, auch die Leitfähigkeit etwas quantitativer zu berechnen, als das Bloch getan hat. Aber all das ist nur Programm, und mit einem rohen Ansatz brachte ich bisher z[um] B[eispiel] für das mittlere Potential im Metall immer das Doppelte des beobachteten Wertes heraus. Nebenbei bemerkt, habe ich selbst schon daran gedacht, den Berechnungsindex für Elektronen durch Absorption zu erklären. Das gibt aber immer Indizes < 1, und die Abweichung von eins beträgt ungefähr ein Prozent des beobachteten Betrages. Ich glaube kaum, dass Herr Aalpern damit glücklicher wird.

Mit dem wissenschaftlichen Niedergang ist auch ein persönlicher verknüpft: Ich amüsiere mich neuerdings auf Maskenbällen. Allerdings stand mir ein Aufgebot verschiedener guttanzender und netter Mädchen aus dem Institut meines Vaters zur Verfügung, zu denen ich immer flüchten konnte, wenn es im Gewühl nicht mehr ging. Aber ich hatte den immerhin erstaunlichen Mut, fremde Mädchen aufzufordern, und sogar die bisher immer fehlende Fähigkeit, sie für einige Zeit an meine liebenswürdige Persönlichkeit zu fesseln, sodass ich am Schluss der Unternehmung prophezeien konnte, dass ich mit 30 Jahren die Wissenschaft des Maskenballs beherrschen würde. Im übrigen betätige ich mich nicht viel gesellschaftlich, sondern meist nur im "trauten Kreise" von beiderlei Geschlechts, was Frau Neuberger, die Ihnen ja dem Namen nach bekannt ist, dazu veranlasst, ihren Sohn und mich als unnormal zu bezeichnen, weil wir in der Faschingszeit zu Hause schmusen. Diese Tätigkeit ist mir aber – trotzdem ich heruntergekommen bin – doch immer noch lieber als Faschingsbälle.

Im Sommer gehe ich, wie ich Ihnen vielleicht schon schrieb, doch nach Stuttgart zu Ewald – als Assistant. Werden Sie in diesem Semester fertig? Und was gedenken Sie dann anzufangen? Eigentlich haben Sie ja im Vergleich zu mir noch viel zu kurz studiert, um schon den Doktor machen zu dürfen. In diesem Sinne mit herzlichen Grüssen.
Ihr

Hans Bethe

Frankfurt, 17.2.[19]29

Dear Peierls!
I haven't been to Göttingen, as you might have noticed.[94] I would probably not have gone there, even if I had not had a cold and struggled with a flu that I am really trying to avoid. Without this struggle I would probably have taken my skis and climbed the rough mountains of the Taunus, which have turned out to be a good option for lack of better activities, despite dissenting opinions. But the Black Forest is nicer. At least I thought so in the Feldberg shelter, from the first to fifth of January this year, even though Petrus didn't grant me five sunny days. He is delivering them now at −22 deg, in some sense as devalued currency (what do you think, could a revaluation lawsuit be successful?).

Still, Göttingen (i.e., the meetings) was presumably very interesting, I guess, Göttingen itself, probably not. A lot of the lectures were within my field of research, such as the motion of electrons, the work function, etc. But Göttingen is far away, and the gravitation of conferences is directly proportional to the attractions of their locations and inversely proportional to the square of the distance.

---

[94] Peierls had been to a conference in Göttingen, which Bethe also had intended to attend.

My productivity has reached a minimum.[95] I have finished the Fourier analysis of incomplete periodic functions, with no result worth publishing. Now I'm delving into Fermi statistics, and unfortunately I must confess that while other people are writing good papers (Bloch, for example), I am not finding out anything.[96] I especially want to put my middle potential in order, which Mr. Aalpern doesn't want to believe. The simple reason why it is wrong, for high ordinal numbers, is that Pauling's shielding numbers are wrong. So first I'm dealing with the shielding numbers, which I want to use Fermi statistics for. After that I'm planning to calculate the conductivity in a more quantitative way than Bloch has done. But these are just my plans; for example, on this first attempt my results for the middle potential of metal have always been twice as high as the observed value. By the way, I have thought about explaining the calculation index for electrons as a result of their absorption. But I always come up with indices < 1, and their deviation is about one per cent of the observed value. I don't think that Mr. Aalpern will be any happier with this.

With my scientific decline also comes a personal one: I have started to like masked balls these days. But I had a choice of many different nice and well-dancing girls from my father's institute at any time. I could go and join them whenever I couldn't stand the crowd anymore.

[95] Bethe had taken up a post at the institute of his former teacher Madelung in Frankfurt. He found the research atmosphere stale and unstimulating, and used the time to learn about current developments in physics. (See Bernstein, *Hans Bethe, Prophet of Energy*, pp. 18–19. See also interview by Lillian Hoddeson with Hans Bethe, 29.4.1981, Niels Bohr Library, AIP.) Nevertheless, his own assessment of his work in the letter to Peierls is difficult to sustain. In this supposedly unproductive phase he published a paper on the passage of cathode rays through electric fields formed by grids ('Über den Durchgang von Kathodenstrahlen durch gitterförmige elektische Felder', *Z. Phys.* **54**, 703–10 (1929)), a paper on the comparison of the distribution of electrons in the helium ground state ('Vergleich der Elektronenverteilung im Heliumgrundzustand nach verschiedenen Methoden', *Z. Phys.* **55**, 431–36 (1929)) and a substantial article on the splitting of terms in crystals ('Termaufspaltung in Kristallen', *Ann. Phys.* **3**, 133–208 (1929)) which was widely used by physical chemists.

[96] Felix Bloch (1905–1983) at the time was working with Heisenberg in Leipzig. He had just completed his doctoral thesis which promulgated a quantum theory of solids that provided the basis for understanding electric conduction.

But surprisingly, I displayed the courage to ask girls I didn't know for a dance and even demonstrated an ability to fascinate them with my lovable personality. By the end I could predict that at the age of 30 I would be a master of the science of masked balls. Besides that I don't go to social functions very often and mostly stay in my "familiar circle of acquaintances" of both sexes. This induced Mrs Neuberger[97] (whom you know) to say that her son and I are a bit abnormal, as we stay home schmoozing at carnival time. I still like this activity more than carnival balls — even though I came down here. This summer I'm going to Stuttgart to work with Ewald,[98] as I have probably told you — as an assistant. Will you finish your thesis this semester? And what are you plans? Compared to me, you were actually too fast with your studies to do your doctorate now.[99]
Sincerely yours,

Hans Bethe

[16] Hans Bethe to Rudolf Peierls

Zug Stuttgart-Heidelberg
27.7.[19]29

Lieber Peierls,
also zuerst mal meinen herzlichsten Glückwunsch, dass Sie offenbar Examen und die damit zusammenhängenden Störungen der Behaglichkeit, wie Lernen und so, hinter sich gebracht haben, und jetzt nicht nur berechtigt sind, A-Assistant zu werden, sondern auch alles wieder zu

[97] Mother of a colleague and friend of Hans Bethe's.

[98] Paul Peter Ewald (1888–1985), at the time Professor of Theoretical Physics at Stuttgart University. Hans Bethe's father Albrecht was a good friend and former assistant to Ewald's uncle in Strasburg. In 1939 Hans Bethe married Ewald's daughter Rose.

[99] In the spring of 1929, Heisenberg went on a lecture tour to the United States and Asia, and, at Heisenberg's recommendation, Peierls joined Wolfgang Pauli at the ETH. Here he soon embarked on a study of thermal conductivity in crystals, which was the basis of his Ph.D.

vergessen, was Sie gelernt haben, und das ist mindestens 500 frs pro Monat wert.

Ich scheine gestern meine Tätigkeit als A-Assistant (vielleicht ausser ein paar Tagen im September) abgeschlossen zu haben. Der Abschluss war sehr feierlich und wurde durch Sortieren von einigen hundert Separatabdrucken von Ewald aus den Jahren 1900–1929 festlich begangen, eine Beschäftigung, die lange genug dauerte, um uns ein Bewußtsein äusserster Pflichterfüllung zu verschaffen. Ab 1. Oktober hab ich mich dann an Sommerfeld verkauft – also doch! – und bekam dafür einige Bonbons in den Mund gesteckt, nämlich 1) ein Reisestipendium von einem Jahr, wovon ich nur $\frac{1}{4}$ Jahr nach New York zu gehen brauche – Sommerfeld hat sich also doch meiner Abneigung gegen längeren Aufenthalt in experimenteller Umgebung nicht verschliessen können und im übrigen nach Cambridge und wahrscheinlich Rom. 2) Habilitation zum nächsten Sommer; nachdem ich erklärt hatte, an Stuttgart reizte mich u.a. die Aussicht, mich baldmöglichst habilitieren zu können, d.h. am Ende des SS1930, blieb S[ommerfeld] nichts anderes übrig als das noch ein halbes Jahr früher zu versprechen, 3) die nächste freiwerdende Assistentenstelle in München; das letzte Versprechen ist zwar nichts wert, weil wahrsheinlich so bald nichts frei wird und weil man ausserdem derartige Versprechen vergessen kann. Immerhin war ich ganz zufrieden mit dem, was ich für 130 DM pro Monat erhandelt habe, so sehr, dass ich meine Arbeit über die Austauschenergie bei Ionenkristallen nicht weiter angeschaut habe und infolgedessen auch Ihr Zitat nicht so bald brauche. (Denn jetzt sind Ferien und überhaupt).

Ich habe übrigns Ihrem Mut bewundert, mit so einfachen Annahmen die Ionisierungsspannung zu berechnen! Ich hätte es nie riskiert, die Störungsenergie einfach mit Wasserstofffunktionen zu berechnen, sondern hätte (warum denn einfach, wenns auch kompliziert geht) sicher erst mal versucht, ein self-consistent field auszurechnen usw. Es ist aber eigentlich sehr hübsch, dass man so leicht durchkommt! Ausserdem habe ich mich bei Ihrer Arbeit überzeugt, dass man etwas "leicht mit Gruppentheorie aber noch leichter ohne sie" sehen kann, eben die Koeffizienten der einzelnen Integrale $\mathfrak{J}$ und $\mathfrak{K}$. Ich habe mich übrigens inzwischen auch bemüssigt gesehen, eine "Ionisierungsspannung" auszurechnen, aber eine "negative", nämlich die Elektronen-Affinität des Wasserstoffs nach der Methode von Hylleraas, im Gegensatz von den Resultaten von Brück und Pauling (negative El[ektronen]-Affin[ität] von 80 bezw. 2 kcal.) finde ich +17 mit einem Fehler von ±1 kcal pro Mol H^- (d.h. Grundterm des H^- gleich 1,053 Rh). Haben Sie übrigens gelesen, was der gute Brück neulich für einen, gelinde gesagt Unsinn, über Hyperfeinstruktur fabriziert hat?[100] Es kommt das z.B. das Axiom vor, dass 2 Vektoren vom Betrag $\frac{3}{2}$ und $\frac{1}{2}$ zur Resultierenden 0 zusammensetzen können.

Momentan fahre ich mal auf zwei Tage nach Heidelberg, um den Werner Sachs als neuen Doktor begrüssen zu können. Dann ev[entuell] 1 Tag nach Ludwigshafen zu Mark, der Elektronen beugen und mich zu disem Zweck interviewen will. Ein paar Tage zu meiner Mutter nach Baden-Baden, und von da aus mit ihr in eine sehr verlockend aussehende Gegend, am unteren Ende des Aletsch-Gletschers (Südhang der Jungfraugruppe) wahrscheinlich Hotel Riederalp. (Aber wenn Sie das Bedürfnis nach schriftstellerischer Tätigkeit haben, schreiben Sie nach Baden, Gunzenbachstr. 27). Die Prospekte von da sehen wunderbar aus!

Und was machen Sie bei 308° K (d.h. momentan sind hier höchstens 295) in der mancherlei Reize bietenden Stadt Leipzig? Eben intoniert der Männergesangverein Bruchsal ein sehr lautes Abschiedslied. Da kann ich Ihnen vor Schreck nur noch herzliche Grüsse schicken!

Ihr

H. Bethe

[100]Hermann Alexander Brück studied at Kiel, Bonn and at Munich where Arnold Sommerfeld supervised his doctoral studies on the wave mechanics of crystals. He later turned to astronomy, moved to Cambridge and Dublin before becoming Astronomer Royal of Scotland in 1957.

On the Stuttgart-Heidelberg train
27.7.[19]29

Dear Peierls,
First, congratulations on passing your exams.[101] You seem to have overcome all the disturbances of your comfort, for example learning etc. You are now qualified to become an A-assistant and are also allowed to forget everything you have learned, which is worth at least 500 frs a month.

It seems that yesterday I finished my time as an A-assistant (except probably for a few days in September). The final [task] was very ceremonial: I had to sort about a hundred of Ewald's reprints from the years 1900–1929, an occupation that lasted long enough to strengthen my feelings of deep duty.[102] I have sold myself to Sommerfeld from the first of October onwards[103] — after all! — and I got some sweets for it, which are: 1) funding to travel for one year. I only have to go to New York for three months — Sommerfeld finally accepted that I don't want to stay in experimental surroundings for too long – and after that I'll go to Cambridge and presumably to Rome. 2) A postdoctoral degree (Habilitation) next summer, after I explained to him that the prospect of early habilitation (SS 1930) made Stuttgart particularly attractive, Sommerfeld was left with no choice but to promise it to me half a year earlier still 3) the next available position as an assistant in Munich. The last of these promises counts for nothing, as there probably won't be an opportunity for some time and one can easily forget such promises. After all is said and done, I was so pleased with what I negotiated for 130 M per month that I haven't dealt with my work on the exchange energy of crystal ions[104] and therefore don't need your quotation in the near future. (And we are on vacation now, etc.)

---

[101]On 23 July 1929, Peierls had passed his doctoral examinations.

[102]Bethe had been working as Ewald's assistant in Stuttgart during the summer semester of 1929.

[103]Bethe enjoyed his semester at Ewald's institute, which he later recalled to be 'among the happiest times of his youth' (Bernstein, *Hans Bethe, Prophet of Energy*, p. 19), but Sommerfeld was keen to see him return to Munich and offered him a very attractive package, both financially and in terms of career prospects.

[104]At Stuttgart Bethe met Douglas Hartree who had recently developed the so-called Hartree method which facilitated the approximate determination of properties

I admire your courage in calculating the ionisation voltage! I would never have risked calculating the interference energy simply by using hydrogen functions, and would have instead tried (why use a simple method, if you can do it in a more complicated way) to calculate a self-consistent field, etc. But it is very nice, if you can do it in such an easy way! Reading our work, I was also convinced that you can see something "easily with group theory but even more easily without it",[105] by which I mean the coefficients of the integrals $\mathfrak{J}$ and $\mathfrak{K}$.

I have felt a need in the meantime to calculate a "voltage of ionisation", but a "negative one", that is, the affinity between electrons and hydrogen following the method of Hylleraas.[106] This is in contrast to the results of Brück and Pauling (negative el[ectron]-affin[ity] of -80, or -2 kcal.) I find $+17$, with an error of ± 1 kcal per Mol H^- (i.e., the first H^- term is equal to -1.053 Rh). By the way, have you read the nonsense that our good Brück has recently written about hyper[fine] structures? For example, he claims as an axiom that 2 vectors with values $\frac{3}{2}$ and $\frac{1}{2}$ can be combined to a resultant 0.

I'm travelling to Heidelberg for two days to say hello to the new doctor Werner Sachs. After that I will probably go to Ludwigshafen for one day to see Mark,[107] who wants to diffract electrons and therefore interview me.[108]

---

of atoms. Bethe had already developed an interest in the quantitative description of atomic structures, and now he applied Egil Hylleraas' calculations of electron energy to the negative hydrogen ion.

[105]At the time, theoreticians began to talk about "Gruppenpest", as many were tempted to use group theory more extensively than necessary. Interview by Thomas S. Kuhn and John L. Heilbron with Léon Rosenfeld, 1.7.1963 and 19.7.1963, Niels Bohr Library, AIP.

[106]E.A. Hylleraas, 'Neue Berechnung der Energie des Heliums im Grundzustande, sowie des tiefsten Terms von Ortho-Helium', *Z. Phys.* **54**, 347–66 (1929).

[107]Hermann Francis Mark (1895–1992), "father of polymer science"; at the time director of the research laboratory of high molecular compounds in IG Farben in Ludwigshafen.

[108]Mark's work with Bethe was used in H.F. Mark and R. Wierl, 'Die Ermittlung von Molekularstrukturen durch Beugung von Elektronen an einem Dampfstrahl' *Z. Elektrochemie* **36**, 675–6 (1930); and in H.F. Mark and K.H. Meyer, *Der Aufbau der hochmolekularen organischen Naturstoffe*, Leipzig: Akademische Buchgesellschaft, 1930. Bethe, in turn, used Wierl's and Mark's results on the determination of the

I will spend a few days with my mother in Baden-Baden, and from there go to a highly attractive area near the lower end of the Aletsch glacier (on the southern slopes of the Jungfrau). Presumably Hotel Riederalp. (But if you feel the need to write, you can write to Baden, Gunzenbachstr. 27). The bochures for this area look wonderful!

And what are you doing at 308° K (i.e., at the moment we have at most 295 here) in the attractive city of Leipzig? Right now, the male choir of Bruchsal is starting a goodbye song. Petrified, I can only send you my greetings!
Yours,

H. Bethe

[17] Hans Bethe to Rudolf Peierls

Stuttgart, 11.9.[19]29

Lieber Peierls,
Vielen Dank für Brief und experimentelle Werte der Ionisierungsspannung. Es stimmt ja wirklich recht schön, und wir könnten wohl jetzt einen gemeinsamen Schrieb verfassen — vorausgesetzt dass Sie nicht schon eine zweite Mitteilung losgelassen haben betreffend die Determinantendarstellung der Atom-Eigenfunktion und das Zustandekommen der Koeffizienten der einzelnen Störungsintegrale. Falls Sie also noch nicht publiziert haben, wäre es wohl richtig, wir schreiben jeder unsere Wissenschaft auf und vereinigen sodann beides zu einem unorganischen Ganzen. Die Berechnung der Wechselwirkung der p-Elektronen ist nicht so schlimm; man braucht nur den gegenseitigen Abstand $\frac{1}{r_{12}}$ nach Kugelfunktionen zu entwickleln, durch die Integration über die Winkel fällt dann fast alles weg und es bleibt eine triviale Integration über r_1 und r_2.

atomic form factor with electrons in his article on the theory of the passage of fast corpuscular rays through matter ('Zur Theorie des Durchgangs schneller Korpuskularstrahlen durch Materie', *Ann. Phys.* **5**, 325–400 (1930)).

Meine Bodensee-Pläne sind in den Bodensee gefallen, besser gesagt vom Darm meiner Mutter nicht verdaut worden, der sich zu diesem Zweck in ein Fress-Sanatorium begeben hat und nicht mit mir an den Bodensee reist. Dafür gehe ich Ende September mit 50% Wahrscheinlichkeit nach Berlin, 50% deshalb, weil die Bekannten, zu denen ich will, möglicherweise verreisen. Treffe ich Sie da, oder sind Sie etwa der Ansicht, es gäbe von Prag nach Zürich einen näheren Weg als über Berlin?

Engadin war sehr schön, und ich habe die Genugtuung, hier am Institut am besten auszusehen. Das obwohl ich nur zweimal richtig auf den rauhen Bergen war und sonst nur an diversen hübschen kleinen Seen auf halber = 2600 m ca. Höhe mit schönem Blick auf und durch das Engadin. Aber der kalte Ozon machts wohl. Da die Familie, mit der ich zusammen war, kein Auto besitzt und folglich leidenschaftlich gern Auto fährt, habe ich die Übereinstimmung colossal vieler Gebiete der Schweiz mit dem Baedecker prüfen und bestätigen können — z.B. Oberalppass, Andermatt, Furka, Rhonegletscher, Grimsel, u[nd] a[ndere]. Es hat sich gelohnt.

Viel Spass für Prag!

Herzlichst Ihr

Hans Bethe

Adresse bis 14. Stuttgart, dann Frankfurt/M, Friedrichstr. 42 (bei Frau Dr. Neuberger)

Stuttgart, 11.9.[19]29

Dear Peierls,

Thank you for your letter and the experimental values of the ionisation voltage. The results correspond very nicely, and we could write a paper together — assuming you have not already written a second paper describing the determinants of the atomic eigenfunction and the coefficients of single interference integrals.[109] If you have not published that,

[109] Peierls had worked on ionisation potentials himself. He had written a paper on their dependence on the atomic number 'Über die Abhängigkeit der Ionisierungsspannung von der Ordnungszahl', *Z. Phys.* **55**, 738–43 (1929).

it would be for the best if we each wrote down our results and brought them together in an inorganic ensemble. The calculation for the interaction of p-electrons is not so difficult; you just have to calculate their mutual distance $\frac{1}{r_{12}}$ with the help of spherical harmonics. The angular integration is easy, and then there is just a trivial integration over r_1 and r_2.

My plans to go to Lake Constance have fallen into Lake Constance, or more precisely: they were not digested by my mother's intestine, which prefers to go to a feed-sanatorium instead of travelling with me to Lake Constance. Instead, there is a probability of 50% that I will go to Berlin at the end of September; the probability is only 50% because the friends I want to visit there may probably go traveling themselves. Will you be there, or do you think there is a faster way to travel from Prague to Zurich than via Berlin?[110]

It was very nice in the Engadin, and I felt the satisfaction of looking better than anyone else in the institute. And that's even though I only went to the rough mountains twice! Otherwise I just went to diverse beautiful little lakes at a middling height = 2600 above sea level, with a nice view on and over the Engadin. The reason must be the cold ozone. As the family I stayed with had no car, and therefore had a passion for driving cars, I could test and confirm the agreement of an enormous number of different parts of Switzerland with the Baedeker — for example, the Oberalppass, Andermatt, Furka, the Rhone glacier, Grimsel, a[nd] o[thers]. It was worth it.
Enjoy Prague!
Sincerely yours,

Hans Bethe

address up to the 14th: Stuttgart, then Frankfurt/M, Friedrichstr. 42 (at Dr. Neuberger)

[110]The 1929 Physikertag of the German Physical Society took place in Prague, and Rudolf Peierls took part in the meeting.

[18] Hans Bethe to Rudolf Peierls

München, 15.10.[19]29
(postcard)

Lieber Peierls,
Sie müssen entschuldigen, aber unsere gemeinsame Arbeit, d.h. das, was Sie gedichtet haben, ruht noch sanft in meiner Mappe. Denn ich glaube jetzt den richtigen Weg zur Behandlung der K-Schale zu haben: Ich gehe vom Variationsprinzip aus, gebe für die einzelnen Elektronen die Form der Eigenfunktion vor und lasse nur eine Konstante variabel (Abschirmungszahl) deren Wert sich dann aus dem V[ariations]p[rinzip] ergibt. Aber die numerische Rechnung wird wohl noch 8 Tage dauern. Berlin war sehr hübsch, und ich fühle mich da jetzt schon vollkommen zu Hause. Theater: Zyankali (sehr gut!), Zwei Krawatten, Die beiden Veroneser von Shakespeare in moderner Aufführung, nicht schlecht. Die "andere Seite" schon in Frankfurt. Fabelhaft! Jetzt werden hier aber lauter Sachen gespielt, die ich schon gesehen habe. Trotzdem ist München sehr schön.
Gruss Ihr

Hans Bethe

Munich, 15.10.[19]29
(postcard)

Dear Peierls,
You will have to excuse that our common work (which means what you have composed) is still sleeping in my folder, because I think I have found the right way of dealing with the K-orbital. My solution takes the variational principle as a starting point, where I assign a form for the eigenfunction to every single electron and only one variable is a free parameter (the shielding number). Its value follows from the variational principle. But the numerical calculation will take 8 more days, I'm guessing. Berlin was very nice, and I already feel very much at home

there. Theatre: Zyankali[111] (very good!), Zwei Krawatten,[112] and The Two Gentlemen of Verona by Shakespeare in a modern performance, not bad. The "other side" is already in Frankfurt. Fabulous! Now a lot of things are being played here, which I have already seen. Munich is very nice though.
Greetings. Yours,

Hans Bethe

[19] Hans Bethe to Rudolf Peierls

München, 5.11.[19]29

Lieber Peierls,
Schade, dass Sie aus unserer Allianz ausscheiden mussten! Aber der Ersatz der Gruppentheorie durch einfachere und gleichzeitig elementare Methoden lag wohl in der Luft. Inzwischen bin ich mit der korrigierten Rechnung glücklich zu Ende, nachdem ich mich 10 Tage lang mit Kurbeln an der Rechenmaschine vergnügt habe. Die Abschirmungszahlen der einzelnen Elektronen-Eigenfunktionen werden so definiert, dass das Variationsintegral $\int grad^2\Psi + V\Psi^2$ ein Minimum wird, wobei Ψ nach Heitler aus den Eigenfunktionen der einzelnen Elektronen aufgebaut ist. Dadurch ergeben sich die Abschirmungszahlen ohne jede willkürliche Annahme, gleichzeitig erhält man auch den Eigenwert exakt, und zwar den tiefsten, der bei Annahme von Wasserstoff-Eigenfunktion mit Abschirmung überhaupt möglich ist. Gleichzeitig ergibt sich, dass die London-Heitlersche Störungstheorie dadurch konsequent gemacht wird, dass man einen willkürlichen Parameter z vorweg in die Eigenfunktionen des ungestörten Problems einführt (sodass $\Psi = e^{-z\frac{r}{a}} \cdot \text{const.}$), und dann nicht nur der Abstand der beiden Kerne, sondern auch z so bestimmt, dass der Eigenwert 1. Näherung — der

[111] Friedrich Wolf, *Zyankali* (Potassium Cyanide) (1929).
[112] Felix Basch's production of *Zwei Krawatten* starring Olga Tschechowa had just appeared in the cinemas.

genau mit dem aus dem Variationsprinzip resultierenden Eigenwert identisch ist — ein Minimum wird. Nach dieser Methode hat schon Wang den Grundterm und Kernabstand des H_2 sehr gut erhalten (ich glaube Phys. Rev.). Meine eigenen Rechnungen stimmen dagegen nicht gut mit der Beobachtung (viel schlechter als die wenigen konsequent berechneten früheren, die ich Ihnen seinerzeit schickte), ohne dass meiner Ansicht nach gegen das Verfahren noch etwas einzuwenden wäre. Die Atome sind eben nicht wasserstoffähnlich, und die Näherung durch eine Wasserstoffunktion kann keine gute sein. Für Li (und noch mehr für Be^+ und Be_{++}) stimmt die neue Rechnung erfreulicherweise recht gut, noch besser für den tiefsten Term des Ortho-He und Ortho-Li^+; je weiter man zum Ende der Periode kommt, umso grösser wird die Abweichung (bei Ne 6 Volt berechnet gegen 21 beobachtet!), man hat also quasi eine Berechnung der Wasserstoffunähnlichkeit vor sich. Inzwischen habe ich mich wieder etwas in die Leitfähigkeit versenkt und einen Einwand des Berner Physikers Grunert gegen die Sommerfeldsche Theorie verfolgt. Bei Sommerfeld ist nämlich die freie Weglänge der Elektronen als reine Funktion der Geschwindigkeit and als unabhängig von der Temperatur betrachtet. Infolgedessen darf man eigentlich das Resultat z.B. der Houstonschen Rechnung (vorausgesetzt dass sie richtig wäre) nicht einfach in die Sommerfeldsche Formeln einsetzen, weil bei Houston $l = \frac{f(v)}{\phi(T)}$.

Dies der Grunertsche Einwand. Setzt man nun etwa die Houstonsche Formel bei Sommerfeld ein, so stimmt zwar die Ableitung für die elektronische Leitfähigkeit, wäre ich Ihnen sehr dankbar für das Resultat und, wenn möglich, Ableitung (Wenn nicht, und Sie auch nicht die Absicht haben, würde ich es gern selbst rechnen). Vielleicht kann man von dieser Seite her den Sommerfeldschen Ansatz im fraglichen Punkt korrigieren, oder zeigen, dass man von vorneherein wellenmechanisch rechnen muss, und keine gemischte Houston-Sommerfeld-Rechnung machen darf.

Persönlich habe ich weniger erlebt, obwohl ein paar Bekannte von mir vorhanden sind, sowohl privater wie institutlicher Natur. Am meisten bin ich momentan mit Schur zusammen und politisiere. Morgen macht Unsöld seinen Privatdozenten; er lief in letzter Zeit dauernd im Zylinder herum, der ihm besonders gut steht. Das Institut ist dicht

bevölkert (im Gegensatz zum Sommer), auch Bechert ist wieder gesund und erschienen. Es wird doch intensiver gearbeitet als in Stuttgart! In diesem Sinne mit bayrischem Gruss
Ihr

Hans Bethe

Munich, 5.11.[19]29

Dear Peierls,
What a pity that you had to leave our alliance! But I guess the replacement of group theory by easier, and at the same time more fundamental, methods was in the air. In the meantime I finished the corrected calculation, after having had the pleasure of cranking the calculation machine for 10 days. The shielding numbers of the single electron eigenfunctions are defined in such a way that the variation integral $\int grad^2\Psi + V\Psi^2$ is minimized, while Ψ is set up following Heitler,[113] by combining the eigenfunctions of single electrons.

By this you get the shielding numbers without any arbitrary assumptions and at the same time you find an exact eigenvalue, namely the lowest possible [eigenvalue] of the hydrogen eigenfunction with shielding. The London-Heitler perturbation theory is made consequential if you first insert an arbitrary parameter z into the eigenfunction of an undisturbed problem (so that $\Psi = e^{-z\frac{r}{a}} \cdot \text{const.}$), and then calculate not just the distance of both nuclei but also z in such a way that the eigenvalue's first approximation — which is identical to the eigenvalue resulting from the variational principle — will be minimized. By this method Wang was able to calculate both the basic term and the distance of H_2 nuclei (I guess in *Phys. Rev.*).[114] My own little calculations don't fit this observation very well (much worse than the few more rigorously calculated calculations which I sent you earlier), but I still think this

[113]Walter Heitler (1904–1981), studied at Karlsruhe, Berlin and Munich, where he obtained his doctorate in 1926. After research at Copenhagen, Zurich and Göttingen, he emigrated to Bristol in 1933. In 1941 he became Professor at the Dublin Institute of Advanced Study, and in 1949 he took up the Professorship for Theoretical Physics at Zurich where he remained until 1974.

[114]S.C. Wang, 'The Problem of the Normal Hydrogen Molecule in the New Quantum Mechanics', *Phys. Rev.* **31**, 579–86 (1928).

method is basically all right. The other atoms are not similar to hydrogen, and the hydrogen function approximation can't be a good one. Fortunately, the new calculation is quite good for Li (and even more so for Be^+ and Be_{++}), and even better for the lowest terms of Ortho -He and Ortho -Li^+; the closer you come to the end of the period, the bigger the deviation gets (as observed for Ne, with 6 Volts calculated against 21 observed!), so you have a sort of calculation for an atom's dissimilarity to hydrogen. I have gone back to conductivity again, trying to investigate the Bernese physicist Grunert's objection to Sommerfeld's theories. According to Sommerfeld the range of free electrons is a function only of their speed, and independent of temperature. You therefore cannot just insert the results of the Houston[115] calculation (provided that they are right), for example, into Sommerfeld's formulas because according to Houston $l = \frac{f(v)}{\phi(T)}$.

This is Grunert's objection. If you insert the Houston formula into Sommerfeld's formulas, the derivation for electric conductivity is still correct, as there is no temperature gradient. In the case of heat conduction, however, a catastrophe occurs: the heat conductivity appears in the first approximation, but not in the second, so the conductivity is 1000 times too big (at room temperature). The Wiedemann-Franz law doesn't work at all anymore. The question now becomes, how does this relate to Bloch's theory? Do you know something about this? Or have you just calculated the crystal's heat conductivity? If you have calculated something on the electron heat conductivity, I would be very grateful for this result, and if possible for its derivation (If not, and if you don't want to, I would like to do the calculation myself). You could probably correct Sommerfeld's method at this weak point, or you could show that you have to calculate everything in terms of wave mechanics from the beginning, so that one shouldn't do a mixed Houston-Sommerfeld calculation.

[115]William Verillion Houston (1900–1968), studied at Ohio State University where he obtained his Ph.D. in 1925. He took up a position at Cal.Tech After some research in Munich and Leipzig, he returned to CalTech where he eventually became Professor of Physics. In 1946 he became President of Rice University in Houston.

I haven't experienced much recently even though some of my friends are here, both personal friends and people I know from the institute. I'm spending most of my time with Schur[116] discussing politics. Tomorrow Unsöld will give his lecture to become *Privatdozent*; recently he has been wearing a stovepipe hat, which suits him very well. The institute is rather crowded (in comparison with the summer), and Bechert[117] has also recovered and is back. They work more intensively here than in Stuttgart.
In this spirit and with Bavarian greetings.
Yours,

Hans Bethe

[20] Hans Bethe to Rudolf Peierls

Munich, 17.12.[19]29
(postcard)

Lieber Peierls,
Meinen herzlichsten Dank für Ihr Ski-Angebot, aber leider bin ich schon vergeben und gehe mit meiner Frankfurter Clique nach Sedrun (Oberrheintal). Es scheint mir bei Ihren Plänen nicht ausgeschlossen, dass wir uns mal begegnen. Nach Sedrun will ich noch ein paar Tage nach Arosa. Meine Adresse ab 21. Baden-Baden, Gunzenbachstr. 27, ab 26. Sedrun, Hotel Oberalp, ab 5. Arosa, Hotel? — Der Zeemaneffekt der Seltenen Erden hat sich sehr befriedigend entwickelt, eine Auswahlregel gilt nur insofern als nur Terme mit bestimmter Darstellung der Kristallgruppe kombinieren — sobald aber mehrere Terme zur gleichen Darstellung

[116] G. Schur, student of Sommerfeld's in Munich.

[117] Karl Bechert (1901–1981); after studying mathematics, physics and chemistry at Munich, he completed his Ph.D. there under Sommerfeld in 1925. Following a year's research in Spain under the Rockefeller scheme, he returned to Munich as Sommerfeld's assistant. He later became professor of physics at Giessen (1933–1946) and Mainz (1946–1969). From 1957 to 1972 he served as member of parliament for the SPD.

gehören, und das ist i[m] allg[emeinen] der Fall, gilt für das Impulsmoment um die opt[ische] Axe keine Auswahlregel, woraus sich die grossen Aufspaltungen ohne besondere Annahme über g ergeben. Auch sonst scheint alles in Ordnung. Später mehr! Herzliche Grüsse, auch an Ihren Chef, dem ich für seine danke.
Ihr

Hans Bethe

Munich, 17.12.[19]29
(postcard)

Dear Peierls,
Thank you for your invitation to go skiing, but unfortunately I have already promised to go with my friends in Frankfurt to Sedrun (Upper Rhine Valley). Based on your plans, I suppose it can't be ruled out that we will meet each other one day. After Sedrun I will go to Arosa for a few days. My address from the 21st: Baden-Baden, Gunzenbachstr. 27; from the 26th: Sedrun, Hotel Oberalp; from the 5th: Arosa, Hotel? The Zeeman effect in rare earths has developed in a satisfactory way, a selection rule is needed only if terms with a specific representation of the crystal group are combined — but if several terms belong to the same representation, which is normally the case, then the selection rule is not needed for the angular momentum around the optical axis, which explains the large disparities without the need for any special assumptions regarding g. Aside from this, everything seems to be all right[118] — More later! Greetings also to your boss, I am thankful for his.
Yours,

Hans Bethe

[118] Bethe published these results in H.A. Bethe, 'Zur Theorie des Zeemaneffektes an den Salzen der Seltenen Erden', *Z. Phys.* **60**, 218–33 (1930).

[21] Hans Bethe to Rudolf Peierls

Korrektur folgt baldigst! (Inliegend zum Trost zwei Marken, mit denen ich nichts anfangen kann)

Zug Baden-Baden – München 30.1.[19]30

Lieber Peierls,
Sie werden schon auf mich geschimpft haben, weil ich Ihre Korrektur so lange nicht zurückgeschickt habe. Dafür habe ich Sie populär gemacht, d.h. in München und Stuttgart über Ihre Arbeit im Seminar vorgetragen, die mir durch besondere Freundlichkeit des Verfassers bereits in der Korrektur zugänglich gemacht wurde. Mein hoher Chef ist nach anfänglichem Kopfschütteln von der Wichtigkeit der Umklapp-Prozesse überzeugt und trauert mit dem ganzen Institut dem Wiedemann-Franzschen Gesetz nach. Dass das nicht herauskommt, ist der einzige Fehler Ihrer Theorie, die sonst sehr schön ist, dass man sie sogar beinahe glauben könnte. Aber wo ein Fehler stecken soll, sehe ich nicht. Im übrigen hab ich "for Spass" überall, wo mir ein Druckfehler zu sein schien, eine entsprechende Korrektur vorgenommen – ohne Garantie, aber vielleicht können Sie es doch für die zweite Korrektur brauchen!

In Arosa war es noch sehr hübsch, obwohl ich am Tag Ihrer Abreise bei der Pasern-Tour, die Ihnen vielleicht ein Begriff ist, (Langwies-Watterscheide (2600m) Küblis (400m)) Minkos bekam. Die Abfahrt war nämlich dermassen von laufenden Leuten bevölkert, dass alle zwei Minuten mindestens drei Leute an einem vorbeifuhren, und zwar alle so schnell, dass man schon alleine vom Luftdruck hinfiel, und dann, wenn man sich mühsam erhoben hatte, gleich nochmal vor Schreck über diesen Vorgang sich hinsetzte. Wenn man davon aber abstrahiert, war diese berühmte Tour sehr schön – eine zweite Tour aufs Matlieshorn auf ganz anderem Weg aber noch schöner, während das Parpaner Rothorn uns infolge des Nebels und Schneefalls versagt blieb.

Eben komme ich von einer Tour ohne Skier nach Stuttgart zu meinem früheren Chef, bei dem ich drei Tage lang ungefähr so ununterbrochen geredet habe, wie Sie neulich in Arosa. Hauptsächlich sollte

ich Ewald eine Arbeit von Rosenfeld über optische Aktivität erklären, wobei ich auch wissen sollte, was R[osenfeld] eigentlich für ein Mann sei und ob er irgendeine besondere Funktion in Zürich hätte. Da ich es nicht wusste, soll ich es eruieren. Anschliessend Abstecher nach Frankfurt und Baden-Baden, sowie einen moralischen Kater wegen 5 Wochen Ferien.
Herzliche Grüsse Ihr

Hans Bethe

Grüssen Sie Ihren neuvermählten Chef!

The correction comes soon! (Inside you will find two stamps I can't do anything with.)

On the Baden-Baden – Munich train
30.1.[19]30

Dear Peierls,
You are probably a bit angry with me, as I haven't sent back your corrections for such a long time.[119] But instead I have made you popular; i.e., I reported on your work in seminars at Munich and Stuttgart, which I was allowed to read at its proof stage in advance thanks to the friendliness of the author.

My boss is now convinced of the importance of Umklapp-processes,[120] after first shaking his head, and is now regretting, together with the whole institute, the demise of the Wiedemann-Franz laws. That these no longer hold, is the only mistake in your theory, but it is very nice anyway — so nice that one might nearly think it true. But where its mistake should lie, I haven't found out yet. Incidentally, I corrected all the probable typos "for fun" — with no guarantees, but you might be able to use it for a second correction!

---

[119]Peierls had asked Bethe to comment on a paper on the kinetic theory of thermal conduction in crystals based on his Ph.D.

[120]In his paper, Peierls looks at thermal resistance in crystals as a result of the anharmonic part of the force between atoms. He refined Debye's earlier results (P. Debye, *Vorträge über die kinetische Theorie*, Teubner, Leipzig/Berlin, 1914).

Arosa was very nice at the end,[121] even though I started to get an inferiority complex on the day of your departure, when I did the Pasern tour which you probably know (Langwies-Watterscheide (2600 m above sea level) — Küblis (400m above sea level)). The slopes were so crowded that every two minutes you were passed by at least three people, all of them so fast that you were knocked down by the air pressure alone, and after getting up again you would immediately have to sit down because of the shock. If you don't take that too seriously, this famous tour was very nice — and a second tour to the Matlieshorn by a totally different route was even nicer, but we couldn't go to the Parpaner Rothorn because it was snowing and became too foggy.

I have just come back from a trip without skis to my former boss in Stuttgart where I talked for three days nonstop just as you recently did in Arosa. I basically had to explain Rosenfeld's work[122] on optical activity to Ewald.[123] He also expected me to know what kind of man R[osenfeld] was, and whether he had a special function in Zurich. Since I didn't know, I now have to find that out. After that I briefly went to Frankfurt and Baden-Baden, and have a guilty conscience because of this 5-week long vacation.

Sincerely yours,

Hans Bethe

Greetings to your newly-wed boss![124]

---

[121]Bethe had joined Thorner and Rudolf Peierls on January 5th in Arosa. Peierls had to leave earlier than the other two, because of his work commitments in Zurich.

[122]Léon Rosenfeld (1904–1974), had completed his Ph.D. in Liège in 1926. After studying in Paris (1926–7) and Göttingen (1927–9), he came to Zurich in the summer of 1929. Kart von Meyenn *et al.* (eds.), *Wolfgang Pauli, Briefwechsel mit Bohr, Einstein, Heisenberg u.a.*, 4 volumes, Berlin, Heidelberg, New York, Tokyo: Springer Verlag, 1979 ff (hereinafter cited as *Pauli Briefwechsel* and volume number). Later, he taught in Liège, Utrecht, Manchester and Copenhagen, where he was also a close collaborator and personal assistant of Niels Bohr.

[123]L. Rosenfeld, 'Über die Gravitationswirkungen des Lichts', *Z. Phys.* **65**, 589–99 (1930).

[124]On 23 December 1929, Pauli had married the Leipzig performer Margarete Käthe Deppner.

[22] Hans Bethe to Rudolf Peierls

München, 11.7.1930

Lieber Peierls!
Schönsten Dank dafür dass Sie Herrn Frank eine falsche Theorie geschrieben haben! Ich bin nämlich dadurch auf die Erklärung der anormalen Grössenordnung der magnetischen Widerstandsänderung (Koeffizient B von Frank) gekommen.

Die vollständig exakte Durchrechnung Ihres Falles – Elektron im elektrischen Feld F_x und magnet[ischen] Feld H_z ohne Beeinflussung durch das Gitterfeld, aber mit endlicher freier Weglänge l – finden Sie anliegend; es ergibt sich ein Widerstand, der vollkommen unabhängig ist von Magnetfeld. Dabei ist angenommen, dass alle Elektronen die gleiche Geschwindigkeit v_0 besitzen, und dass nach jedem Stoss alle Bewegungsrichtungen des Elektrons gleich wahrscheinlich und unabhängig von der Bewegung vor dem Stoss sind (Mittelung über ϕ). Der Fehler bei Ihnen war offenbar, dass Sie den Stoss als gleichwahrscheinlich an jedem Punkt der Bahn des Elektrons betrachteten, das bedeutet aber Vernachlässigung von Gliedern der Ordnung $\frac{\beta^2}{h^2} = \left(\frac{\text{Krümmungsradius im Magnetfeld}}{\text{freie Weglänge}}\right)^2$; infolgedessen müssten Sie in Ihrer Formel $\rho = \frac{\frac{1}{\alpha}H^2}{1+\text{const.}\cdot h^2}$ konsequenterweise auch die 1 im Nenner gegen das Glied mit H^2 streichen, und bekommen demnach sicher keine Aussage über den Widerstand in der Nähe der Wendetangente (d.h. $\beta \approx h$).

Dass der Widerstand bei fester Elektronengeschwindigkeit streng unabhängig vom Magnetfeld wird, entspricht der Tatsache, dass die magnetische Widerstandsänderung in der "ersten Näherung" von Sommerfeld verschwindet. Eine Widerstandsänderung ist erst dann zu erwarten, wenn mehrere Elektronensorten verschiedener Geschwindigkeiten sich an der Leitung beteiligen, man bekommt dann in meinen Bezeichnungen für kleine Felder ($h \ll \beta$)

$$\mathfrak{F}_x = enf \int \frac{\beta(v)}{\beta^2(v) + h^2} w(v)dv = enf \cdot \left(\int \frac{w(v)dv}{\beta(v)} - h^2 \int \frac{w(v)dv}{\beta^3(v)} \right)$$

$$\mathfrak{F}_y = enf \left(h \int \frac{w(v)}{\beta^2(v)} dv - h^3 \int \frac{w(v)}{\beta^4(v)} dv \right)$$

wobei $w(v)$ die Anzahl der Leitungselektronen von der Geschwindigkeit v ist. Rechnet man dasselbe etwas strenger nach Sommerfeld durch, so kommt man zum gleichen Resultat; und es wird

$$dv \cdot w(v) = v\frac{df_0}{dv}d\Omega$$

in den Bezeichnungen von Sommerfeld. Wenn nun

$$\overline{\beta^n} = \int \beta^n w(v)dv$$

ist, so folgt in unserer Näherung (d.h. solange man nach H entwickeln darf!):

$$\sigma = \frac{\mathfrak{F}_{x^2} + \mathfrak{F}_{y^2}}{F\mathfrak{F}_x} = \frac{enf}{F}\frac{\left(\overline{\beta^{-1}} - h^2\overline{\beta^{-3}}\right)^2 + \left(h\overline{\beta^{-2}}\right)^2}{\overline{\beta^{-1}} - h^2\overline{\beta^{-3}}}$$

$$= \frac{e^2}{m}n\frac{\overline{\beta^{-1}}^2 + h^2\left(\overline{\beta^{-2}}^2 - 2\overline{\beta^{-1}}\,\overline{\beta^{-3}}\right)}{\overline{\beta^{-1}}\left(1 - h^2\frac{\overline{\beta^{-3}}}{\overline{\beta^{-1}}}\right)}$$

$$= \frac{e^2}{m}n\overline{\beta^{-1}} \cdot \left(1 + h^2 \cdot \frac{\overline{\beta^{-2}}^2 - \overline{\beta^{-1}}\,\overline{\beta^{-3}}}{\overline{\beta^{-1}}^2}\right)$$

Der mit $H^2 \sim h^2$ proportionale Teil enthält also eine typische Schwankungsgrösse: $\overline{\beta^{-2}}^2 - \overline{\beta^{-1}} \cdot \overline{\beta^{-3}}$. Bisher hat man nun die Schwankungen von $\beta^{-1} = \frac{l}{v}$ immer nur soweit in Betracht gezogen, als sie durch die Fermische Verteilung bedingt sind, d.h. durch die kleine Variation der Energie der Leitungselektronen. In Wirklichkeit wird aber, weil die Elektronen sich im Gitterfeld bewegen, die freie Weglänge bezw. β nicht eine reine Funktion der Energie der Elektronen sein, sondern von der Bewegungsrichtung abhängen (von C in der Bezeichnung von Bloch). Das macht nichts aus für die Berechnung des Widerstandes selbst ($\overline{\beta^{-1}}$), aber alle Schwankungsgrössen werden natürlich enorm vergrössert — und das ist gerade das, was wir zur Erklärung der magnetischen Widerstandsänderung brauchen: β schwankt bei fester Energie viel stärker als vermöge der kleinen Energiedifferenzen unter den

Leitungselektronen. Eine quantitative Ausrechnung scheint mir ziemlich hoffnungslos, aber man kann leicht sehen, dass man bei Annahme von vernünftigen Schwankungen der β-Werte für die Leitungselektronen auch zu vernünftigen Grössenordnungen für die Widerstandsänderung kommt. Vielleicht hilft die neue Arbeit von Morse (Phys. Rev. Juni) zu einer etwas quantitativeren Behandlung solcher Fragen. Der Punkt, wo das quadratische Gesetz zu gelten aufhört, wird durch die Nicht-Freiheit der Elektronen nicht berührt, weil er nur davon abhängt, dass die freie Weglänge selbst vergleichbar wird mit der Bahnkrümmung im Magnetfeld, und nicht den Schwankungen der mittleren freien Weglänge bei den verschiedenen Leitungselektronen. Infolgedessen stimmt Franks C und nicht sein B.

Es tut mir leid, dass auf diese Weise nicht nur Frank, sondern auch ich Ihnen ins Handwerk pfusche. Aber man kann nichts dagegen machen, dass einem bisweilen eine Idee kommt!

Übrigens war ich in Leipzig schon ziemlich gespannt auf Landaus Theorie des Diamagnetismus im Elektronen-Kasten. Könnten Sie ihn wohl veranlassen, uns hier etwas darüber zu schreiben? Oder können Sie das selbst tun? Mein Chef interessiert sich glaube ich auch sehr dafür.

In Leipzig verschwand ich ohnehin so sang- und klanglos in Hunds Arbeitszimmer, sodass Sie mir dort nicht bloss den Diamagnetismus sondern auch irgendeinen herrlichen Witz schuldig blieben, den Hund nicht hören sollte. Was machen Sie sonst? Hier ist augenblicklich eine erträgliche Temperatur ausgebrochen, sodass sich leben lässt.
Herzliche Grüsse
Ihr

Hans Bethe

Munich, 11.7.1930

Dear Peierls!
Thank you very much for having sent a wrong theory to Mister Frank![125]
Through this I have found an explanation for the abnormal order of

[125]Nathaniel Hermann Frank (1903–84), postdoctoral fellow from MIT, who later returned to Princeton. Frank was working on electron theory of metals.

magnitude for the change in resistance due to a magnetic field (Frank's coefficient B).[126]

You can find attached the complete and exact calculation of your case: an electron in an electric field F_x and magnet[ic] field H_z, which is uninfluenced by the boundary field and has a mean free path l. The result is a resistance which is totally independent of the magnetic field. This is based on the hypothesis that all electrons have the same speed v_0, and that after a collision all directions of motion are equally probable and independent of the electron's velocity before the impact (average over ϕ). Your mistake was obviously that you thought the collisions should be equally probable at every point along the path of the electron, which means that you neglected all terms of order $\frac{\beta^2}{h^2} = \left(\frac{\text{radius of curvature in the magnetic field}}{\text{mean free path}}\right)^2$. Therefore, in your formula $\rho = \frac{\frac{1}{\alpha}H^2}{1+\text{const.}\cdot h^2}$ you would have to cancel the 1 in the denominator against the term containing the H^2 and then you certainly won't get a result about resistence near the inflection point (i.e., $\beta \approx h$).

The result that resistance is strictly independent of magnetic field strength for electrons with a fixed speed has to do with the fact that the change of resistance due to a magnetic field disappears in Sommerfeld's "first approximation". You only expect a change in the resistance if electrons are moving at several different speeds, or better if electrons

---

Peierls had been working on the increase of electric resistivity of metals caused by a magnetic field. Experiments showed that after an initial gradual rise, resistivity became proportional to the strength of the magnetic field, but all the theories determined the additional resistivity as proportional to the square of the field.

[126] Peierls believed to have found a way out by arguing that in strong magnetic fields the electrons which follow spiral orbits in the magnetic field, would complete several turns and therefore the quadratic law would cease to exist. Shortly before the Leipziger Festspiele, where he was to present his ideas, Peierls realised that, mathematicaly, the different parts cancelled each other and would not serve to explain the situation. See R. Peierls, 'Das Verhalten metallischer Leiter in starken Magnetfeldern', *Leipziger Vorträge: Elektronen Interferenzen*, Leipzig: Hirzel, 1930, pp. 75–85. In writing up the paper, Peierls could insert a note with the correct explanation which he had found, independently of Bethe's calculations in this letter.

with different $\beta = \frac{v}{l}$ take part in the conduction.[127] According to my description, for small fields ($h \ll \beta$) you will find

$$\mathfrak{F}_x = enf \int \frac{\beta(v)}{\beta^2(v) + h^2} w(v)dv = enf \cdot \left(\int \frac{w(v)dv}{\beta(v)} - h^2 \int \frac{w(v)dv}{\beta^3(v)} \right)$$

$$\mathfrak{F}_y = enf \left(h \int \frac{w(v)}{\beta^2(v)} dv - h^3 \int \frac{w(v)}{\beta^4(v)} dv \right)$$

where $w(v)$ is the number of conduction electrons having speed v. If you calculate this more strictly following Sommerfeld's approach, you find the same result:

$$dv \cdot w(v) = v \frac{df_0}{dv} d\Omega$$

in Sommerfeld's notation. If

$$\overline{\beta^n} = \int \beta^n w(v)dv,$$

in our approximation (i.e., as long as you can expand in H!)

$$\sigma = \frac{\mathfrak{F}_{x^2} + \mathfrak{F}_{y^2}}{F \mathfrak{F}_x} = \frac{enf}{F} \frac{\left(\overline{\beta^{-1}} - h^2 \overline{\beta^{-3}}\right)^2 + \left(h \overline{\beta^{-2}}\right)^2}{\overline{\beta^{-1}} - h^2 \overline{\beta^{-3}}}$$

$$= \frac{e^2}{m} n \frac{\overline{\beta^{-1}}^2 + h^2 \left(\overline{\beta^{-2}}^2 - 2\overline{\beta^{-1}}\,\overline{\beta^{-3}}\right)}{\overline{\beta^{-1}} \left(1 - h^2 \frac{\overline{\beta^{-3}}}{\overline{\beta^{-1}}}\right)}$$

$$= \frac{e^2}{m} n \overline{\beta^{-1}} \cdot \left(1 + h^2 \cdot \frac{\overline{\beta^{-2}}^2 - \overline{\beta^{-1}}\,\overline{\beta^{-3}}}{\overline{\beta^{-1}}^2}\right)$$

The part proportional to $H^2 \sim h^2$ has therefore a typical variation: $\overline{\beta^{-2}}^2 - \overline{\beta^{-1}} \cdot \overline{\beta^{-3}}$. Until now you have only looked at variations in $\beta^{-1} = \frac{l}{v}$, insofar as these are contingent on a Fermi distribution; i.e., on small variations in the energy of conduction electrons. In fact, neither the free wavelength nor β will be a pure function of the electron energies because

[127] Handwritten addition: more precisely with different $\beta = \frac{v}{l}$.

the electrons are moving within the boundary field. Rather, they will also depend on the direction of motion (or of C, in Bloch's notation).[128] This has no effect on the calculation of resistance itself, $\left(\overline{\beta^{-1}}\right)$, but all the variations will be enormously enlarged — and that is exactly what we need to explain the variations in the magnetic resistance. β varies much more at high energies than when there is little variation in the energies of conduction electrons. Calculating this quantitatively seems hopeless to me, but you can easily see that if you assume reasonable variations in the β-values of conduction electrons then you also arrive at reasonable orders of magnitude for the change in resistance. For a quantitative solution to your question the latest work of Morse[129] may prove quite helpful (*Phys. Rev.*, June).[130] The point where the square law ends is untouched by the non-freedom of the electrons. The reason for that is simply that the mean free path itself can only be compared to diffraction of the trajectory through the magnetic field, and not to the variations in the average mean free path of different conduction electrons. Franks' C is therefore right, and not his B.

I am sorry that this way I am not only treading on Frank's toes but also on yours. But sometimes one just happens to have an idea!

Incidentally, when I was in Leipzig I grew curious about Landau's theory of diamagnetism in an electron box.[131] Could you possibly ask

---

[128]For Bloch's notation see: F. Bloch, *Über die Quantenmechanik der Elektronen in Kristallgittern*, Dissertation Leipzig, July 1928. A published version can be found in *Z. Phys.* **52**, 555–600 (1928).

[129]Philip M. Morse, (1903–1985) obtained his Ph.D. from Princeton in 1929; he studied in Munich and Cambridge before obtaining a faculty position at M.I.T. in 1931. He conducted distinguished research in physics and scientific administration and was best known for his contributions to operations research.

[130]Philip M. Morse, 'The Quantum Mechanics of Electrons in Crystals', *Phys. Rev.* **35**, 1310–24 (1930).

[131]Lev Davidovich Landau (1908–1968), graduated form Leningrad University in 1927 and obtained his doctorate from Leningrad in 1929. After research at Göttingen, Leipzig, Copenhagen, Cambridge and Zurich, he headed the department of theoretical physics at Kharkov. From 1937 until his death in 1962 he was head of the Theoretical Division at the Institute for Physical Problems in Moscow.

Lev Landau spent 18 months travelling working among others in Germany, Switzerland, and with Niels Bohr in Copenhagen. When he arrived in Zurich in 1930 he had him to write something about that for us? Or could you write something yourself? I'm guessing that my boss is also very interested in this.

In Leipzig I went to Hund's office so quickly that you couldn't tell me anything about diamagnetism, or any great joke that Hund shouldn't hear. What are you doing now? The temperature has become quite acceptable, so life is pleasant.
Sincerely yours,

Hans Bethe

[23] Hans Bethe to Rudolf Peierls

[ohne Ort], 24.7.1930

Lieber Peierls,
Entschuldigen Sie, dass ich Ihren Sommerfrischen-Seelenfrieden in Karersee störe — aber da Herr Guth mir im Verlaufe einer unendlichen, nicht konvergenten, Reihe von Mitteilungen u[nter] a[nderem] auch sagte, dass Sie sowieso Material zum ixen mitgenommen haben, ists vielleicht doch nicht so schlimm.

Also herzlichen Dank für Ihren Brief, aus dem ich entnehme, dass Sie im gleichen Moment die gleiche Erklärung für die Widerstandsänderung im Magnetfeld gefunden haben wie ich. Anscheinend hatte ich mich in meinem mehr langen als durchdachten Brief so unklar wie irgend möglich ausgedrückt, sodass ich nochmal den Versuch mache, Ihnen zu sagen, was ich damals sagen wollte und was Ihnen sowieo bekannt ist, also:

Bei gebundenen Elektronen ist die Energie nicht mehr eine Funktion der Geschwindigkeit allein, sondern hängt ausserdem von der Richtung der Bewegung des Elektrons ab. Infolgedessen lässt sich auch die Geschwindigkeitsverteilung der Elektronen im elektrischen Feld nicht mehr in der Form $f_o(\varepsilon) + \xi \cdot \chi(\varepsilon)$ (ε = Energie, ξ = Komponente der

just completed his theory on diamagnetism which was of great interest to Pauli who himself had developed a theory of paramagnetism due to electron spins.

Geschwindigkeit in der Feldrichtung) schreiben, sondern χ hängt auch wieder von allen drei Geschwindigkeitskomponenten einzeln ab. Dasselbe tut die Sommerfeldsche freie Wellenlänge, die ja direkt proportional zu χ ist (nach Definition). Sie wird also bei gebundenen Elektronen "richtungsabhängig", d.h. hängt ausser von ε von der Bewegungsrichtung ab. Sie sehen, dass meine Erklärung mit Ihrer identisch ist.

Da Sie sich schon immer mit magnetischer Widerstandsänderung befasst haben, während ich die Sache eigentlich bloss per Zufall fand, ist es unbedingt richtig, wenn Sie die Sache publizieren. Andererseits möchte allerdings Sommerfeld gern, dass in der Z[eitschrift] f[ür] Physik im Anschluss an die Franksche Widerstandsänderung bei grossem H auch gleich die Richtigsstellung der absoluten Grösse der W[iderstands] Ä[nderung] bei kleinen Feldern erscheint, entweder von Ihnen oder von mir oder eventuell gemeinsam. Was denken Sie dazu? Es sollte nur eine kurze Notiz sein, die die Sache qualitativ erklärt, ohne tiefe Theorien. Franks Arbeit ist fertig.

Ich reise am 1. VIII nach Frankfurt, Blumenstr. 12, und am 7. mit meiner Mutter in die Schweiz, vermutlich nach Fetan (Adresse Baden-Baden, Gunzenbachstr. 27). Gute Erholung und herzliche Grüsse
Ihr

Hans Bethe

[location unspecified], 24.7.1930

Dear Peierls,
Sorry for disturbing the peace of your vacation in Carezza — but Mr. Guth[132] told me in one of his endless, non-convergent messages that you took some material for ixing[133] with you anyway, so it's probably not too bad of me.

[132]Eugène Guth (1905–), Viennese physicist who was working on the application of the Thomas–Fermi model to positive ions with Peierls. They later published a paper on this problem; R.E. Peierls, E. Guth, 'Application of the Thomas-Fermi model to positive ions', *Phys. Rev.* **37**, 217 (1931). Guth later moved to the University of Notre Dame and worked mainly on polymer physics.

[133]A term used by the young physicists for tedious calculations.

Thank you very much for your letter. I see that you found an explanation for the variation of resistance in a magnetic field at the same moment as I did.[134] It seems as if I was as unclear as possible in my letter, which was more long rather than clever. So I will try again to tell you what I wanted to tell you before, and what you already know anyway:

In the case of bound electrons, energy is no longer just a function of speed but also depends on their direction of motion. The velocity distribution of electrons in an electric field therefore cannot be described as $f_o(\varepsilon)+\xi\cdot\chi(\varepsilon)$ (ε = energy, ξ = component of the speed in the direction of the field); rather, χ depends on all three components of the velocity. Sommerfeld's free wavelength, which is proportional to χ (by definition), has the same behaviour. In the case of bound electrons the wavelength will be "dependent on direction", meaning that it is dependent on both ε and the direction of motion. You can see that my explanation is identical to yours.

As you have always dealt with the effect of magnetic fields on motion, while I just came across this result by chance, it would be absolutely right for you to publish it. On the other hand, Sommerfeld really wants Frank's variation of resistance with big H, followed by a correction to the absolute value of v[ariation] of r[esistance] for small fields, to appear in the *Z[eitschrift] f[ür] Physik*. He wants this to be written either by you or by me, or eventually together. What do you think? It is supposed to be just a short note, which explains the thing quantitatively but without a more profound theory.[135] Frank's work is finished.[136]

---

[134]See Peierls, *Bird of Passage*, p. 59.

[135]Peierls published a paper, though not in *Zeitschrift für Physik*, as Sommerfeld had wished. R. Peierls, 'Zur Theorie der magnetischen Widerstandsänderung', *Ann. Phys.* **10**, 97–110 (1931).

[136]Frank was working on a joint paper with Arnold Sommerfeld. See Correspondence Sommerfeld-Bethe-Frank, 2.11.1930, 27.11.1930 and 15.12.1930, Eckert and Märker, *Sommerfeld Briefwechsel*, II, pp. 313–16; Arnold Sommerfeld and Nathaniel H. Frank, 'The statistical theory of thermoelectric, galvano- and thermo-magnetic phenomena in metals', *Rev. Mod. Phys.* **3**, 1–42 (1930) and N.H. Frank, 'Über die metallische Widerstandsänderung in starken Magnetfeldern', *Z. Phys.* **64**, 650–56 (1930), and he published a paper independently on the same topic; N.H. Frank, 'Bemerkungen zur Theorie der metallischen Widerstandsänderung in einem Magnetfeld', *Z. Phys.* **60**, 682 (1930).

On August 1st I will go to Frankfurt, Blumenstr. 12, then on the 7th to Switzerland with my mother, most likely to Fetan (address Baden-Baden, Gunzenbachstr. 27). Enjoy your time!
Sincerely Yours,

Hans Bethe

[24] Hans Bethe to Rudolf Peierls

Zuoz, 12.8.[19]30
(post card)

Lieber Peierls,
Vielen Dank für Ihren Brief + Manuskript. Ich bin sehr damit einverstanden, und werde in der Arbeit von Frank, deren Korrektur ich bekomme, einen entsprechenden Hinweis zufügen. Ich nehme an, Sie haben noch ein zweites Korrektur-Exemplar und schicke meines an Sommerfeld. Ich hoffe, das ist Ihnen recht.
Herzlichst Ihr

Hans Bethe

Viel Spass in Odessa. Hier kann man auf mehr oder weniger rauhe Berge steigen und sich dabei aufs Abendessen freuen. Adresse bis zum 21. VIII hier, dann postlagernd Meran.

Zuoz, 12.8.[19]30
(postcard)

Dear Peierls,
Thank you for your letter + manuscript. I totally agree, and I will add a note to Frank's work, of which I have received the corrections.[137]

[137]See letter [23]. Bethe communicated with Sommerfeld about the calculations. See letter Hans Bethe to Arnold Sommerfeld, 2.11.1930, Eckert and Märker, *Sommerfeld Briefwechsel*, II, pp. 313–16.

I assume you have the second proof, so I will send mine to Sommerfeld. I hope you are happy with that.
Sincerely yours,

Hans Bethe

Have fun in Odessa.[138]
Here you can climb more or less rough mountains and look forward to dinner at the same time. My address is here up to August 21st, then poste restante Meran.

[25] Hans Bethe to Rudolf Peierls

Rome, 23.4.[19]31

Lieber Peierls,
Sie haben es aber eilig! Meinen allerherzlichsten Glückwunsch! D.h. vorerst hat mir meine Mutter Ihren Brief noch gar nicht nachgeschickt, sondern nur den Inhalt mitgeteilt. Nicht nur verlobt, sondern gleich geheiratet ist eine etwas zu grosse Überraschung. Russin? Ich bin jedenfalls sehr auf "sie" gespannt.

Inzwischen habe ich meinen Wohnort, wie Sie sehen, in den Frühling verlegt, und ich bin begeistert von Rom. Die Stadt kann man immer wieder von neuem ansehen, und findet immer noch etwas Schönes. Ich entwickele dabei einen gewissen Sinn für das Altertum und sehr wenig für Renaissance und Barock-Kirchen, die gar nicht dem entsprechen, was ich mir unter einer Kirche vorstelle. Herrlich (und sehr wesentlich) das Essen. Auch auf der Speisekarte kann man Entdeckungsreisen machen, und 53 Abarten von Maccaroni, Tintenfische, Artischoken schmecken entschieden besser als mutton and cabbage in Cambridge. Der Clou von Rom ist aber unbedingt Fermi, der meiner Ansicht nach in der

[138]The 7th All-Union Congress, which took place in Odessa in August 1930, attracted a great number of scientists from the Soviet Union and other countries.

physikalischen Welt noch lange nicht genug geschätzt wird. Bewundernswert sein Überblick über jedes x-beliebige Problem: er sagt unmittelbar, ob eine Arbeit, die er eben zum ersten Mal gesehen hat, sinnvoll ist oder nicht, und sein Urteil über theoretische wie experimentelle (!) Literatur ist eigentlich unfehlbar. Ausserdem ist er der erste meiner Chefs, die sich richtig um mich bzw. meine Arbeit kümmert, Sie haben es bei Heisenberg und Pauli vielleicht besser gehabt. Jedenfalls habe ich hier wieder Lust an der Arbeit bekommen, während mir in Cambridge vorkam, als ob die noch ungelösten Probleme der Physik uninteressant oder unlösbar wären. Es hat mich gefreut, dass Sie die 2·137 – 1 – Notiz richtig fanden. Manche Leute, vor allem Sommerfeld, fanden, solche Scherze gehörten nicht in ernste Zeitschriften, und es sei eine Herabwürdigung der Naturwissenschaften. Nichtphysiker, denen ich davon erzählte, sind meist derselben Meinung. Immerhin hat es den meisten Physikern gefallen; nur fürchte ich, dass der Erfolg gleich Null sein wird.

Bei der Widerstands-Änderung habe ich in Cambridge ein paar sehr formale Rechnungen gemacht, aus denen ich sah, dass die Sache sehr vom speziellen Modell abhängen muss. Darauf gab ich die Geschichte auf, nachdem ich mich überzeugt hatte, dass aus der Blochschen Theorie bei Entwicklung nach Potenzen von H die Temperaturabhängigkeit

$$\frac{\Delta R}{R} \sim \frac{1}{R^2}$$

folgt, die merkwürdig gut mit den Experimenten stimmt. Sogar Kapitza gab das zu, obwohl er die Ableitung für verkehrt hielt und die ganze Leitfähigkeitstheorie für Unsinn erklärt. Sie sind offnbar doch einiges weiter gekommen als ich. Ich habe, wie gesagt, keinerlei Ambitionen, spezielle Modelle zu rechnen. Momentan habe ich die Termaufspaltung in Kristallen wieder aufgenommen und zunächst die quantitative Berechnung der Terme von Kristall von dem Fall eines Elektrons ohne Berücksichtigung des Spins auf mehrere Elektronen mit Spin ausgedehnt. Dabei kamen nebenbei einige komische Beziehungen zwischen den Multipletts von Atomen der gleichen Übergangsgruppe, z.B. der seltenen Erden, heraus, die einige Punkte der Salpeterschen Multiplettarbeit noch klarer machen. Dann fand ich dass das von Becquerel

beobachtete Absorptionsspektrum aus verbotenen Linien besteht, weil nur Elektronen einer Schale sich umgruppieren; die Übergänge kommen erst durch das Kristallfeld zustande, und die Intensität, die man mit dieser Annahme abschätzt, stimmt mit dem Experiment. Dann fand ich, dass die Zeeman-Aufspaltung bei kubischen Kristallen für gewisse (4-fach entartete) Terme von der Richtung des Magnetfeldes abhängt, ebenso die Polarisatioen aller Zeemankomponenten, was mit neuen Experimenten in Jena stimmt. Jetzt sehe ich zu, wieweit die Annäherung berechtigt ist, die Ionen des Kristalls einzeln zu behandeln, und finde, dass die Ionen mit unabgeschlossenen Schalen nicht zu nahe aneinander liegen dürfen. Dadurch scheiden Metalle von vornherein aus, was vernünftig ist. Ausserdem mache ich in Paramagnetismus der Kristalle. Aber nochmals herzlichen Glückwunsch und beste Grüsse
Ihr

Hans Bethe

Rome, 23.4.[19]31

Dear Peierls,
You are in a hurry! My congratulations! I.e. my mother still hasn't sent me your letter, but just told me what it is about. Not just engaged but even married, which is nearly too big a surprise.[139] Russian? I am very curious to get to know "her" in any case.

In the meantime I have moved my domicile into the spring, as you can see, and I am excited about Rome.[140] You can go to the city over and over again, and always find something beautiful. I am more and

[139] In the summer of 1930, Peierls had attended the 7th All Union Congress at Odessa, where he had met, among others, the young Russian physicist Eugenia (Genia) Nikolaevna Kannegiser. The following spring, Peierls returned to Russia to give some lectures, and on that occasion married Genia in Leningrad. See Peierls, *Bird of Passage*, pp. 63–64, 67–70.

[140] Hans Bethe spent six months of a Rockefeller fellowship in Cambridge, and the subsequent six months in Rome. Two years later, Rudolf Peierls chose the same two locations for his year abroad as part of the same fellowship, but he judged that he improved significantly on the arrangement by swapping the order and staying in Rome in the winter months and moving to Cambridge in the spring.

more enthusiastic about antiquity and less about the Renaissance and baroque churches, which do not correspond to my idea of what a church should look like. Great (and very important) is the food. You can make a discovery tour just through the menu, and the 53 different kinds of macaroni, calamari, and artichokes taste much better than the mutton and cabbage at Cambridge.

But the *Clou* of Rome is definitely Fermi, who I think hasn't been appreciated enough in the world of physics. His ability to summarize any problem is amazing; he can immediately tell whether or not a paper makes sense, even if he has only just seen it for the first time, and his judgement of the theoretical and experimental (!) literature is virtually infallible. Besides that, he is the first boss who's really interested in me and my work; you were probably luckier with Heisenberg and Pauli. At least I feel like working again, whereas in Cambridge it appeared as if the unsolved problems of physics were uninteresting and unsolvable. I was really happy that you enjoyed the 2·137 – 1 – note.[141] Some people, especially Sommerfeld, thought that such jokes do not belong in a serious journal and that the sciences would be denigrated. The non-physicists that I told about it mostly think the same way. But most physicists have liked it. I'm just afraid that it won't make any difference at all.

[141]Hans Bethe, together with Guido Beck and W. Riezler, who had been research fellows at Cambridge together, had listened to Eddington's talk about the 'number 137'. Eddington claimed that he could derive the value of the fine structure constant to be $1/\alpha = \hbar c/e^2 = 137$, starting from Dirac's theory which used a 4×4 matrix for an electron. Eddington assumed that there should be a 4×4 matrix pertaining a proton, and on the basis of the assumption that the two then known elementary particles that, electron and proton, one would need a 16×16 matrix which, if symmetric, would have 136 distinct matrix elements to which 1 was added for the orbital motion of the electron around the proton. The resulting number 137 was close to the then known value of $1/\alpha$. Bethe, Beck and Riezler published 'On the Quantum Theory of the Temperature of Absolute Zero', *Die Naturwissenschaften* **19**, 39 (1931), which made fun of the talk and Eddington's papers on the subject. In a later issue of *Naturwissenschaften*, (6.3.1931) the journal editors published the following 'correction': The note ··· was not meant to be taken seriously. It was intended to characterize a certain class of papers in theoretical physics of recent years which are purely speculative and based on spurious numerical agreements. In a letter received by the editors from these gentlemen they express regret that the formulation they gave to the idea was suited to produce misunderstanding.'

In Cambridge I did some very formal calculations on the variation of resistance, and found that it has very much to do with the specific model. So I gave up the whole thing, after having been convinced that according to Bloch's theory the power series of H follows from the temperature dependence

$$\frac{\Delta R}{R} \sim \frac{1}{R^2},$$

which corresponds surprisingly well to the experimental data.[142] Even Kapitza[143] agreed with this, although he thought that the derivative was wrong and the whole theory of conductivity was nonsense. You obviously seem to have advanced much further.

As I said, I have no ambition to calculate specific models. For the moment I have gone back to the splitting of terms in crystals, and I have progressed in my quantitative calculations from a crystal having one spinless electron to a crystal having several electrons with spin. In doing so I have found some strange relationships between the multiplets of atoms in the same group of transitions, e.g. of rare earths, which make some aspects of Salpeter's work on multiplets even clearer.[144] Then I realised that the atomic absorption spectrum observed by Becquerel consists of forbidden lines, since only electrons from one shell can regroup; the allowed transitions depend only on the crystal's field and its intensity, which you can estimate by supposing that the model agrees with experiment. Then I found that for certain (4-times degenerate) terms the Zeeman splitting in cubic crystals depends only on the direction of the magnetic field. The same is true of the polarisation for all Zeeman components, which agrees with the new experimental results

---

[142]Bethe returned to the problem and published his results in H. Bethe, 'Change of resistance in magnetic fields', *Nature* **127**, 336 (1931).

[143]Pyotr Leonidovich Kapitza (1894-1984), Russian physicist who worked in Cambridge between 1921 and 1934, when after his annual visit to the Soviet Union he was prevented from returning to England. He later founded the Institute for Physical Problems of the Russian Academy of Sciences, Moscow. His refusal to join the Soviet nuclear weapons programme led to his dismissal in 1946; in 1955 he was reinstated as director of the Moscow institute.

[144]See earlier paper of H. Bethe, 'Zur Theorie des Zeemaneffektes an den Salzen der Seltenen Erden', *Z. Phys.* **71**, 205–26 (1930).

from Jena. Now I want to find out how accurate the approximation is that treats the ions of a crystal individually; I am finding that ions with unclosed shells cannot be too close together. Metals are therefore excluded generally, which is reasonable. I'm also dealing with the paramagnetism of crystals.

Congratulations again, and my best greetings.

Yours,

Hans Bethe

[26] Hans Bethe to Rudolf Peierls

München, 12.11.[19]31

Lieber Peierls,

Man soll doch immer alle Annahmen, die man macht, in eine Arbeit schreiben, und nicht, wie ich diesmal, die Hälfte weglassen.

Ich habe nämlich stillschweigend angenommen, dass die Formel (5) nur dann gilt, wenn,

$$m_1 < m_2 < m_1 + N \tag{I}$$

ist, d.h. wenn der Abstand der beiden Rechtsspins kleiner als die Länge der Periode ist. Eine solche einschränkende Bestimmung zu treffen, ist offenbar notwendig, sonst entwickelt man sich mit der Periodizitätsbedingun in Widersprüche – Kostproben dafür haben Sie mir ja genug gegeben. Macht man die Einschränkung (I), so ist alles in Butter: Für $a(m_1, m_2 + N)$ gilt dann die Formel (5) nicht mehr, weil $m_2 + N > m_1 + N$, und es bleibt ausser $a(m_1, m_2)$ selbst nur $a(m_2, m_1 + N)$ als einziges a übrig, dessen Argumente die Bedingung I erfüllen. Damit kommt man zwangsläufig auf meine Randbedingung (10).

Wenn meine Periodizitätsbedingung (10) erfüllt ist, gilt offenbar auch die Gl[eichung] (4b) auch dann, wenn $m_2 - m_1 = N - 1$, d.h. falls der zweite Spin unserer Periode neben dem ersten Spin der nächsten

Periode der Kette liegt. Die Gültigkeit von (4b) in diesem Fall zu verlangen, erscheint doch notwendig, gleichgültig ob man einen geschlossenen Ring von Atomen hat oder eine Kette mit aufgezwungener Periodizitätsbedingung – warum Sie einen Unterschied zwischen diesen beiden Fällen konstruieren, habe ich nicht recht verstanden.

Die Einschränkung (I) für die m's zu treffen, scheint mir nicht nur notwendig, sondern auch vernünftig: Man darf an jeder beliebigen Stelle der Ketten anfangen, die Rechtsspins aufzuzählen, muss sie dann aber in der Reihenfolge aufzählen, wie sie aufeinanderfolgen, und darf nicht zwischendurch einen auslassen. Allerdings muss ich nach alledem zugeben, dass meine Grenzbedingung (3) irreführend oder sogar direct verkehrt ist, man darf tatsächlich nur fordern

$$a(m_1 m_2 \ldots m_r) = a(m_2 \ldots m_r m_1 + N)$$

wenigstens, wenn man will, dass beide a's durch die geschlossene Formel (5) bezw. (25) gegeben sind. Ich glaube nicht, dass Ihnen meine Randbedingung (10) jetzt noch so künstlich erscheint. Ausserdem ist es durchaus notwendig, eine scharfe Randbedingung zu stellen und nicht zu sagen, dass es auf die Abzählung erst in zweiter Linie ankommt: Wenn man z.B. den tiefsten Eigenwert im nicht ferromagnetischen Fall $\mathfrak{F}$ ausrechnet (der ja einem Gesamtspin Null entspricht), so gehen die Phasen ϕ_{ik}, die ja eben durch die Periodizitätsbedingung hereinkommen, ganz wesentlich in das Resultat ein. Sie bewirken nämlich, dass die Wellenzahlen f_i der Spinwellen sich alle um $f_i = \pi$ herum anhäufen, und da eine Spinwelle im Fall $\mathfrak{F} < 0$ umso weniger Energie hat, je näher ihre Wellenzahl an π liegt ($E = \mathfrak{F}\ (1 - \cos f_i)$) so wird dadurch die Energie wesentlich herabgesetzt. Es kleben also nicht nur die Spins zusammen, sondern auch die Wellenzahlen der Spinwellen! (Übrigens ist der tiefste Eigenwert für $\mathfrak{F} < 0$

$$e = E_0 + N\mathfrak{F}\,(\lg 4 - 1) = E_0 + 0{,}386N\mathfrak{F} \qquad \text{(II)}$$

Slater fand $e = E_0 + 0{,}293N\mathfrak{F}$, also erheblich zuviel. (II) ist exakt bis auf Grössen der Ordnung $\mathfrak{F}$.)

Noch ein paar Kleinigkeiten zu Ihren "Angriffen" gegen meine Periodizitätsbedingung: Die komplexen Wellen befriedigen selbstverständlich auch die Periodizitätsbedingung. Denn so ist ja gerade das $\varphi = \psi + i\chi$ nach (16) gewält. Die a's sehen etwa so aus (siehe (23))

$$a(m, m_2) = e^{\frac{1}{2}ik(m_1+m_2)} \cdot \cos(\frac{1}{2}N - (m_2 - m_1)) \qquad \text{(III)}$$

Die Amplitude a fällt also exponentiell ab, wenn m_2 sich von m_1 entfernt, erreicht ein Minimum für $m_2 - m_1 = \frac{1}{2}N$ und steigt dann wieder wenn m_2 sich dem ersten Spin der nächsten Periode der Kette $m_1 + N$, nähert. Für $m_2 = m_1 + N - 1$ bekommt man genau dieselbe Amplitude wie für $m_2 = m_1 + 1$. (III) gilt übrigens nur für gerade k, für ungerade tritt ein hyperbolischer Sinus an die Stelle des Cosinus. Man kann das alles leicht nachrechnen, wenn man (16) in (5) einsetzt.

Warum ich in (11) $\lambda_1 = \lambda_2$ ausschliesse? Weil sich (11), (8) dann überhaupt nicht lösen lässt. Wenn $f_1 = f_2$, so ist unweigerlich $\phi = \pm\pi$, d.h. sowohl für $\lambda_1 = \lambda_2 + 1$ wie $\lambda_1 = \lambda_2 - 1$ führen zu $f_1 = f_2$. Ist $f_1 > f_2$ so muss $\lambda_1 > \lambda_2 + 1$ sein, und für $f_1 < f_2$ muss $\lambda_1 < \lambda_2 - 1$ sein (ausgenommen für die ganz kleinen $f < \frac{1}{\sqrt{N}}$, für die die komplexen Lösungen in reelle degenerieren). Jedenfalls gibt es überhaupt keine reellen Werte $f_1 f_2$, für die $\lambda_1 = \lambda_2$ werden kann — bei den Komplexlösungen kommt dann ja (für gerade k) $\lambda_1 = \lambda_2$ vor. Schliesslich noch die Feststellung, dass man die komplexen Lösungen bei der Abzählung mitnehmen muss — ich bin sogar nur dadurch darauf gekommen, dass die reellen Lösungen nicht die nötige Anzahl gaben. Für Gesamtspin Null ($r = \frac{N}{2}$) bleibt sogar nur eine einzige Lösung mit reellen Wellenzahlen, wegen des Verbots, dass nie zwei Wellenzahlen gleich sein dürfen (diese Lösung führt bei $\mathfrak{J} < 0$ auf den tiefsten Eigenwert), praktisch alle Lösungen sind also komplex. Ihren Einwand, meine komplexen Lösungen befriedigten die Perioditätsbedingung (10) nicht, habe ich ja schon oben besprochen. Ihre Umschreibung meiner Rechnungen gefällt mir sehr gut. Besonders erkennt man viel besser als bei mir, warum es bei gegebenem k nur einen diskreten Eigenwert gibt und warum $f_1 = f_2$ auszuschliessen ist — dass das eben an der "Randbedingung" für $l = m_2 - m_1 = 0$ liegt und nicht etwa an der Perioditätsbedingung, wie es bei mir aussehen könnte. "Aneinenderkleben" tun die Spins übrigens natürlich nur bei der komplexen Lösung, im allgemeinen (bei den reellen Wellen) ist sogar eine grössere Entfernung wahrscheinlich, weil im Mittel über alle reellen und komplexen

Lösungen jede Entfernung zwischen m_1 und m_2 gleich wahrscheinlich sein muss.

Ihre Ersatzperiodizitätsbedingung $u(l_0) = 0$ geben Sie vielleicht mir zuliebe auf, weil Sie sich von der mathematischen Eindeutigkeit meiner Bedingung überzeugt haben. Wenn nicht, stets gern zu weiteren Diensten.

Ob man mit Ihrer Schrödingergleichung beim dreidimensionalen Fall etwas anfangen kann, weiss ich nicht. Die grosse Schweinerei ist nämlich, dass (schon bei zwei Dimensionen) sich die Definition der a's für den Fall zweier Rechtsspins am gleichen Atom nicht mehr willkürfrei machen lässt: Definiert man $a(m_1 m_2)$ einmal so, dass die Gleichung, die (4b) entspricht, befriedigt ist, wenn m_2 rechts neben m_1 liegt, das andere mal für m_2 über m_1, so widersprechen sich die beiden Definitionen. Dann kriegen Sie auch keine Grenzbedingung für Ihre Schrödingerfunktion mehr. Aber ich muss mir die Sache nochmal überlegen.

Hoffentlich rechnen Sie bei der Lichtabsorption in festen Körpern nicht zuviele Sachen, die ich in Rom gerechnet habe. Aber ich habe mich ja immer auf die Atome und Ionen mit unabgeschlossenen Schalen kapriziert, sodass wir uns wohl nicht ins Gehege kommen. Natürlich würde es mich interessieren, was an nicht-trivialen Dingen herauskommt. Ich schwitze an einem Handbuchartikel über H und He und komme garnicht vorwärts, teils aus eigener Insuffizienz, teils weil mich alle fünf Minuten jemand stört. Ausserdem merke ich jetzt nach meiner "Weltreise" erst richtig, dass hier im Institut meist nur Mathematik gemacht wird und keine Physik.

Kann denn irgend jemand von uns, d.h. von Ihnen "selbst wirtschaften"? Ist es möglich, in St. Antonien Lebensmittel zu kriegen, oder steht eine Hungersnot zu befürchten? Glauben Sie dass das billiger ist als eine Pension? Kurz gesagt: An sich gerne, aber ich mag mich noch nicht binden. Ich habe auch nebenbei noch ein zweites "Projekt" Sachs laufen, dessen Entwicklung ich noch abwarten muss. Ihre Dispositionen sind ja wohl auch unabhängig von der genauen Anzahl der Beteiligten. Es wäre aber nett, uns auf die Weise endlich wiederzusehen. Grüssen Sie Ihre Frau, die Sie doch überraschend schnell herausgekriegt haben, Ihren Chef und sich selbst von Ihrem

H[ans] Bethe

Munich, 12.11.[19]31

Dear Peierls,
In any work one must write down every single assumption, and not leave half of it out as I did this time. I have implicitly assumed that formula (5) applies only if

$$m_1 < m_2 < m_1 + N, \tag{I}$$

in other words if the difference between the two right spins is smaller than the period. Such a restricted definition is necessary so as not to run into problems with the periodicity condition — you have given me enough examples of that. If you adopt this condition (I), then everything is all right: For $a(m_1, m_2 + N)$ formula (5) is no longer correct because $m_2+N > m_1+N$, and except for $a(m_1, m_2)$ only $a(m_2, m_1+n)$ remains as the only a, whose arguments fulfil condition I. By this you come to my basic condition (10).

If my condition of periodicity (10) is fulfilled, then equation (4b) is evidently correct, even if $m_2 - m_1 = N - 1$, i.e. if the second spin of our period lies next to the first spin of the next period in the chain. The validity of (4b) is necessary in this case; it makes no difference whether you have a closed ring of atoms or a periodic chain — and why you construct a difference between these two cases, I haven't really understood.

To me it seems not just necessary but reasonable to adopt condition (I) for the m's; You can start counting the right spins at any point along the chain, but you have to count them one after the other and you are not allowed to leave out a single one. But I have to confess, after all, that my limiting condition (3) is misleading or even wrong; you can only demand

$$a(m_1 m_2 \cdots m_r) = a(m_2 \cdots m_r m_1 + N),$$

at least if you want both a's to be given by the closed formulas (5) and (25) respectively. I don't think that my other limiting condition (10) will appear so artificial any more. Furthermore, it is absolutely necessary to adopt a strict limiting condition, and not to claim that the counting is of secondary importance. If you calculate the lowest eigenvalue $\mathfrak{F}$ in a non-ferromagnetic case (which corresponds to a total spin of zero), for example, then the phases ϕ_{ik} which are taken into account because of the periodicity conditions, enter into the result in a significant way. They have the effect that the wave numbers f_i of the spin waves all accumulate around $f_i = \pi$, and as the energy of a spin wave in the case $\mathfrak{F} < 0$ decreases as its wavelength approaches π ($E = \mathfrak{F}$ $(1 - \cos f_i)$), so the energy is substantially reduced. Not only do the spins stick together, but so do the wave numbers of the spin! (By the way, the lowest eigenvalue with $\mathfrak{F} < 0$ is

$$e = E_0 + N\mathfrak{F}(\lg 4 - 1) = E_0 + 0,386N\mathfrak{F} \qquad \text{(II)}$$

Slater[145] found $e = E_0 + 0,293N\mathfrak{F}$, which is much too large. (II) is correct, except for eigenvalues of order $\mathfrak{F}$.)

Some comments on your "attacks" on my periodicity condition: of course the complex wavefunctions satisfy the periodicity conditions, as $\varphi = \psi + i\chi$ is chosen according to (16). The a's look like this (see (23))

$$a(m, m_2) = e^{\frac{1}{2}ik(m_1+m_2)} \cdot \cos\left(\frac{1}{2}N - (m_2 - m_1)\right) \qquad \text{(III)}$$

The amplitude of a declines exponentially as the difference between m_2 and m_1 increases, reaches its minimum at $m_2 - m_1 = \frac{1}{2}N$, then increases again as m_2 approaches the first spin of the next period in the chain $m_1 + N$. For $m_2 = m_1 + N - 1$, the amplitude is exactly the same as for $m_2 = m_1 + 1$. (III) is correct only for even values of k, and for odd k the cosine is replaced by a hyperbolic sine. You can easily calculate this for yourself by using (16) in (5).

Why do I exclude $\lambda_1 = \lambda_2$ in (11)? Because it would then be impossible to solve (11), (8). If $f_1 = f_2$ then $\phi = \pm\pi$, which also means that $\lambda_1 = \lambda_2 + 1$; $\lambda_1 = \lambda_2 - 1$ similarly leads to $f_1 = f_2$. If $f_1 > f_2$,

[145] John Clarke Slater (1900–1976) received his Ph.D. from Harvard in 1923, he worked with Niels Bohr and Werner Heisenberg. In 1930 he became head of physics at MIT, where, among others, he worked on the electromagnetic theory of microwaves. Slater had just published a paper with some tentative suggestions about the links between the superconducting states of metallic electrons and the perturbation theory of Bloch's theory. J.C. Slater, 'The Nature of the Superconducting State', *Phys. Rev.* **51**, 195–202 (1937).

so λ_1 has to be $> \lambda_2 + 1$, then for $f_1 < f_2$ we must have $\lambda_1 < \lambda_2 - 1$ (except for very small $f < \frac{1}{\sqrt{N}}$, in which case the complex solutions degenerate into real ones). In any case, there are no real values f_1, f_2 for which $\lambda_1 = \lambda_2$. For complex solutions (in the case of even k) we can have $\lambda_1 = \lambda_2$. Finally the comment that you have to take the complex solutions into account when you do the calculation — I found this out only because there were not enough real solutions. For a total spin of zero ($r = \frac{N}{2}$) there is just one solution with real wave numbers, because of the requirement that two wave numbers can never be the same (this solution corresponds to the lowest eigenvalue in the case of $\mathfrak{F}$), so nearly all the solutions are complex. I have already discussed your objection that the complex solutions wouldn't satisfy the periodicity condition (10). I really like your transcription of my calculations. It makes it easier to understand why there is only one discrete eigenvalue for any given k, and why $f_1 = f_2$ has to be excluded — it has to do with the "limiting condition" for $l = m_1 - m_2 = 0$, and not with the periodicity condition as my description makes it seem. The spins only "stick together" in the case of a complex solution. In general (as is the case in real waves) greater distances are more probable, because on average, considering all real and complex solutions, every value for the distance between m_1 and m_2 has to be equally probable.

You will probably give up your extra periodicity condition $u(l_0) = 0$, now that you have been convinced by the mathematical uniqueness of my condition. If not, I am always at your service.

I don't know whether one can do something with your Schrödinger equation in the case of a three-dimensional field. The big mess is that one is not allowed (even in the case of two dimensions) to arbitrarily define the a's for two right spins in the same atom. Once you define $a(m_1 m_2)$ in such a way that the equation corresponding to (4b) is satisfied, if m_2 lies to the right of m_1, the other time for m_2 above m_1, then the two definitions are contradictory. And then you no longer have any limiting condition for your Schrödinger function. But I have to think about this some more.

I hope that when working on the absorption of light by solid bodies you won't spend too much time calculating things which I have already done in Rome. But I concentrated on atoms and ions with unclosed shells, so there shouldn't be any problem. I'm interested to know whatever non-trivial things you've found out. I am busy with an article on H and He that I'm writing for a *Handbuch*, on which I am not making any progress at all,[146] partly because of my own insufficiency and partly because I am disturbed every five minutes. After my "world tour"[147] I now realise that here at the institute they are only doing mathematics, not physics.

Are any of us, by which I mean of you, capable of "keeping house on his own"? Is it possible to get some food in St. Antonien, or do we have to fear famine? Do you think this is less expensive than a B& B? In short: I would like to, but I don't want to commit myself yet. I am working on another "project" Sachs and I have to see how it develops. Your arrangements are independent of the exact number of guests, I hope? It would be nice if we finally met this way. Remember me to your wife, whom you managed to get out [of Russia] surprisingly fast, and to your boss, and to yourself.
Yours,

H[ans] Bethe

[27] Hans Bethe to Rudolf Peierls

München, 30.11.[19]31

Lieber Peierls,
Ich komme also mit nach St. Antonien, Familie Thorner wird dann wohl auch einverstanden sein. Ist es Ihnen recht, wenn ich Erwin Strauss noch auffordere? Oder haben Sie keinen Platz mehr? Wie sind die Einzelheiten Ihrer Pläne — ich möchte Weihnachten zu Hause, d.h. in

---

[146] Initiated by his teacher Sommerfeld, Hans Bethe was writing the core of a substantial review article to be published under joint authorship of Sommerfeld and himself. Sommerfeld, A. and H. Bethe, 'Electron Theory of Metals', *Handbuch der Physik* **24**, 2 (1933).

[147] Refers to his year abroad as part of his Rockefeller fellowship, when he spent six months in Cambridge and six months with Fermi in Rome.

Baden-Baden sein, und schlage 26. als Reisetermin vor (was nicht hindert, dass Sie schon frueher). Endpunkt der Unternehmung von meiner Seite aus ziemlich unbegrenzt. Dass Morse-Allies "grundfalsch" ist wissen wir hier tatsächlich nicht. Es wundert einen zwar, dass es so gut stimmt, obwohl Austausch und Polarisation vernachlässigt sind (die Polarisation ist jedenfalls noch wichtiger), aber die Schematisierung des Potentialfeldes des Atoms ist doch wohl ganz vernünftig, und die Rechnung als solche scheint mir in Ordnung. Bitte belehren Sie uns! Sonst rechnet hier alles drauf los, ohne irgendeine moralische Berechtigung dafür zu haben. Schade dass bei der Verallgemeinerung der Metalltheorie auf zwei Dimensionen nichts rauskommt.
Herzlichen Gruss
Ihr

H[ans] Bethe

Munich, 30.11.[19]31

Dear Peierls,
So, I will come to St. Antonien. The Thorner family will agree to that, I presume. Would it be all right if I also ask Erwin Strauss? Or is there no more room? What are the details of your plans — I would like to be at home for Christmas, which means in Baden-Baden, and I would propose leaving on the 26th (but you could also leave earlier). For me the end of this little trip is quite open. Here, we really don't know that the Morse-Allies result[148] is "totally wrong". One wonders why it corresponds so perfectly, even though exchange and polarisation have been neglected (in any case, the polarisation is more important), but the schematisation of the atomic potential is quite reasonable and the calculation itself seems to be all right. Please teach us! If not, everybody will start calculating without any moral justification. What a pity that the generalisation of the metal theory to two dimensions hasn't brought any result.
Sincerely yours,

H[ans] Bethe

---

[148] W.P. Allis and P.M. Morse, 'Theory of the scattering of slow electrons by atoms', *Z. Phys.* **70**, 567–82 (1931).

[28] Hans Bethe to Rudolf Peierls

München, 16.12.[19]31
(postcard)

Lieber Peierls,
Erwin Strauss kommt wahrscheinlich nach Neujahr zu uns. Thorner will noch ein Mädchen aus F[rankfurt] mitbringen, nebenbei eine früher heftige Freundin von mir. Mein Chef hat die Vor-Ferien-Arbeitswut, unter der ich auch zu leiden habe.
Ihr

Hans Bethe

Das Wichtigste vergessen: Er lässt mich erst zwei Tage später fort als ich wollte, sodass ich meiner Mutter versprechen musste, erst am 27. (Sonntag) zu Ihnen zu fahren.

Munich, 16.12.[19]31
(postcard)

Dear Peierls,
Erwin Straus will probably join us after New Year's Day. Thorner wants to bring along a girl from F[rankfurt], who used to be a passionate friend of mine by the way. My boss is working like crazy before the vacations, which makes me suffer too.
Yours,

Hans Bethe

I've forgotten the most important thing: He just asked me to leave two days later than I wanted, so I had to promise my mother that I wouldn't join you before the 27th (Sunday).

[29] Rudolf Peierls to Hans Bethe

Cambridge, 14.6.[19]33

Lieber Bethe!
Ich habe Ihnen immer noch nicht auf Ihren Brief geantwortet, hauptsächlich in der Annahme, Ihnen demnächst irgendwelche offiziösen Mitteilungen in der lecturer-Angelegenheit machen zu können. Aber die Entscheidung hierin geht erstens beliebig adiabatisch vor sich und zweitens ist hier nichts herauszubringen, sodass Sie vermutlich mehr darüber wissen als ich. Dem Vernehmen nach soll "einer von Ihnen" eine Stelle bekommen haben, aber ob Sie das sind, oder ob das eine lectureship ist oder eine sehr schofel bezahlte assistant l[ectureship] weiss ich nicht. Ich stehe jedenfalls nicht in der engeren Wahl.

Was haben Sie sonst fuer Ideen? Ich habe ziemliche Aussichten auf ein Stipendium von Langevin, und mache Ihnen ausserdem Konkurrenz in Belfast. In letzterem scheinen allerdings die Aussichten fuer Sie und mich ziemlich infinitesimal zu sein, da es sich um eine Stelle mit viel Verwaltungsarbeit zu handeln scheint, sodass man wohl vermutlich einen der hiesigen Leute hinsetzen wird. Ich liebäugele auch ein bisschen mit Manchester, wo bei Bragg sogar vermutlich zwei Stellen frei sind — wenn nämlich Williams sein Rockefeller bekommt. Bragg scheint auch Sie nach hier umlaufenden Gerüchten sehr stark in Betracht zu ziehen.

Ich habe jetzt meine Metalle ein bisschen aufgesteckt und ixe mit den Dirac-Löchern=Blackett-Elektronen herum. Vielleicht hat diese Diracsche Theorie doch in irgendeiner Näherung einen vernünftigen Sinn.

Sonst gibt es von hier nicht viel Neues, denn dass es in Cambridge sehr nett ist, dass Wilson noch immer nicht seinen Fehler einsieht, dass Frau Blackett allen Leuten erfrischende Frechheiten sagt, und dass Fowler jedes zweite Kollegium vergisst, dürfte wohl für Sie nicht neu sein.
Mit besten Grüssen Ihr

R. Peierls

Gruss auch von Genia

Cambridge, 14.6.[19]33

Dear Bethe!
I haven't answered your letter yet, as I hoped to give you some official news about the lecturer issue.[149] But first the decision is totally adiabatic, and second it has been impossible to get to know anything. So I suspect you might know more about this than I do. It has been said that "one of you" will get a job, but whether that's you, or whether it is a lectureship or a badly paid assistant l[ectureship] I don't know. I am not on the short list, anyway.

What are your other plans? My chances of getting a grant at Langevin's[150] are good, and aside from that I am competing with you in Belfast. Neither of us has a real chance there, as the post includes a lot of administrative work, so they will presumably take someone from here. I would also like to go to Manchester, where there might be two openings with Bragg[151] — if Williams[152] gets his Rockefeller's. Bragg is also seriously considering you; at least, that's what people are saying here.

I have given up my metals for a while and am now ixing with Dirac-holes=Blackett-electrons. Maybe this theory of Dirac's, in some approximation, will make some sense after all.[153]

---

[149] The younger scientists were certain that Cambridge needed a theoretical physicist in the autumn of 1933, but it was not clear what form the post would take. Many letters exchanged between Peierls, Bethe and their friends contained references to these posts and to others that were opening up in various British universities. See below letters [30–32].

[150] Paul Langevin (1872–1946) had been director of the École municipale de Physique et Chimie in Paris since 1929.

[151] William Lawrence Bragg (1890–1971) studied mathematics at Adelaide before completing a physics degree at Cambridge. In 1919 be was appointed Langworthy Professor of Physics at Manchester University. In 1938 he became Cavendish Professor of Experimental Physics at Cambridge.

[152] Evan James Williams (1903–1945), at the time reader in physics at Manchester, was appointed professor of Physics at Aberystwyth in 1939 where, interrupted by his service for the ment of Naval Operational Research, he stayed until his premature death in 1945.

[153] See also letter Wolfgang Pauli to Rudolf Peierls, 22.5.1933, in *Pauli Briefwechsel*, II, pp. 163–66.

Other than that there is not much to report from here. Cambridge is very nice, Wilson[154] still doesn't see his error, Mrs. Blackett is still very impertinent to everybody in a refreshing way and Fowler forgets every other colloquium, but that's nothing new to you.
Best wishes yours,

R. Peierls

Greetings also from Genia

[30] Rudolf Peierls to Hans Bethe

Cambridge, 28.6.[19]33

Lieber Bethe!
Vielen Dank für Ihre prompte Antwort. Was den Lorentzfonds betrifft, so habe ich von Fokker eine Absage bekommen, das sie Ihre Mittel jetzt für dringendere Aufgaben (d.h. offenbar für kurzfristige Einladungen) brauchen und ich ja ausserdem immer nach Zürich zurückkönnte (Ob das richtig ist, hängt allerdings von der Definition des Wortes "können" ab). Ich weiss übrigens nicht, ob ich Ihnen geschrieben habe, dass für mich merkliche Aussichten auf ein Pariser Stipendium zu bestehen scheinen.

Bezüglich Cambridge hat sich die Situation insofern wieder geändert als plötzlich wieder zwei Stellen frei sind. Erstens hat man Bloch eine Stelle mit 120-200£ angeboten, die er wohl mit grosser Sicherheit ablehnen dürfte, da er nicht auf das Rockefeller Stipendium verzichten wird. Ausserdem hat man eine zweite Stelle — ich glaubte fälschlich es wäre dieselbe — Born angeboten, der inzwischen auch abgelehnt hat. Für mindestens eine von diesen beiden Stellen haben Sie erhebliche Chancen, als erster berufen zu werden, was nämlich erstens aus Aeusserungen des — allerdings nicht immer korrekt informierten — Mott hervorgeht und mir auch a priori plausibel erscheint. Allerdings

[154] Alan Herries Wilson (1906–1976) had taken issue with some aspects of Peierls' doctoral thesis, and Peierls engaged in a dispute with him in which he successfully defended the view taken in his thesis.

kann man bei den hiesigen Bonzen nie wissen; ich war z.B. sehr erstaunt, dass man Bloch vor Ihnen berufen hat, zumal man sie hier sehr zu schätzen scheint. Haben Sie sich etwa hier mit jemand verkracht? Das ärgerlichste ist nun, dass das Komitee, das die Stellen neu besetzen soll, von der Leistung Bloch und Born zu berufen, so erschöpft war, dass es schleunigst in die Ferien ging, und vor Oktober nicht wieder zusammentritt. Trotz dieser unbegrenzten Adiabasie sollte man ja nun eigentlich schon vorher unoffiziell feststellen können, wer eingeladen wird, denn da es ziemlich fest zu stehen scheint dass man Physiker berufen wird, so wird es wohl ganz von Fowler abhängen. Der ist aber so ein Geheimniskrämer, dass ich nicht weiss, ob man vor Oktober irgendwas erfahren wird.

Mott liebäugelt übrigens mit einem Tauschhandel. Wie Sie wissen, hat er Stobbe gegen Heitler eingetauscht, um den ersteren loszuwerden, und scheint dies nun schrecklich zu bereuen, da ja Stobbe jetzt auch ohne Tausch beliebig viele Stellen bekommen kann. Er wird nun weiter tauschen, und jedenfalls wenn Sie in Manchester sind, wird er sicher nach einem halben Jahr versuchen, Heitler gegen Sie einzutauschen. Allerdings weiss ich nicht, ob Bragg sich dies gefallen lässt, und ob Sie das Risiko eingehen werden, ev[entuell] nach einem halben Jahr gegen Heisenberg eingetauscht zu werden. All dies ist übrigens nicht offiziell.

Die zweite Stelle in Manchester hat sich inzwischen auch auf Null reduziert, d.h. wird mit einem Engländer besetzt werden.

Wissen Sie übrigens dass auch die Royal Society Stipendien zu vergeben hat? Über die allerdings noch keine näheren Einzelheiten bekannt sind. Wenn Sie sich dafür interessieren, so schicken Sie doch Ihre Personalien an das Academic Assistance Council, Burlington House, Piccadily, London.

Wir haben mit Ihnen herzliches Mitleid, nachdem uns Mott erzählt hat, wie Sie in Manchester gehungert haben. Genia macht die ganze Zeit Pläne, wie man es einrichten kann, dass Sie bei uns wohnen und bei uns anständig zu essen kriegen (in Quantitäten wie etwa in St. Antönien), aber das geht natürlich nur dann, wenn wir in einer Stadt wohnen, wofür die Wahrscheinlichkeit sehr gering ist.

Sehen werden wir Sie ja vermutlich auf dem Weg nach Manchester, denn wir bleiben bis Anfang Oktober hier und Sie kommen doch wohl

sicher über Cambridge, aber dann wird Genia keine Zeit haben, Sie zu füttern da sie dann damit beschäftigt sein dürfte, einen Mitbürger zu füttern, der 27 Jahre und zwei Monate jünger ist als Sie.
Sonst nichts Neues, ich ixe Diraclöcher.
Beste Grüsse
Ihr

R. Peierls

Cambridge, 28.6.[19]33

Dear Bethe!
Thank you for your prompt reply. Concerning the Lorentz fund, I received a negative answer from Fokker,[155] because they needed their funding for more urgent tasks (i.e., evidently for *ad hoc* invitations) and because I could always go back to Zurich (whether this statement is correct, however, depends on how you define the word "could").

I don't know whether I told you, but it seems as if I have some chance of getting a grant in Paris.[156]

As for Cambridge, the situation has changed again. Suddenly there are two openings. First Bloch was offered a job for 120-200£, which he will surely reject as he won't pass up his Rockefeller grant.[157] In addition, a second job — which at first I thought was the same one — was offered to Born, who has also rejected it in the meantime.[158] You have a considerable chance of being appointed first for at least one of these jobs; this can be concluded from statements of Mott[159] (who, however, is not always properly informed) and seems to be reasonable a priori. On the other hand, you never know what will happen with

[155] Adriaan Fokker was professor of physics in Leiden between 1928 and 1955.

[156] One of Peierls' options of a fellowship at Langevin's institute; see letter [29].

[157] Bloch had not rejected the offer. See letter [31].

[158] Max Born (1882–1970), studied at Breslau, Heidelberg, Zurich, Göttingen, where he obtained his doctorate in 1907. He later taught at Göttingen, Berlin, Frankfurt and again as professor in Göttingen. Being forced to flee Germany in 1933 Born initially went to India and in 1936 accepted the Tait professorship in Applied Mathematics at Edinburgh.

[159] Nevill Mott (1905–1996), received his Ph.D. from Cambridge in 1926, and taught at Manchester, Cambridge and Bristol before returning to Cambridge as Cavendish professor of Physics in 1954, where he stayed until his retirement in 1971.

these big shots; I was very surprised for example that Bloch had been appointed before you, especially as they seem to like you a lot here. Are you in trouble with someone here? The most annoying thing is that the committee which is supposed to fill vacancies was so exhausted, as a result of appointing of Bloch and Born, that they went on vacation right away and won't meet again until October. Despite this open-ended adiabatic process, it should be possible to learn in advance unofficially who will be invited. It seems to be pretty clear that in the appointment of physicists, everything depends on Fowler. But he is such a mystery-monger that I don't know whether you can find out anything before October. Mott fancies a deal, by the way. As you know, he exchanged Stobbe[160] for Heitler to get rid of the former. He seems to regret that terribly now, because Stobbe can now find a new job easily without an exchange. He will keep on trading, and if you are going to be in Manchester, he will surely try to exchange Heitler for you after six months. But I don't know whether Bragg would accept this, or whether you would risk being exchanged for Heisenberg six months later. All this news is unofficial, by the way.

The second job in Manchester has been nullified;[161] i.e., an Englishman is going to be appointed.

Did you know that the Royal Society is offering grants as well? But we don't know the details yet. If you are interested, send your application to the Academic Assistance Council,[162] Burlington House, Piccadilly, London.

We felt very sorry for you after Mott had told us how you were starving in Manchester. Genia is always thinking about how to arrange

---

[160]Martin Stobbe (1903–1944), physicst in Göttingen; he had resigned from his post in protest of the Nazis, spent some time in England and later Oslo where he died in 1944. See Alan D. Beyerchen, *Scientists under Hitler*, Yale University Press, 1977, p. 30.

[161]See interview by Charles Weiner with Rudolf Peierls, 11.–13.8.1963, Niels Bohr Library, AIP. Bethe was appointed to a lectureship at Bragg's institute at Manchester, but it became clear that it would not have been acceptable to appoint another German to a second lectureship, should this become available. Peierls was then appointed at Manchester to a fellowship funded by the Academic Assistance Council.

[162]The Academic Assistance Council was formed in May 1933 as a British response to help central European scholars persecuted by the Nazis. It provided, among others, scholarships to distinguished academics.

things so that you could stay here with us and get enough to eat (in St. Antönien quantities), but that would only be possible if we lived in the same city, which is not very likely.

We will meet you on the way to Manchester, as we are staying here until the beginning of October and I'm sure you will come via Cambridge. Genia won't have any time to feed you then, as she will have to feed a citizen 27 years and two months younger than you.
Besides that nothing new, I'm ixing Dirac holes.
Best wishes,
Yours,

R. Peierls

[31] Rudolf Peierls to Hans Bethe

Cambridge, 6.7.[19]33

Lieber Bethe!
Der Mensch denkt und das Faculty Board of mathematics lenkt. Und daraus ergeben sich dann alle möglichen Falschmeldungen. Z.B. stimmt es nicht, dass B[loch?] abgeleht hat, im Gegenteil, er kommt bestimmt her, (nur weiss ich nicht, wie weit das offiziell ist). Trotzdem gibt es ausserdem noch zwei Stellen, von denen eine wohl ein hiesiger bekommen wird, während die andere mit grosser Wahrscheinlichkeit — nach vertraulicher Aussage von Newman — einer der drei Nordheim, Sie oder ich bekommen werden. Da die Entscheidung hierüber nicht vor September zu erwarten ist, so muss man anfangen, Eventualprogramme zu machen. Können Sie jetzt schon feststellen, was Sie machen würden, wenn Sie hier eingeladen würden, ob Sie annehmen oder nach Manchester gehen würden? Das ist mir deswegen von Interesse, weil ich im ersteren Fall zu Bragg gehen würde. Hängt diese Frage mit der Höhe des hier gezahlten Gehalts empfindlich zusammen?
Mit besten Grüssen
Ihr

R. Peierls

Warum sitzen Sie eigentlich in München und kommen nicht her? Ist es nicht in dem Münchener Institut jetzt sehr langweilig?

Cambridge, 6.7.[19]33

Dear Bethe!
Human beings think and the Faculty Board of mathematics rules.[163] This results in many different false reports. For example, it isn't true that B[loch] has rejected the offer; on the contrary, he will certainly come here (I just don't know yet how official this is). Still, in addition there are another two jobs, one of which will probably go to someone from here. The other one might be offered — according to a confidential message from Newman[164] — either to Nordheim,[165] you, or me. Since there won't be a decision before September, you have to start preparing yourself for all eventualities. Could you decide now what you would do, if you were invited here? Would you accept, or would you go to Manchester? I ask because in the first case, I would go to Bragg. Is what they are willing to pay very important to your decision?
Sincerely yours,

R. Peierls

Why are you still sitting in Munich and not coming here? Isn't it very boring in the Munich institute these days?

---

[163]This is a play on words, as the German terms for 'think' and 'rule' rhyme.
[164]Max Newman (1897–1984), mathematician at the time lecturer of mathematics at Cambridge.
[165]Lothar W. Nordheim (1899–1985), studied at Hamburg, Munich, and Göttingen, where he obtained his doctorate under Max Born. In 1927 he took up a post at the Cambridge Cavendish Laboratory, and in 1835 he moved to the US, taking up positions at Purdue University (1935), Duke University (1937) and at the John Hopkins Laboratory (1956).

[32] Rudolf Peierls to Hans Bethe

Cambridge, 24.7.[19]33

Lieber Bethe!
Das Laufende ist nicht viel weiter gelaufen, ich möchte Sie aber trotzdem darauf halten. Bragg war jetzt hier ich hatte ihn vorher nur flüchtig gesehen und der reizende Eindruck, den er schon damals auf mich gemacht hat, ist so vestärkt worden, dass das fast alle Unannehmlichkeiten der Stadt Manchester aufwiegen dürfte. Ich bin daher fast ganz sicher entschlossen, das Angebot von Bragg anzunehmen, es sei denn dass sich Fowler doch noch in den nächsten Tagen entschliesst, mir etwas hier anzubieten, wofür die Wahrscheinlichkeit nicht höher als 5,73% anzusetzen ist. In diesem Fall würden wir in Manchester zusammen sein, was natürlich sehr erfreulich wäre. Armstrongs Taktlosigkeit nimmt natürlich hier niemand ernst, zumal er ja in seinem Brief auch die Tatsachen entstellt hatte, eine Richtigstellung ist bisher nur in der Zeitung veröffentlicht worden. Im übrigen stimmt man aber insofern mit ihm überein, als die allgemeine Praxis die ist, uns keine etatmässigen Stellen zu geben, sondern nur Stipendien oder ad hoc geschaffene Stellen. Ob das noch in einem oder zwei Jahren genau so sein wird, kann man natürlich nicht sagen. Vielleicht hat man sich bis dahin mit unserer Existenz schon abgefunden.

Noch eine technische Frage: Wir werden uns in Manchester natürlich eine Wohnung nehmen, und es würde ein relativ geringer Aufwand an Mühe sein, zu dieser Wohnung noch ein Zimmer mehr zu nehmen und das Ihnen zu geben. Da Sie das englische Essen nicht lieben, und wie Mott erzählte, sogar in Manchester besonders zu leiden hatten, so könnte ich mir denken, dass Ihnen diese Lösung angenehm wäre, zumal Sie dadurch in den Stand gesetzt würden, "langsam und reichlich" zu essen! Was die Kostenfrage betrifft, so würden Sie natürlich nur zu bezahlen haben, was Sie uns kosten, was wohl sicher weniger ist als in jeder Pension. Da wir schon sehr bald nach Manchester fahren wollen, um die Wohnungsmöglichkeiten zu studieren, wäre es mir lieb, wenn Sie bald antworten würden, wie Sie sich zu diesem Vorschlag stellen würden — natürlich vorbehaltlich der Entscheidung, ob Sie und ob wir überhaupt in Manchester sein werden.

Die positiven Elektronen werden sicher durch Zusammenstoss des Lichtquants mit einem Kern erzeugt, also einem Vorgang, der sich in dem Diracschen Modell durch Photoeffekt aus dem negativen ins positive Energiespektrum beschreiben lässt. Die Wahrscheinlichkeit hierfür habe ich noch immer nicht fertig geixt. Oppenheimer hat eine Formel hierfür in der letzten Phys.Rev. veröffentlicht, die, soweit ich sehen kann, für Oppenheimer ausserordentlich korrekt ist, es sind nämlich anscheinend nur die Zahlfaktoren falsch, jedenfalls für $E \ll mc^2$. Im entgegengesetzten Fall kinetischer Energien scheint mir seine Formel überhaupt verdächtig, ohne dass ich allerdings beweisen kann, dass sie falsch ist.

Ueber die Frage, ob man Ihnen hier noch eine Stelle anbieten wird, und wenn was für eine, ist mir nichts Neues bekannt. Mit besten Grüssen Ihr

R. Peierls

Genia lässt grüssen.

Cambridge, 24.7.[19]33

Dear Bethe!
Nothing really new has happened, but I wanted to inform you of the latest developments anyway.[166] Bragg has been here; I had only seen him once before in passing, but that intial good impression has now been reinforced in a way that it should balance out all the inconveniences of Manchester. So I'm almost certain that I will accept Bragg's offer, unless Fowler decides within the next few days to offer me something here. But the probability of this is no higher than 5.73%. In this case we would be together in Manchester, which would certainly be very nice.

Nobody really cares about Armstrong's indiscretion, especially since he manipulated the facts in his letter. So far, a correction has only been published in the newspaper. Apart from that, we can agree with him that it is now common practice not to give us the jobs accounted for in the budget, but just to offer grants or jobs created *ad hoc.* Whether

[166]The German is a pun and does not translate literally into English.

this will remain the case beyond the next year or two, nobody knows. We will probably have gotten used to it by then.

A technical question: we will of course rent an apartment in Manchester, and it wouldn't be much trouble to rent an additional room for you. As you don't like English food and had to suffer so much in Manchester, as Mott has told us, I am guessing that you might like this solution. Then you would be able to eat "slowly and plenty"! Regarding the costs, you would of course have to pay simply what you cost us, which would surely be less than any bed and breakfast. Since we want to go to Manchester soon to study the housing situation, I would appreciate, if you replied soon to tell us what you think of our proposal — contingent, of course, on whether you and I will be in Manchester at all.[167]

The positive electrons are certainly produced by the impact of a photon with the nucleus, in other words by a process which according to Dirac's model can be described by a photoelectric effect from the negative energy spectrum to the positive. I haven't yet ixed the probability for this. Oppenheimer[168] published a formula for it in the last *Phys. Rev.*,[169] which is surprisingly correct for Oppenheimer, as far as I can see. It seems as if only the numerical factors were wrong, at least for $E \ll mc^2$. In the opposite case of kinetic energies, his formula generally seems to be suspicious even though I can't prove that it is really wrong.

I still don't know whether you will be offered a job, nor do I know of what kind.

Sincerely yours,

R. Peierls

Greetings from Genia.

[167]Bethe did indeed live with the Peierls family in Manchester. As he wrote to Arnold Sommerfeld at the end of the year, this arrangement had 'disadvantages for the English' of the two physicists, but 'great advantages for their general well being and for their physics'. See Eckert and Märker, *Sommerfeld Briefwechsel*, vol. 2, p. 196.

[168]Robert Oppenheimer (1904–1967) graduated from Harvard before embarking on research in Cambridge and Göttingen where he obtained his Ph.D. in 1927. He took up a professorship at Berkeley. He headed the Manhattan Project and after the war became Chairman of the General Advisory Committee of the Atomic Energy Commission. In 1947, Oppenheimer became Director of the Institute of Advanced Studies at Princeton.

[169]J.R. Oppenheimer and M.S. Plesset, 'On the Production of the positive Electron', *Phys. Rev.* **44**, 53–55 (1933).

2. Kellnerinnen — A Slight Amount of Respectability into the Theory

The years between 1935 and 1940 were a period of consolidation in the careers of both Bethe and Peierls. Hans Bethe had already made a name for himself as one of the masters of theoretical physics. He had written what he considered his best paper, on the stopping power of charged particles as they go through matter, the paper which earned him the habiliation allowing him to teach at German universities.[170] He had written two Handbuch articles on the quantum mechanics of one- and two-electron problems, and (nominally together with Sommerfeld) on the electron theory of metals.[171] And together with Wolfgang Heitler, he had written an influential paper on bremsstrahlung and pair production which became a classic.[172]

When Bethe came to Cornell in 1935, he felt at home immediately among the then around 15 faculty members and around 40 students.[173] Shortly after his arrival, Robert Bacher joined the department, and together with him and M. Stanley Livingstone, Bethe turned Cornell into one of the leading centres of nuclear physics. He found that many of his American colleagues and advanced students were full of enthusiasm and eager to learn about nuclear theory, but that they were not very

---

[170]H.A. Bethe, 'Zur Theorie des Durchgangs schneller Korpuskularstrahlen durch Materie', *Ann. Phys.* **5**, 443-49 (1930).

[171]H.A. Bethe, 'Quantenmechanik der Ein- und Zwei-Elektronenprobleme', *Handbuch der Physics*, 24, part 1, Springer, Berlin, 1933; H.A. Bethe and A. Sommerfeld, 'Elektronentheorie der Metalle', *Handbuch der Physik*, 24, part 2, Springer, Berlin, 1933, pp. 333–622.

[172]H.A. Bethe and W. Heitler, 'On the Stopping of Fast Particles and on the Creation of Positive Electrons', *Proc. Roy. Soc.* **A146**, 83–112 (1934).

[173]Bethe conversation with Gottfried. http:ifup.cit.cornell.edu/bethe/bethe-weisskopf240.mov.

knowledgeable[174] and Bethe was invited to tour the US to explain some of the basic concepts;[175] in addition he decided to write down these basics, and create what was to become known as the 'Bethe Bible', a series of review articles (together with Bacher and Livingstone). The reviews which summarised everything Bethe knew on nuclear physics, or as others would put it, everything there was to know about nuclear physics in 1936/7,[176] were written at his own initiative to 'give the nuclear-physics community the theoretical perspectives that would direct their research'.[177] And then followed, in 1938/9, the work that would earn him the Nobel Prize in 1967, the solution of the problem of energy generation in stars.[178]

For Peierls, similarly, the thirties were a prolific period of consolidation and a phase of making his influence felt. Two productive years at Cambridge saw the publication of his now famous paper on the Ising model,[179] work based on collaborative efforts with David Shoenberg,[180] and an important statistical mechanics paper which had been initiated by some earlier work of Ralph Fowler's.[181] Furthermore, the networking at Cambridge was important for shaping Peierls' future career. In

[174]Bernstein, *Hans Bethe, Prophet of Energy*, p. 45.

[175]Hans A. Bethe, interviewed by Judith R. Goldstein, 17.2.1993, Archives CalTech, Pasadena, California, http://resolver.CalTech.edu/CaltechOH:OH_Bethe_H.

[176]H.A. Bethe and R.F. Bacher, 'Nuclear Physics. Part A, Stationary States of Nuclei', *Rev. Mod. Phys.* **8**, 82–229 (1936); H.A. Bethe, 'Nuclear Physics. Part B, Nuclear Dynamics, Theoretical', *Rev. Mod. Phys.* **9**, 69–244 (1936); M.S. Livingstone and H.A. Bethe, 'Nuclear Physics. Part C, Nuclear Dynamics. Experimental', *Rev. Mod. Phys.* **9**, 245–390 (1936).

[177]S.S. Schweber, 'The Happy Thirties', in Brown and Lee, *Hans Bethe and His Physics*, pp. 131–146, here p. 141.

[178]H.A. Bethe, 'Energy Production in Stars', *Phys. Rev.* **55**, 434–456 (1939).

[179]R.E. Peierls, 'On Ising's model of Ferromagnetism', *Proc. Camb. Phil. Soc.* **32**, 477–81 (1936).

[180]David Shoenberg (1911–2004), Russian-born British physicist and close friend of the Peierls couple. In 1932 Shoenberg had graduated from Cambridge where he remained for the rest of his working life, working on low-temperature physics and, in particular, superconductivity.

R.E. Peierls, 'Magnetic Superconductors', *Proc. Soc. Roy. Lond.* **A155**, 613–28 (1936).

[181]R.E. Peierls, 'Statistical Theory of Adsorption with Interaction between the Absorbed Atoms', *Proc. Camb. Phil. Soc.* **32**, 471–81 (1936).

the mid-1930s the Cavendish Laboratory at Cambridge was in flux with many excellent men moving to and fro. In a letter to Bethe,[182] Peierls mentioned the disintegration of the Cavendish. Chadwick had moved to Liverpool, Blackett to London, Kapitza had not been allowed to return to the UK from Russia, and Oliphant, in 1936, was appointed to the chair of physics at Birmingham University. The connections with Oliphant, fostered during their time at Cambridge, encouraged Peierls to apply for the chair in mathematical physics at Birmingham, a post to be created in 1937. The networking was therefore crucial for Peierls' succeeding in what many other refugee scientists in the UK and the US failed to achieve, namely an early permanent academic position.

Bethe's move across the Atlantic did not end the collaboration between him and Peierls. The correspondence in the latter half of the 1930s demonstrates an intense exchange of ideas and results. Many of the letters dealt with nuclear physics and concerned publications in the making.[183] Despite this focus on nuclear matters, Bethe and Peierls still engaged in cutting-edge research in solid state physics. One such example is prominent in the correspondence soon after Bethe's arrival at Cornell. While still at Manchester, Lawrence Bragg had interested him in the problem of order/disorder transition in binary alloys. Bragg and Williams had just published the simplest mean-field theory of this transition analogous to Weiss's molecular field theory of ferromagnetism.[184] Bethe wanted to go beyond this simple Bragg-Williams theory and in his paper on the statistical theory of superlattices,[185] he added the important concept of 'short range order'. In addition, he introduced some

---

[182]See references in letter [36].

[183]H.A. Bethe, 'Nuclear Radius and Many-Body Problem', *Phys. Rev.* **50**, 977–79 (1936); H.A. Bethe and M.E. Rose, 'Nuclear Spins and Magnetic Moments in the Hartree Model', *Phys. Rev.* **51**, 205-13 (1937); H.A. Bethe and G. Placzek, 'Resonance Effects in Nuclear Processes', *Phys. Rev.* **51**, 450–84 (1937); R.E. Peierls, 'Interpretation of Shankland's Experiment', *Nature* **137**, 904–905 (1936); R.E. Peierls, 'The dispersion formula for nuclear reactions', *Proc. Roy. Soc.* **A166**, 277–95 (1938).

[184]W.L. Bragg and E.J. Williams, 'Effect of thermal agitation on atomic arrangement in alloys', *Proc. Roy. Soc.* **A145**, 699–730 (1934).

[185]H.A. Bethe, 'Statistical Theory of Super-lattices', *Proc. Roy. Soc.* **A150** 552–75 (1934).

remarkably accurate approximations which were consistent with Bragg-Williams in some cases, improved on the Bragg-Williams results in other cases and, in Bethe's words, 'put just a slight amount of respectability into the theory of Bragg and Williams on the way order and disorder came about.'[186]

Bethe's work was then followed up by Peierls,[187] who further developed these approximations and generalised the Bethe method for superlattices to cases where the concentrations of the two constituents are unequal.[188] This has been referred to as the 'Bethe-Peierls method' or, more commonly the 'Bethe-Peierls approximation'. Genia Peierls, who followed the genesis of her husbands and Bethe's work, referred to Bethe's method as 'Kellnerinnen' (waitresses), because it involved the labourious counting of configurations,[189] a term which Peierls and Bethe adopted in their correspondence around that time.[190]

Among Peierls' many scientific achievements in the subsequent years, two deserve particular mention, although only the former features in the correspondence with Bethe. Following his participation in the Copenhagen Conference on Nuclear Physics in September 1937 and the subsequent All Union Conference in Moscow later that month, Peierls engaged in active exchanges with two of his mentors, Wolfgang Pauli and Niels Bohr.[191] It was the collaboration with the latter and

[186]Hans Bethe in *From a Life in Physics: Evening Lectures at the International Centre for Theoretical Physics*, Trieste, Italy, Special supplement of the International Atomic Energy Agency (IAEA) Bulletin, Vienna, 1969.

[187]R.E. Peierls, 'Statistical Theory of Superlattices with Unequal Concentration of Components', *Proc. Roy. Soc.* **A154**, 207–222 (1936).

[188]Many of the thought processes are evident in the correspondence between them. See e.g. letters [35, 40].

[189]Letters [212, 214].

[190]See e.g. letters [33–34].

[191]Niels Bohr (1985–1962), studied at Trinity College, Cambridge and received his Ph.D. from Copenhagen in 1911. After research under J.J. Thomson at Cambridge and Ernest Rutherford in Manchester he became Professor of Physics at the University of Copenhagen and Director of the newly-created Institute of Theoretical Physics there in 1920, arguably the most important research institution in this discipline in the first half of the 20th century. Bohr contributed to the Los Alamos Project. After his return to Copenhagen he continued research and teaching.

with his friend and colleague George Placzek,[192] that took up much research time in the years leading up to the war. An investigation of the compound nucleus, and in particular nuclear reactions in the continuous energy region, occupied the three scientists for much of the period between 1937 and 1939, and they drafted and redrafted a paper which gained considerable fame as the 'most-frequently-cited unpublished paper'. The tripartite work was based, among others, on research which Bethe and Placzek had carried out jointly,[193] and therefore it is not surprising that it features in the correspondence between Bethe and Peierls.[194]

In early 1940 followed what was arguably the most influential of Peierls' papers, influential not as a path-breaking scientific achievement but as a result with far-reaching political consequences and global political impact. After the discovery of nuclear fission in 1938, Lise Meitner[195] and her nephew Otto Frisch had developed a qualitative theoretical explanation of the process.[196] It was understood that the relatively rare

[192] George Placzek (1905–1955), studied physics in Vienna and Prague, where he obtained his Ph.D. in 1928. After research at Utrecht, Leipzig, Rome, Copenhagen, Jerusalem, Kharkov and Paris, he moved to the US in 1939. Initially professor at Cornell, Placzek went to Montreal and Los Alamos as part of the Manhattan Project. In 1948 he obtained a permanent position at Princeton in 1948.

[193] H.A. Bethe and G. Placzek, 'Resonance Effects in Nuclear Processes', *Phys. Rev.* **51**, 450–84 (1937).

[194] A short sketch of the solution to the problem was published in 1939. N. Bohr, R. Peierls and G. Placzek, 'Nuclear Reactions in the Continuous Energy Region', *Nature* **144**, 200–201 (1939). The longer complete paper became a victim of Bohr's notorious labouring over the formulation of his research. It was not finalised before the outbreak of the war, and after the war the three physicists decided that, as the knowledge was already in the public domain, there was not need to revisit the problem for publication purposes. One of the surviving drafts of the paper has now been published in Niels Bohr's Collected Works. R. E. Peierls (ed.), *Niels Bohr. Collected Works*, vol. 9, Nuclear Physics (1929–1952), Amsterdam: North Holland, 1986, p. 49, note 86 and p. 50, notes 87–89.

[195] Lise Meitner (1878–1968), Austrian-Swedish physicist, close collaborator of Otto Hahn, with whom she studied radioactivity. In 1918, they discovered the element protactinium, and in the late 1930s their work would provide the evidence for nuclear fission.

[196] L. Meitner and O. Frisch, 'Disintegration of Uranium by Neutrons: A New Type of Nuclear Reaction', *Nature* **143**, 239 (1939).

uranium isotope U235 was the most promising candidate for fissionable material that would allow a self-sustaining chain reaction. Otto Frisch,[197] who had joined the University of Birmingham to work with Marc Oliphant, and Rudolf Peierls thus tackled the question of the critical mass of U235, which was generally believed to be in the region of tons rather than kilos. They calculated that a chain reaction was not only theoretically feasible, but also that the amount of fissionable material needed for such a self-sustaining chain reaction was far less than previously assumed.[198] This discovery, duly communicated to the government committee concerned with the feasibility of a nuclear chain reaction, eventually resulted in the establishment of the so-called MAUD Committee, a government committee which sought to clarify the implications of the Frisch-Peierls findings and led to the kick-starting of the Anglo-American nuclear programme.[199]

Many of these significant developments do not find expression in the communications between Bethe and Peierls. War work was top secret, so much so that Peierls and Frisch themselves, as enemy aliens, were not allowed to access the secret memorandum they had composed. Indeed, it was war-related issues that took priority in the letters exchanged between Bethe and Peierls in the late 1930s and early 1940s. The correspondence was taken up by concerns about the deteriorating situation in Germany. Bethe's mother and Peierls' parents were still in Germany, and both Bethe and Peierls tried to persuade them to emigrate. Hans Bethe, as usual during the summer, visited Europe in the

[197]Robert Otto Frisch (1904–1979), received his doctorate from the University of Vienna in 1926; he continued research in Berlin, Hamburg and Copenhagen before moving to Birmingham in 1939. Here, Frisch and Peierls calculated that the fissioning of uranium had the fissioning of uranium had the potential to create a chain reaction which could be used to develop an atomic weapon. Frisch later joined the Manhattan Project. After the war he worked for the Atomic Energy research Establishment at Harwell before taking up the Jackson Chair of Physics at Cambridge in 1947.

[198]'The Frisch-Peierls Memorandum of 1940', R.H. Dalitz and R.E. Peierls (eds.), *Selected Scientific Papers of Sir Rudolf Peierls*, Imperial College Press, London, 1997, 277–282.

[199]See S. Lee, " 'In no sense vital and actually not even important'? Reality and Perception of Britain's Contribution to the Development of Nuclear Weapons", *Contemporary British History* **20**, 159–186 (2006).

summer of 1938. After spending a short holiday in Cornwall with the Peierls family (by then Rudi and Genia had a daughter of almost 5 and a son of almost 3), he went to Germany where he met not only his own family but also Rudolf Peierls' parents. As the latter could not communicate freely with his parents due to censorship, Bethe's description of the situation in Germany and particularly the circumstances of his parents were of utmost interest to him.[200] The situation of both the Bethe and the Peierls families was typical of Jewish middle-class families. The younger generation often recognised the signs of the time earlier, they had less to leave behind and found it easier to make a fresh start in emigration. By the time the older generation had realised the precariousness of the political situation for them in Germany, it had become very difficult to leave, both because of the increasingly repressive regulations in Germany affecting those who chose to leave, and because of the limitations posed by the potential recipient countries.[201] Bethe's mother and Peierls' father and stepmother could eventually leave, if under difficult circumstances.[202]

Another common concern remained the fate of colleagues who had to leave Germany or Central Europe because of the political situation. One example of this is George Placzek with whom both Bethe and Peierls had collaborated very closely during the preceding years and who contemplated taking up a position at Cornell.[203]

After the outbreak of war, communications about work-related issues ceased, but during the year 1940, the Bethe and Peierls couples[204] exchanged some letters which are indicative of the depth of friendship between Hans Bethe and Rudolf Peierls. With the prospect of a German

---

[200] Letter [47].

[201] See letters [47, 49, 53–56].

[202] Hans Bethe had submitted his calculations about the energy production in stars to the New York Academy of Sciences, which had a competition for the best original unpublished research in this field. Bethe used half the money he was awarded for his paper, $250 (which would later earn him the Nobel Prize) to pay for the release of his mother's belongings.

[203] See letters [48–53].

[204] Hans Bethe had married Rose Ewald, his former mentor's daughter, in 1939. See letter [57].

invasion of Britain looming, Hans and Rose Bethe had offered to put up the Peierls children Gaby, who was seven at the time, and Ronnie who was five. By 1940, Rudolf Peierls' sister had settled in the US, as had his father and some other family, but the correspondence is unequivocal in the Peierls' preference for their children to live with the Bethes and not their own family. In the event, Gaby and Ronnie were evacuated to Canada as part of a university programme organised by Toronto and Birmingham University. Nevertheless, Hans Bethe became guardian for the Peierls children, a sign of the trust and friendship between the two scientists.[205]

Although initially their German background was a hindrance to much of the war work, both Bethe and Peierls soon became involved in research with war-related applications. Peierls had been naturalised in March 1940, and this facilitated his endeavours to contribute to the British war effort in a way that went beyond fire watching.[206] But he was prevented from work on Birmingham physics' war-related flagship project, the radar. Thus he focused on nuclear work, which at the time was thought to be of less immediate impact on the war effort.

Bethe became an American citizen in March 1941, and accordingly had to wait even longer than his friend for clearance to work on classified projects. His solution to the problem was typical in its pragmatism and corresponded to his philosophy that, as a scientist, he would only tackle problems for which he had an unfair advantage.[207] After the fall of France, 'desperate to do something — to make some contribution to the war effort'[208], Bethe began studying the penetration of armour by projectiles, and with the help of an engineer friend, George Winter, he developed a theory of armour penetration. Furthermore, together

[205] Lee, *Selected Correspondence*, 1, chapter 4.

[206] Lee, *Selected Correspondence*, 1, chapter 5.

[207] John N. Bahcall and Edwin E. Salpeter, 'Stellar Energy Generation and Solar Neutrinos', in Brown and Lee, *Hans Bethe and His Physics*, pp. 147–156, here pp. 153–4.

[208] Bernstein, *Hans Bethe, Prophet of Energy*, p. 61.

with Teller,[209] who had similar problems of clearance for war work, he studied shock waves.[210]

Most significantly, however, once Bethe and Peierls had been cleared and once the Anglo-American nuclear project, the Manhattan Project got under way, both contributed to this in prominent positions: Bethe as head of the theoretical division and Peierls as leader of the implosion group at Los Alamos. After Munich and Manchester, they now had the opportunity to engage in yet another period of intense collaboration; one which naturally does not find expression in the correspondence.

[209]Edward Teller (1908–2003), studied at Karlsruhe, Munich and Leipzig where he obtained his doctorate under Heisenberg in 1930. After working at Göttingen, Copenhagen and London, he accepted a position at Washington in 1935. Teller made important contributions to the development of the atomic bomb and was among the architects of the hydrogen bomb.

[210]H.A. Bethe and E. Teller, 'Deviations from Thermal Equilibrium in Shock Waves', in H.A. Bethe, *Selected Works of Hans A. Bethe*, World Scientific, 1997, pp. 295–345; H.A. Bethe, 'The Theory of Shock Waves for an Arbitrary Equation of State', Office of Scientific Research and Development, Report 545 (1942).

[33] Rudolf Peierls to Hans Bethe

[Manchester], 12.2.1935

Lieber Bethe,

1. Mit den Kellnerinnen hast Du ja in ein Wespennest gestochen. Genauer gesagt in den Williams, bei [dem] sich offenbar ganz tiefsitzende Komplexe aufgestört fanden, wie immer wenn jemand etwas besser weiss als er. So war er z.B. über den Anfang schwer gekränkt, in dem seiner Meinung nach dort nämlich behauptet war, dass er und Bragg der Meinung seien, dass tatsächlich Wechselwirkungskräfte auf grosse Entfernung vorhanden wären. Er legte grossen Wert darauf, mir an Hand seines unveröffentlichten Manuskripts zu beweisen, dass er wohl gewusst hätte, dass dies nicht der Fall sei. Da man wohl dieses Punktes wegen nicht gut nach Amerika schreiben konnte, wurde ich als Dein Agent aufgefasst und um die Genehmigung angegangen, diese und verschiedene andere Stellen leicht zu ändern. Diese Genehmigung gab ich dann schliesslich auch her, indem diese Stelle nämlich erstens wirklich nicht zu verstehen war, und indem Du ja zweitens alle diese Veränderungen in der Korrektur wieder rückgängig machen kannst, falls sie Dir nicht passen. Ferner legte Bragg Wert darauf, die Arbeit schon so weit stilistisch in Ordnung zu haben, dass sie Aussicht hat, gleich gedruckt zu werden, ohne vom Referenten noch einmal zurückgeschickt zu werden. Zu diesem Zwecke musste man im wesentlichen überall dort partition function einsetzen, wo Wahrscheinlichkeit stand, etc. Ferner legte Williams Wert auf eine aufklärende Fussnote, dass Dein s mit dem von BW in der Definition identisch sei. Er beantragte noch eine Fussnote, dass bei $z = \infty$ sich die BW-Theorie nicht nur in erster Näherung sondern sogar streng ergibt, was ich ihm an sich glaube, eine solche Bemerkung wollte ich Dir doch aber selbst überlassen. Schliesslich kostete es mich noch einen Nachmittag, Williams davon zu überzeugen, dass es berechtigt ist, im Summary zu sagen, die BW-Theorie gebe eine fair first approximation,

er wollte entweder nur fair oder nur first, beides konnte ich ihm aber nicht konzidieren.

Ich finde jetzt übrigens Deine Methode doch sehr schön, und habe sie gleich mit Begeisterung auf den Fall $p \neq \frac{1}{2}$ angewandt. Das sieht zunächst sehr schrecklich aus, denn man bekommt zunächst drei Parameter (Wirkung der äusseren Schalen auf die Ordnung der a-Atome, Wirkung der äusseren Schalen auf die Ordnung der b-Atome und partition function dass das Zentralatom b ist). Die Rechnung geht dann aber doch sehr leicht, und man bekommt das lustige Resultat, dass der Umwandlungspunkt bei einer bestimmten Konzentration (nämlich $1/z$) mit senkrechter Tangente zu Null geht. Die Ordnung bei $T = 0$ geht dementsprechend gleichfalls bei $p = 1/z$ plötzlich zu Null, von da steigt sie in einem sehr kleinen Intervall auf 1.

2. Beiliegend ein Fragebogen des Academic Assistance Council bezüglich Fröhlich, den sie diesem nicht nach Russland schicken wollen (ich halte dies für übertriebene Vorsicht). Ich nehme an, Du weisst über Fröhlich besser Bescheid als ich. Fülle also bitte alles so aus, was Du weisst und schicke ihn an den AAC zurück.

3. Gleichfalls beiliegend ein Brief von den Spring Mines, den wir versehentlich aufgemacht haben.*

4. Die Rechnung für die Diplonseparata betrug 18/3, ich bitte mir also 9 sh. gutzuschreiben.

5. Mit der Neutronenstreuung bin ich noch nicht viel weiter. Dagegen habe ich mit Leipunski einen Versuch überlegt, der anscheinend die beste Klärung der Frage ermöglicht, wie langsam eigentlich die wirksamen Neutronen sind (d.h. ob sie wirklich die Energie kT haben). Zu diesem Zwecke soll man Neutronen mit homogener Energie nehmen, die man durch Chadwick-Goldhabereffekt produzieren kann, und sie durch Paraffinschicht gehen lassen, die klein gegen die freie Weglänge rascher Neutronen ist. Aus der Zunahme der Wirksamkeit mit der Schichtdicke kann man dann ohne viel Rechnung (d.h. mit so viel Rechnung, wie sich

noch gut beherrschen lässt) auf die Grössenordnung der "wirksamen Energie" schliessen. Das merkwürdige ist, dass Leipunski sogar behauptet, dass dieser Versuch ausführbar ist, und dass er ihn machen wird.

Wir haben eine neue Jüdin und ab April neues (altes) Haus in Kingston Road. Wie gefällt es Dir in der Wildnis?
Beste Grüsse

Peierls

(handwritten addition)
Heisenberg legt Wert darauf, dass unsere Überlegung über mögliche Geschwindigkeitsabhängigkeit der Fermiglieder im Zusammenhang mit der Grössenordnungsschwierigkeit bei den Austauschkräften zitierfähig publiziert wird. Hast Du noch irgendeine Korrektur oder MS von der diesbezüglichen Diskussionsbemerkung in London? Das ist eigentlich öffentlich genug.

Bragg schreibt einen zweiten Teil seiner Arbeit und möchte daher gerne Deine Kurven für AuCu3 (kubisch-flächenzentriertes Gitter) haben, da dieser Fall häufig vorkommt. Hast Du die Absicht, das zu rechnen? Sonst bin ich gern dazu bereit, nur da die Rechnung wegen der Nachbarschaft zu 1. Ring nicht völlig trivial ist, lohnt es nicht, sie doppelt zu machen. Bragg möchte es aber bald haben, also lass mich bitte umgehend wissen, ob Du es machst, oder schon gemacht hast. Separata vom Diplon gehen demnächst an Dich ab, später folgt Liste an wen ich schon von hier aus geschickt habe. Mit der Massenversendung denke ich bis zum Erscheinen der Streuarbeit zu warten.
Mit Gruss auch von Genia

R.P.

*Im Augenblick identisch verschwunden, folgt im nächsten Brief.

[Manchester], 12.2.1935

Dear Bethe,

1. You stirred up a hornet's nest with the waitresses.[211] Or rather, you stirred up Williams.[212] His deep complexes seem to have been woken up, which always happens when someone knows something better than he does. For example, he felt hurt about the beginning of the paper.[213] According to him, it said that he and Bragg thought there really were interaction forces over long distances.[214] He thought it was very important to show me his unpublished manuscript, to prove that he knew this wasn't the case. As he couldn't send a letter about this to America, he asked me whether I, as your agent, would accept some slight changes in this and other passages. I finally gave my permission, first because this passage wasn't really understandable, and second because you can cancel all these changes in the corrections if you don't like them. Furthermore, Bragg insisted that the paper should be so well written that it would have a chance of being published as is, without being sent back by the reviewer. This basically means that we had to change 'partition function' for 'probability', etc. Moreover, Williams wanted an explanatory footnote to emphasize that your definition of s is identical to BW's. He also asked for a footnote to the effect that if $z = \infty$, the BW theory arrives at the same

[211]Genia Peierls had coined the phrase 'Kellnerinnen', waitresses, for Bethe's method of working out order/disorder transitions in binary alloys.

[212]Evan James Williams (1903–1945), studied at the University of Wales, Swansea and at Manchester, Cambridge from where he received his Ph.D. in 1929. After research in Copenhagen, he taught at Manchester, Liverpool and, from 1938 at Aberystwyth, where he remained until his death.

[213]H.A.Bethe, 'Statistical Theory of Superlattices', *Proc. Roy. Soc.* **A150**, 552–575 (1935); (communicated December 1934).

[214]The relevant papers by Bragg and Williams are: W.L. Bragg and E.J. Williams, 'The Effect of Thermal Agitation on Atomic Arrangement in Alloys', *Proc. Roy. Soc.* **A145**, 699–730 (1935), W.L. Bragg and E.J. Williams, 'The Effect of Thermal Agitation on Atomic Arrangement in Alloys. II', *Proc. Roy. Soc.* **A151**, 540–566 (1935), E.J. Williams, 'The Effect of Thermal Agitation on Atomic Arrangement in Alloys. III', *Proc. Roy. Soc.* **A152**, 231–252 (1935).

answer not just in the first approximation but even in the exact solution. This might be true, but I wanted you to make the decision on such a note. It took me one more afternoon to convince Williams that we were justified in stating in the summary that the BW theory would give a fair first approximation. He wanted to accept only 'fair' or 'first'; I could not concede him both.

I now like your method very much, by the way, and have immediately applied it to the case $p \neq \frac{1}{2}$. This looks terrible at first, since you get three parameters (the effect of the outer shell on the order of the a-atoms, the effect of the outer shell on the order of the b-atoms, and the effect of the partition function if the atom in the centre is of type b). The calculation turns out to be rather easy anyway, and you get the funny result that at one specific concentration (which is $1/z$) there is a singularity and the conversion point goes to zero. The order at $T = 0$ therefore goes to zero as well when $p = 1/z$, then over a very small interval increases to 1.

2. Enclosed you will find a questionnaire from the Academic Assistance Council[215] concerning Fröhlich,[216] which they don't want to send to him in Russia (I think this is too cautious!). I presume that you know more about Fröhlich than I. Please fill in what you know, and send it back to the AAC.

3. I have also enclosed a letter from Spring Mines that we accidentally opened.*

4. The total cost for the dipole reprints was 18/3, so I would ask you to credit me 9 sh.

---

[215]The Academic Assistance Council, later the Society for the Protection of Science and Learning, was formed in May 1933 to help central European scholars persecuted by the Nazis. Its Academic Assistance Fund provided research fellowships to distinguished academics.

[216]Herbert Fröhlich (1905–1991), studied in Munich under Arnold Sommerfeld, where he obtained his Ph.D. in 1930. After being dismissed from his post at Freiburg in 1933 he joined Joffe's Physio-Technical Institute in Leningrad as a 'foreign expert'; in 1935 he was forced to flee again. He took up a position in Bristol, where he stayed until 1948 when he took up the first Chair of Theoretical Physics at Liverpool which he held until his retirement in 1973.

5. I haven't really made progress with the neutron diffraction.[217] I and Leipunski have figured out an experiment, which might be the best way to clarify the question of how slow the effective neutrons are (i.e., whether they really have energy kT). For this purpose you need a source of neutrons with uniform energy, which you could produce with the Chadwick-Goldhaber effect.[218] Then you let them pass through a sheet of paraffin, whose thickness has to be small relative to the mean free path of fast neutrons. If you find that the effect increases with the thickness of the sheet then you will learn the order of magnitude of the "effective energy" without too much calculation (meaning just as much calculation as you can easily handle). Strangely, Leipunski is even claiming that the experiment could be done and that he will do it.

We have a new Jew here, and from April a new (old) house on Kingston Road. How do you like it in the wilderness?
Sincerely,

Peierls

(handwritten addition)
Heisenberg has emphasized that our thoughts on the possible velocity dependence of Fermi elements, and its relation to the difficulty of predicting the right order of magnitude for the exchange force, should be published in such a way that it can be easily quoted. Do you have any corrections or a transcript of your comments in the discussion, which can address this? I'd say this is enough for the public.

Bragg is working on writing the second part of his paper,[219] and would therefore like to have your curves for $AuCu_3$ (face-centered cubic lattice) as this case is quite common. Are you planning to calculate this? I'm ready to do this myself, but as the calculation isn't trivial, due to its proximity to the first ring, it isn't worth doing twice. But

[217] H.A. Bethe and R.E. Peierls, 'The Scattering of Neutrons by Protons', *Proc. Roy. Soc.*, **A149**, 176–183 (1935).
[218] J. Chadwick and M. Goldhaber, 'The Nuclear Photoelectric Effect', *Proc. Roy. Soc.* **A151**, 479–493 (1935).
[219] See above letter [33], note 214.

Bragg would like to have it soon, so please let me know immediately whether you are planning to do it or have already done it. I will send you reprints of the dipole first, and then follow up with a list of other people to whom I have sent them. I think I will delay the mass sending until the scattering paper has appeared.
Greetings also from Genia

R.P.

*This seems to have disappeared for the moment, and will follow with the next letter.

[34] Rudolf Peierls to Hans Bethe

[Manchester], 25.4.1935

Lieber Bethe,
Kennst Du die Geschichte von dem Amerikaner, der in der Inflationszeit in einem deutschen Restaurant für einen Dollar ein Essen bestellte, und der nie fertig wurde, weil der Dollar inzwischen immer noch gestiegen war? Genauso geht es mir mit diesem Brief, denn immer wenn ich gerade genügend Zeit habe, um zu schreiben, kommt wieder etwas neues dazu, was mit in den Brief soll, und dann reicht die Zeit nicht aus, und es lohnt sich folglich nicht anzufangen. Einmal muss man aber doch irgendwo anfangen. Deine Theorie der Capture und Streuung hat mir sehr gut gefallen, übrigens scheint das Fermi wieder mal schon lange gewusst zu haben, jedenfalls wenn ich seine Arbeit in den Proceedings richtig verstanden habe. Die Notwendigkeit von Neutronen thermischer Energie scheint mir aber doch eine ernsthafte Schwierigkeit zu sein. Da nämlich meine Rechnungen über die Steuung noch immer nicht konvergiert haben (die Schwierigkeit liegt darin, dass man Deinen hübschen Witz nicht anwenden kann, da die Unterschiede zwischen den Energien verschiedener Neutronen, die eine bestimmte Anzahl Stösse erlitten haben, von derselben Grösse sind, wie diese Energien selbst) so habe ich die Cambridger dazu überredet, doch einmal Experimente

zu machen, aus denen man mit weniger Rechnung direktere Schlüsse ziehen kann. Das scheint nun geschehen zu sein, besonders Leipunsky und Goldhaber haben verschiedene Messreihen gemacht, indem sie die Zunahme der Anzahl der "langsamen" Neutronen mit der Dicke der Paraffinschicht bei sehr kleiner Dicke gemessen haben. Dabei stellt sich z.B. heraus, dass bei 1 cm Paraffinschicht schon etwa 2% aller Neutronen verlangsamt werden. Nun stossen in einer so dünnen Schicht nur etwa 10% überhaupt erst einmal. Die Anzahl derer, die zweimal stossen, wird also schon von der Grössenordnung 2% sein, wenn nicht der Stossquerschnitt für die einmal gestossenen schon weit grösser als der für rasche Neutronen ist. Es scheint also zu folgen, dass entweder die zweimal gestossenen Neutronen schon langsam sind, oder dass die einmal gestossenen einen sehr grossen Wirkungsquerschnitt haben. Beides wäre wohl mit Deinem Modell nicht in Einklang zu bringen. Allerdings sind diese Versuche, die übrigens nicht abgeschlossen sind, nicht ganz einwandfrei. Erstens kommen aus der Quelle (Be + γ-Strahlen) doch noch ein paar langsame Neutronen, die von den experimentellen Werten abgezogen sind. Es könnte also sein, dass das erste Anwachsen bei dünnen Paraffinschichten auf die weitere Verlangsamung dieser Neutronen zurückzuführen ist.

Zweitens sind diese Versuche mit einem Lithiumzähler gemacht. Nun hat aber Li einen wesentlich kleineren Wirkungsquerschnitt als etwa Silber oder gar Cd. Man könnte also vermuten, dass Li nicht so sehr selektiv für ganz langsame Neutronen ist. Dann sollte die Kurve mit einem anderen Zähler langsamer mit der Paraffindicke ansteigen. Das ist in der Tat auch der Fall, denn z. B. die Kurve von Bjerge und Westcott in den Cambridge Proc. steigt langsamer als die, die mir Goldhaber zeigte. Leider haben Bjerge und Westcott in dem interessanten Gebiet dünner Schichten nicht genau genug gemessen, man kann also aus ihren Werten nicht viel schliessen. Trotz aller dieser "Abers" scheint es doch jedenfalls sehr schwer zu glauben, dass die Neutronen wirklich nur kT haben. Ganz Cambridge besteht auf mindestens 104 Volt.

Diese Experimente würden übrigens noch viel sauberer werden, wenn man ähnliche Messungen, (insbesondere solche wo die Neutronen durch Photoeffekt bekannter γ-Strahlen entstehen und daher kein Kontinuum bilden) mit schwerem Wasser statt Paraffin vornehmen würde. Dann ist nämlich der Energieverlust, den ein Neutron in einem Stoss erleiden kann, begrenzt, und aus der Anzahl der Stösse lässt sich streng eine untere Grenze für die Endenergie angeben. Die Cambridger können das nicht machen, weil sie nicht genügend schweres Wasser haben. Kannst Du nicht jemand dazu überreden? Chadwick behauptet übrigens weiter, dass der Photoeffekt im Deutron eine richtungsunabhängige Emission von Neutronen ergibt, insbesondere dass nach vorn und unter 90° gleich viel emittiert wird. Wenn das kein Dreckeffekt ist, (was mit vorläufig noch plausibel scheint) und wenn die verwendete γ-Strahlung nicht zufällig eine Linie dicht an der Kante enthält, (dann würden nämlich Proton und Neutron beide langsam und würden beide nach vorn gehen, um den Impuls zu erhalten, die Linie müsste aber, um nach vorn und nach der Seite gleich viel zu bekommen bis auf 3000 Volt mit der Kante übereinstimmen die kleine Wahrscheinlichkeit des Effekts in der Nähe der Kante könnte durch die hohe Wirksamkeit langsamer Neutronen allenfalls kompensiert werden) wenn das alles also nichts hilft, dann ist das doch sehr sehr interessant! Denn dann heisst es, entweder kommen die Neutronen auf einer s-Eigenfunktion des Kontinuums heraus — dann kann der Grundterm kein s-Term sein und ist vermutlich p. Dies würde eindeutig gegen Wigner und gegen Heisenberg sprechen und die Majoranasche Theorie direkt beweisen. Diese allerdings mit dem umgekehrten Vorzeichen der Austauschfunktion, sodass die Majoranasche Begründung seines Ansatzes, sowie die ganze schöne Wignersche Diskussion (wie auch nota bene unsere beiden Arbeiten) wegschwimmen. Oder aber, zweite Möglichkeit, die Neutronen kommen auf einer d-Bahn mit einem Maximum bei 45° und 135° heraus, dann ist zwar der Grundterm noch ein s, aber der Dipolübergang fällt wegen schlechten Wetters aus, und es passiert nur der Quadru. Grund unverständlich. Oder schliesslich die Linie ist eine Interkombinationslinie mit starker Mitwirkung des Spins, was auch nicht viel leichter zu verdauen ist. Wie dem auch sei, sollte Chadwick recht haben, so ist an allem, was bisher über die Wechselwirkung von Neutron und Proton gesagt worden ist, kein wahres Wort.

Was die Kellnerinnen betrifft, so glaube ich jetzt, dass Bragg und Williams doch damit recht haben, dass im Falle AuCu3 eine Umwandlungswärme existiert. Ein erster Ansatz mit Deiner Methode hat sich allerdings als zu grob erwiesen. Ich nahm dort nur die Wechselwirkung der Atome zwischen Zentrum und erster Schale mit, während die Wechselwirkung der Atome in der ersten Schale untereinander mit in das e (genauer, in die drei Parameter, die dort leider an die Stelle von Deinem e treten) hineingesteckt werden sollte. Das führt aber zu einer viel zu hohen Ordnung insbesondere zu einer zu hohen "flüssigen Ordnung" über dem Umwandlungspunkt. Ich bin jetzt gerade dabei, die Wechselwirkung in der ersten Schale mitzunehmen, vermutlich braucht man das nur für den Zustand ohne long distance order zu tun. Es ist aber auch eine ziemliche Viecherei.

Fowler hat übrigens einen jungen Mann, der sehr keen darauf ist, sich mit Kellnerinnen einzulassen. Da Fowler selbst Deine Arbeit nur zur Hälfte gelesen hat, und eine Fortsetzung der Lektüre ihm sein immer vorhandenes Refereegewissen nicht erlaubte, so wird er mir den Jüngling herschicken, und ich werde ihn als willkommenen Rechenknecht ausbeuten.

Williams hat sich jetzt mit der Existenz Deiner Arbeit ausgesöhnt. Er beschäftigt sich jetzt damit, die verschiedenen Resultate verschiedener Methoden in verschiedene Variable umzuschreiben, und will hierüber auch etwas publizieren. Ab und zu kommt übrigens dabei auch etwas heraus, z.B. hat er zu Recht festgestellt, dass Dein Integralkriterium nur dann etwas über die Güte der Theorie aussagt, wenn man von der Theorie weiss, dass sie sicher eine zu grosse, oder sicher eine zu kleine Ordnung ergibt. Es besagt ja nichts anderes als dass die gesamte Entropie, d.h. die Anzahl der Konfigurationen stimmt. Daher ist es z.B. in der Theorie von Bragg und Williams streng erfüllt. Er hat auch verstanden, was eigentlich die Ordinate und Abszisse in den Braggschen Kurven sind (sowas ist immer sehr instruktiv) nämlich sie bedeuten Ableitung der Energie und der Entropie nach der Ordnung. Der Schnittpunkt bedeutet also, dass die freie Energie ein Extremum ist. Auf diese Weise kann man dann auch verstehen, welcher von mehreren Schnittpunkten der richtige ist. Dies war der einzige nicht triviale Punkt in Fowler's Rotationsarbeit, und den hat er dort prompt falsch gemacht. Die Arbeit von Heitler und Nordheim, (da wir schon mal bei falschen Arbeiten sind) die Du im Zentralblatt mit so gefurchter Stirn referiert hast, ist natürlich auch falsch. Es ist natürlich keine Rede davon, dass zwei Kerne vom gleichen e/m beim Stoss keine Paare erzeugen, und ebensowenig ist es richtig, dass der Effekt mit M^2 geht. Vielmehr haben H.und N. nur die niedrigste auftretende Potenz der Geschwindigkeit genommen, ohne sich zu überlegen, wie gross der Faktor ist. Das ist natürlich leicht richtigzustellen, und ich hatte Herrn Finkelstein aus Dnepropetrowsk an diese Frage gesetzt, er ist aber schon wieder weg, und ob er das alleine fertig bringt, weiss der Himmel.

Dass Chadwick Professor in Liverpool wird, hast Du vermutlich gehört. Ich spare auf diese Weise viel Fahrgeld, aber für die Wissenschaft ist es schade, denn er wird dort kein Radium haben, sondern nur mit einer 10^5-Volt-Anlage (nebbich!) arbeiten.

Beiliegend ein Quartal gemischte Post, das eigentlich schon lange hätte nachgeschickt werden sollen, es war aber in tiefere geologische Schichten versunken und kam erst beim Umzug wieder zum Vorschein.

Der Umzug ging gut, und zur Erschütterung der Nachbarn an einem Sonntag vor sich. Du hast recht, Kingston Road ist die Strasse neben Didsbury Park, in der Du mal Kindermädchen spielen musstest während wir ein Haus ansahen. Unser jetziges Haus ist zwar nicht dasselbe aber sehr ähnlich. Es ist vor allem gross, und erregt den Neid aller Besucher.

Die deutsche Jüdin No. 2 wurde plötzlich abgerufen und verschwand identisch innerhalb von 6 Stunden. Am nächsten Tag legte sich Genia mit Grippe und einer Lebergeschichte (diesmal wirklich Leber!) ins Bett. Natürlich war dann auch noch MacGregor verreist und Mrs Higgins hatte eine verstauchte Hand. Jetzt sind wir ernsthaft auf der Suche nach einem Dienstmädchen. Anlässlich der Hochzeit von Tolansky mit der Tochter des Kantors Pinkasovitch waren wir alle in der Synagoge, was, wie Du Dir vorstellen kannst ein reines Vergnügen war. U.a. ergab sich eine völlig unerwartete Entdeckung: Die alten frommen Juden (je älter und je frömmer desto mehr) vollführen dort beim Beten rythmische Schwingungen des Oberkörpers, die völlig mit den sprichwörtlichen Schwingungen von Pauli identisch sind. Ein ungemein interessanter Fall von "inheritance of acquired characters" unter erschwerenden Umständen. Pauli spielt übrigens Tennis und hat sich einen Flügel gekauft, auf dem er spielt.(!!)

Ich habe übrigens, wie ich sehe, noch eine Frage Deines Briefes nicht beantwortet: Die Umwandlungstemperatur im Falle eines einfach-kubischen Gitters geht für z=∞ wie $p(1-p)$ (wo $p =$ Konzentration einer Komponente). Im Falle von endlichem z geht sie sehr ähnlich, solange p wesentlich grösser als $1/z$ ist, dann fällt sie natürlich steiler ab. Das Bradleysche Resultat kann man also auf diese Weise nicht verstehen, jedenfalls kann man nicht verstehen, warum die Linien von FeAl gerade bei 25% exakt verschwinden. Ich habe aber auch schon vermutet, dass das ein Zufall ist, und dass für die von Bradley gerade benutzte Wärmebehandlung die Umwandlungstemperatur gerade bei 25% nicht mehr erreicht wird. Er hat auch versprochen, Messungen mit variierender Wärmebehandlung zu machen.

Sykes macht jetzt auch Röntgenaufnahmen und bekommt dabei endlich das heraus, was wir immer haben wollten: Wenn man eine Legierung (in seinem Falle allerdings Cu3Au) auf eine nicht zu tiefe Temperatur abschreckt und dort hält, so kann man sehen, wie die Ordnung allmählich entsteht. Dabei sind die Ueberstrukturlinien wenn sie erscheinen, zunächst sehr breit, erst allmählich werden sie schmaler. Die Zeit, die hierfür erforderlich ist (Bildung von grossen Gebieten einheitlicher Ordnung) ist wesentlich länger und stärker temperaturabhängig als die, die für das Auftreten der Linien erforderlich ist, und die auch für die Zeitabhängigkeit des elektrischen Widerstandes charakteristisch ist.

Fankuchen fragt mich jeden Tag, was ich von Dir höre und ob Du seine Voraussagen über Ithaca bestätigt gefunden hättest. Bergs und Deutsche haben je einen Sohn. Bell ist melancholisch und schimpft auf "this married life", Hartree ist so rot und so dick wie noch nie, denn seine Maschine ist fertig und geht sogar, und sie ist wirklich reizend. Ueber die Note von Born und Schrödinger habe ich mich sehr geärgert, denn ich hatte Born schon immer mit Briefen bombardiert, um ihm klarzumachen, dass ein Fehler seiner Theorie der ist, dass sie eben nicht die richtige Formel für die magnetische Selbstenergie liefert, die er jetzt als eine triviale Konsequenz seiner Theorie behandelt. Born geht es übrigens jetzt recht schlecht, er hat ein Gallenleiden und man kann ihn nicht beschimpfen. Bretscher sitzt mit einem Rockefeller in Cambridge, da ihm Scherrer verboten hat, zu Blackett zu gehen.

Das ist aber, glaube ich, alles. Beste Grüsse

R. Peierls

(Handwritten addition)
Könntest Du nicht gelegentlich den Spring Mines verraten, dass Du in den USA bist?

(Handwritten addition by Genia)
Lieber Bethe,
Bist Du schon ganz verwildert? Isst Du nur Quark mit Orangensauce und liest Du nur den Ithaca Herald? Hier ist noch ein Amerikaner King (nicht Kingfish) und er wohnt in Alderley Edge! Ho ho! Wir blühen (augenblicklich). Haus ist unglaublich schön und gross, alle platzen, sogar Deutschs. Wir haben für 2£ Farbe gefunden — ... Schränke und unglaublich hübsch! Und die Wandschränke in der Küche sind orange wie verrückt gewordene Apfelsinen. Bergs haben Sohn mit Kaiserschnitt, und Deutschs ohne. Gaby spricht, läuft sehr schnell, ist zu dick und arbeitet im Garten. Wann kommst Du in das arme kleine Europa? Und zu uns? Grüsse, Grüsse, und Grüsse. Hast Du schon Silberminen geholt? Und wieviel hast Du in boom verdient?

*Im Augenblick identisch verschwunden, folgt im nächsten Brief.

[Manchester], 25.4.1935

Dear Bethe,
Have you heard the story of the American who ordered a meal in a German restaurant for one dollar during Hyperinflation,[220] and could never finish it since the dollar kept rising higher and higher? I feel the same with this letter, since whenever I find sufficient time to write, something new is happening that I also want to write about. Then there is not enough time anymore, so it's not worth starting the letter.

At some point you just have to start anyway. I liked your theory of capture and scattering very much.[221] As usual Fermi seems to have

[220]This refers to hyperinflation in Germany in the second half of 1923, when the Reichsmark lost value against the dollar by 381,700,000,000% between July and November of that year.

[221]The research was about to be published (see below, letter [35], note 239) and Peierls had read a draft of the work.

known about it for a long time, by the way, if I understand his paper in the Proceedings properly[222] The need for thermal neutrons still seems to be a serious difficulty. My calculations on the scattering still haven't converged (the difficulty is that we can't apply your nice trick, because the variation in neutron energies after colliding a certain number of times is as large as their initial energy, so I have convinced the scientists at Cambridge to do some experiments which will give more direct results with fewer calculations. This now seems to have happened; Leipunsky[223] and Goldhaber[224] have performed different series of measurements, by measuring the increase of the number of "slow" neutrons with increasing thickness of the paraffin sheets in very thin sheets.[225] It turns out, for example, that for a paraffin coating 1 cm thick 2% of all the neutrons are already slowed down. In a coating this thin, only about 10% of them hit each other at all. The number of neutrons colliding twice will be in the order of 2%, if the cross-section for a neutron that has already scattered once isn't much bigger than that of the faster neutrons. It seems to follow that either the neutrons colliding twice were already slow, or the neutrons colliding once have a very large cross-section. Neither fits your model, but these experiments

[222]Fermi had published two papers on a related topic in 1934. E. Fermi, E. Amaldi, O. D'Agostino, *et al.*, 'Artificial Radioactivity Produced by Neutron Bombardment', *Proc. Roy. Soc.* **A146**, 483–500 (1934), E. Fermi, E. Amaldi, O. D'Agostino, *et al.*, 'Artificial Radioactivity Produced by Neutron Bombardment. II', *Proc. Roy. Soc.* **A149**, 522–558 (1935).

[223]Alexander Leipunski, Ukrainian physicist, at the time working in Cambridge.

[224]Maurice Goldhaber (1911–), completed his Ph.D. in Cambridge in 1936. In 1938 he took up a post at the University of Illinois before moving to Brookhaven in 1950 where he stayed until 1973. Since 1961 he has also been Adjunct Professor at Stony Brook.

[225]Several research groups were concerned with these experiments and results were published in subsequent years. For example, M. Goldhaber, 'Scattering of Neutrons by Protons', *Nature* **137**, 824–825 (1936); A. Leipunksi, L. Rosenkewitsch and D. Timoshuk, *Phys. ZS.SU.* **10**, 625 (1936); A. Leipunksi, L. Rosenkewitsch and D. Timoshuk, *Phys. ZS.SU.* **10**, 751 (1936); A.I. Leipunski, 'Determination of Energy Distribution of Recoil Atoms during β-Decay and the Existence of the Neutrino', *Proc. Camb. Phil. Soc.*, **32**, 301–303 (1936); T. Goloborodko and A. Leipunski, 'Scattering of Photoneutrons from Deuterium by the Nuclei of Atoms of Light Elements', *Phys. Rev.* **56**, 891–892 (1939).

haven't been finished yet and are not absolutely accurate. First, there are still some slow neutrons coming from the source (Be + γ-rays) which have been subtracted from the experimental values. It could therefore be that the increase seen in the case of thin paraffin sheets could result from additional slowing of these neutrons.

Second, these experiments were done with a lithium counter. But lithium has a much smaller cross-section than, for example, silver or even Cd. So you might think that lithium is not a very good way of selecting very slow neutrons. With a different counter, the curve should rise more slowly with increasing paraffin thickness. This is actually the case; for example, the curve of Bjerge and Westcott in the Cambridge Proc. rises more slowly than the one Goldhaber showed me.[226] The measurements of Bjerge and Westcott weren't accurate enough in the interesting regime of thin sheets, unfortunately, so their values don't tell us much.

Despite all these "buts", it is still very hard to believe that the neutron energy is really just kT. All of Cambridge insists on at least 10^4 Volts. These experiments would be much more precise, by the way, if you did similar measurements with heavy water instead of paraffin. (Especially if the neutrons were built up using the photo effect of γ-rays with known energy, and therefore didn't form a continuum.) This would limit the neutron's energy loss from a single impact, so the number of impacts would impose a strict lower limit on the final energy. The Cambridge scientists can't do this, as they don't have enough heavy water. Couldn't you try to persuade someone to do this? Chadwick still claims that the photo effect on deuterons creates a neutron radiation which is independent of direction, and in particular that the radiation is the same in forward emission and at angles of less than 90°. If this is not a perturbation effect (which seems plausible to me at the moment), and if the γ-radiation used is not from a line close to the edge by coincidence (Since then the proton and neutron would both be slow and go towards the front to scatter. For the front and sides to be the same,

[226] T. Bjerge and C.H. Westcott, 'On the Slowing down of Neutrons in Various Substances Containing Hydrogen', *Proc. Roy. Soc.* **A150**, 709–728 (1935); C.H. Westcott and T. Bjerge, 'Some Experiments on the Slowing Down of Neutrons by Collisions with Hydrogen Nuclei', *Proc. Camb. Soc.* **31**, 145 (1935).

however, the line would have to near the 3000 Volt edge, and the small probability of an effect next to the edge could at least be compensated for by the high efficiency of slow neutrons), and if that all doesn't help, then it is still very, very interesting! This would mean that the neutrons are coming from an s-eigenfunction of the continuum — but then, the ground state cannot be an s-value and is presumably a p. This would definitely contradict Wigner and Heisenberg, and directly prove Majorana's theory.[227] But Majorana's theory[228] has no inverted signs of the exchange function, so both his argument and the whole of Wigner's nice discussion[229] (just like *nota bene* our work) come to nothing. There's also another possibility: that the neutrons come out on a d-heading with maxima at 45° and 135°. Then the basic term is s again, but the dipole transition is cancelled due to bad weather and only the quadrupole happens. The reason is incomprehensible. Or finally, perhaps the line is an interrelation line with a strong spin effect, which isn't much easier to take. Be that as it may, if Chadwick is right then everything we know about the interaction between neutron and proton is wrong.

Concerning the waitresses, I now think Bragg and Williams were right that the transformation heat can be calculated for $AuCu_3$. A first attempt using your method turned out to be too imprecise. I have just taken into account the interaction between the centre of the atom and the first shell of the atoms. The interaction of atoms in the first shell should be put in terms of ε (or to be more precise, in terms of three separate parameters which unfortunately replace your ε at this point in the calculation). But this leads to a degree of order that is far too high, and especially to a "fluid order" that is well over the conversion point. I'm now trying to take the interactions in the first shell into account, which you probably only have to do for those states without any long-range order. But this is also quite a torture.

---

[227] Ettore Majorana (1906–1938), had joined Fermi's team in Rome as one of the 'Via Panisperna Boys'; he later joined Heisenberg in Leipzig, where he carried out his investigation of exchange forces which further refined Heisenberg's theory of the nucleus.

[228] E. Majorana, 'Über die Kerntheorie', *Z. Phys.* **82**, 137–145 (1933).

[229] E. Wigner, 'Über die Streuung von Neutronen an Protenen', *Z. Phys.* **83**, 253–258 (1933).

Fowler has a young man who is very keen to start working with the waitresses, by the way. Fowler has just read half of your work, and his never-dormant referee's conscience prevents him from reading any more of it, so he is sending this young man to me and I will exploit him as a calculation servant.

Williams has reconciled himself to the existence of your work. He is now working on rewriting the various results of different methods in different variables, and wants to publish something about it. Sometimes he even finds out something, e.g. that he was right when he said that your integral criterion only says something about the quality of a theory if you already know that the theory gives too much or too little order. In other words, it says nothing except for whether the total entropy (i.e., the number of configurations) is correct. He has also understood what the ordinates and abscissa represent in Bragg's curves (which is always very instructive). They are the derivatives of energy and entropy with respect to the order, so an intersection point is where the free energy is an extremum. This way you can also tell which of several intersections is the right one. This was the only non-trivial aspect of Fowler's rotation paper, and he promptly got it wrong.[230]

The paper by Heitler and Nordheim (since we are talking about wrong papers),[231] of which you gave a review in the *Zentralblatt* with many concerns (meaning with a lot of frowning), is also certainly wrong. It is of course impossible that two nuclei which have the same e/m ratio don't produce any pairs in a collision, and it is not right to say that the effect works with M^{-2}. All H. and N. have done is to take the lowest possible power of velocity without thinking about how big the factor needs to be. This is easy to correct, and I asked Mister Finkelstein from Dnepropetrovsk to find the answer to this question. He has already left, however, and heaven knows whether he can do it on his own.

I guess you have heard that Chadwick will be appointed as a professor at Liverpool. I can save a lot in transportation expenses, but it's

[230]R.H. Fowler, 'A Theory of the Rotations of Molecules in Solids and of the Dielectric Constant of Solids and Liquids', *Proc. Roy. Soc.* **A149**, 1–28 (1935).

[231]W. Heitler and L. Nordheim, 'Production of Pairs by Collision of Heavy Particles', *Journal de Physique et le Radium* **5**, 449–454 (1934).

a pity for the sciences as he won't have any radium and their facility is only 10^5 Volts (nebbish!).

Enclosed you will find mixed post from the last quarter, which was supposed to be forwarded much earlier but disappeared into deeper geologic layers and only just turned up again during our removal. The removal went fine, and it was a shock for the neighbours that it happened on a Sunday. You were right, Kingston Road is the street next to Didsbury Park where you had to play nanny while we were looking for a house. Our present house is not the same, but looks quite similar. First and foremost, it's huge and all our visitors are quite envious.

German Jew No. 2 was suddenly called away, and disappeared completely within 6 hours. The next day Genia was lying in bed with the flu and a liver problem (this time, really the liver!). MacGregor was away, of course, and Mrs. Higgins had a sprained hand. Now we are really looking for a maid. We all went to the synagogue for Tolansky's[232] wedding to the daughter of cantor Pinkasovitch, which as you can imagine was a pure pleasure. Among other things, I made a totally unexpected discovery: the pious old Jews (the older they are, the more pious) were performing rhythmic movements with their upper bodies while praying, which were totally identical to Pauli's proverbial movements.[233] A very interesting case of "inheritance of acquired characteristics" under difficult conditions. Pauli is playing tennis now, and he has bought a grand piano which he actually plays. (!!)

I just noticed that I haven't answered one of the questions in your letter: the conversion temperature for a simple cubic lattice is just the same for $z = \infty$ as for $p(1-p)$ (where $p =$ the concentration of one component). The case of finite z is quite similar, but of course as long as p is much bigger than $1/z$ it will fall off more steeply. Bradley's result[234] can't be understood this way; at least, one can't understand

---

[232]Samuel Tolansky (1907–1973), professor of physics at Royal Holloway College, London, 1947–1973.

[233]See comments on Pauli in a publication of his colleagues written during the Odessa Conference in 1930. Lee, *Selected Correspondence*, 1, chapter 2.

[234]A.J. Bradley and R.A.H. Hope, 'The Atomic Scattering Power of Iron for Various X-Ray Wave Lengths', *Proc. Roy. Soc.* **A136**, 272–288 (1932), A.J. Bradley and A.H. Jay, 'The Formation of Superlattices in Alloys of Iron and Aluminium', *Proc.*

why the lines of FeAl disappear exactly at 25%. I suspected that this is a coincidence and that for the heat treatment used by Bradley the conversion temperature isn't reached at just 25% anymore. He has also promised to take some measurements with different heat treatments.

Sykes is also doing X-ray pictures, and he has finally found out what we always wanted to know.[235] If you cool an alloy (in his case Cu_3Au) down to some temperature that is not too low and keep it there, then you can see how order is gradually built up. If any hyperstructure lines can be found, they will be very broad at first and only start getting smaller a while later. The time needed for this (the production of large areas of equal order) is much longer (and more closely related to temperature) than the time needed for the first appearance of the lines. This behaviour is also characteristic of the dependence of the electrical resistance on time. Fankuchen[236] asks me every day what I have heard from you, and whether you could confirm his predictions about Ithaca. Bergs and Deutsche have both had a son. Ball is melancholic, and complains about "this married life". Hartree[237] is redder and fatter than ever before, because his machine is finished and even working, which is really cute.

I was very upset about the note by Born and Schrödinger,[238] as I had bombarded Born with letters over and over again to make him understand that his theory has the error that it doesn't give the right formula for magnetic self-energy. He is now treating it as a trivial con-

Roy. Soc. **A136**, 210–232 (1932); A.J. Bradley, 'The Crystal Structure of the Heusler Alloys', *Proc. Roy. Soc.* **A144**, 340–359 (1934).

[235] C. Sykes, 'Methods for Investigating Thermal Changes Occurring during Transformations in a Solid Solution', *Proc. Roy. Soc.* **A148**, 422–446 (1935).

[236] Isidor Fankuchen (1904–1964), crystallographer; he and his wife became good friends of Genia and Rudolf Peierls after they had emigrated to England. Fankuchen later moved to the US where he was Professor of Physics at the Polytechnic Institute of Brooklyn from 1942 to 1964.

[237] Douglas Rayner Hartree (1897–1958), studied at Cambridge where he obtained his Ph.D. in 1926. In 1929 he became Professor of Applied Mathematics at Manchester; in 1946 he took up the Plummer Professorship of Mathematical Physics at Cambridge.

[238] M. Born and E. Schrödinger, 'The absolute field constant in the new field theory', *Nature* **135**, 342 (1935).

sequence of his theory. Born is not well at all, by the way; he has a bilious complaint and one shouldn't offend him at the moment.

Bretscher is in Cambridge with a Rockefeller grant, since Scherrer hasn't allowed him to go to Blackett. That is all, I think.
Sincerely,

R. Peierls

(Handwritten addition)
Couldn't you tell the Spring Mines occasionally that you are in the USA?

(Handwritten addition by Genia)
Dear Bethe,
Have you become totally uncivilized? Do you just eat curd cheese with orange sauce and read the Ithaca Herald? We have an American here too, called King (not Kingfish), and he is staying at Alderly Edge! Ho ho! We are blossoming (at the moment). The house is incredibly nice and huge, and everyone is very envious, even Deutschs. We have found colour for 2£ — ... wardrobes, and incredibly nice! And the cupboards in the kitchen are as orange as crazy oranges. Bergs had a son by caesarean, and Deutschs without. Gaby is talking, running very fast, too fat, and now working in the garden. When are you coming back to poor little Europe? And to us? Greetings, greetings, and greetings. Have you got any silver mines? And how much have you earned in the boom?

*At the moment it has similarly disappeared, and will come with the next letter.

[35] Rudolf Peierls to Hans Bethe

[Manchester], 8.6.1935

Lieber Bethe,

Deine *Phys. Rev.* Arbeit über "Zertrümmerung von Neutronen" (nach Behauptung der Seitenüberschriften) ist sehr hübsch. Ob ich sie Dir glauben soll, weiss ich eigentlich aber noch nicht. Folgende Dinge scheinen mir dagegenzusprechen: 1. Wie schon in meinem letzten Brief erwähnt, scheint es fast unmöglich, zu verstehen, wie die Neutronen schon so rasch auf thermische Geschwindigkeiten kommen. 2. was ärger ist, scheint es — nach einer Diskussion des empirischen Materials, die ich mit Ehrenberg durchgeführt habe, unmöglich, den scheinbaren Wirkungsquerschnitt von 10^{-24} für capture schneller Neutronen als Wirkung von langsamen Neutronen zu interpretieren. Nämlich erstens tritt dieser Effekt in einigermassen reproduzierbarer Grösse auch dann auf, wenn man als Neutronenquelle Deuton plus Deuton benutzt. Dann ist es aber sehr schwer einsehbar, woher die langsamen Neutronen kommen sollen. Allerdings liegen über diesen Punkt, soweit mir bekannt ist, nur Experimente von Bjerge und Westcott vor, die vor der Entdeckung des Fermieffektes gemacht worden sind, und bei denen möglicherweise Dreck eine Rolle spielen könnte. Ferner würde aber auch der Tatsache, dass z.B. der anscheinende Wirkungsquerschnitt schneller Neutronen in Ag von 10^{-24} in Wirklichkeit davon herrührt, dass 1/1000 der Neutronen einen Wirkungsquerschnitt von, sagen wir 10^{-21} die anderen gar keinen haben, aus dieser Tatsache, wollte ich diesen langen Satz anfangen, würde folgen, dass eine Schicht von ca. 10^{-2} cm Ag diesen Effekt der "schnellen" Neutronen völlig wegabsorbieren sollte (auch ohne jegliches Paraffin). Von einer solchen Wirkung ist jedoch nichts bekannt, zwar scheint das Experiment nie direkt in dieser Form ausgeführt worden zu sein, doch ist es z.B. sicher, dass weder dicke Cd. noch I-Schichten den Effekt in Ag merkbar reduzieren. Ich versuche Ehrenberg dazu zu überreden, das Experiment mit dem Ag Absorber und Ag Indikator zu wiederholen, um ganz sicher zu sein. Es scheint aber sehr unwahrscheinlich, dass sich die Situation retten lässt. Wenn aber der Wirkungsquerschnitt von 10^{-24} für schnelle Neutronen echt ist, so kommt man wohl um langreichweitige Kräfte nicht herum, die sogar noch ziemlich stark sein müssen, um bei schnellen Neutronen etwas auszurichten. Oder aber, es ist hier noch etwas ganz anderes los, wozu wir noch zu dumm sind.

Die letzte Panikmeldung aus Cambridge bezüglich Richtungsverteilung der Photoneutronen hat übrigens Chadwick zurückgezogen, nachdem ich ihm in lebhaften Farben schilderte, was aus diesem Resultat folgen würde. Er sagt jetzt, dass die Geometrie so ungenau war, dass man aus den Experimenten nichts schliessen kann. Neue Messungen in der Wilsonkammer scheinen das $\sin^2$-Gesetz zu bestätigen, sind aber noch nicht zahlreich genug, um Statistik zu machen. Was übrigens die Statistik in Deiner Arbeit betrifft, so konnte ich aus Deinen Angaben nicht entnehmen, ob Du berücksichtigt hast, dass jedes Element mehrere Isotope hat, und man nur den Bruttoquerschnitt beobachtet, ich hoffe aber Du hast diesen tiefliegenden Tatbestand nicht übersehen. Ich habe Dir kürzlich die Separata von Neutron II geschickt und Londoner Konferenz zugeschickt, und lege hier eine Liste der Leute bei, an die ich Separata geschickt habe. Die Liste ist hier und da etwas dünn, da ich Dir gleich zu Anfang die Hälfte von Neutron I geschickt habe, was aber doch eine zu pessimistische Einschätzung des nichtamerikanischen Anteils in der Physik bedeutete. Vielleicht füllt Du die eine oder andere Lücke gelegentlich aus.

Kommst Du im Sommer in das alte senile Europa zurück? Fankuchen äusserte die Ueberzeugung, dass Du sicher in Cornell bleibst, weil Franck im Sommer dort ist, ich bin aber nicht ganz sicher, ob das ein völlig überzeugender Grund zum Dortbleiben ist.
Beste Grüsse

R. Peierls

P.S. Ich erfahre eben, dass Fröhlich stellenlos ist da die Russen seine Aufenthaltsbewilligung ohne ersichtlichen Grund nicht verlängert haben. (Das gleiche ist Wasser und anderen Ausländern dort passiert.) Er reist im Augenblick herum, um wenigstens etwas kurzfristiges zu finden. Kannst Du etwas für ihn drüben tun? Seine Adresse c/o Notgemeinschaft Deutscher Wissenschaftler im Ausland, Löwenstr. 3/IV, Zürich.

Ist Dir ferner bekannt, dass Ehrenbergs Stipendium hier nicht verlängert wird? Blackett scheint keinen gesteigerten Wert auf sein Bleiben zu legen, vielleicht braucht er weniger selbständige Mitarbeiter, vielleicht haben sie sich auch nicht so gut vertragen. Er hat ihm also nahegelegt, eine Stelle in Charkow anzunehmen, und die Situation ist jetzt die, dass Ehrenberg sich verpflichtet nach C. zu gehen, wenn man ihn in Valuta zahlt. Ob das bewilligt wird, ist noch zweifelhaft, aber die Entlassung Fröhlichs lässt das natürlich fraglich erscheinen. Wenn er es bekommt, ist natürlich auch nicht sicher, ob das eine sehr glückliche Lösung ist, denn er würde infolge seiner "Unbegabtheit", wie er es nennt, den grösseren Strapazen des Lebens dort drüben vielleicht nicht gewachsen sein. Er begrüsste an dieser Lösung vor allem, dass er in Charkow eine Stelle haben würde, in der er für eine paar Jahre ruhig sitzen und experimentieren könnte. Kannst Du nicht auch für ihn etwas in USA tun?

[Manchester], 8.6.1935

Dear Bethe,
Your *Phys. Rev.* paper on the "Disintegration of neutrons" (according to the page headings) is very nice.[239] But I don't know yet whether I agree with it. The following points seem to be contradictory: 1. As I mentioned in my last letter, it seems almost impossible to understand how neutrons reach thermal velocities so quickly. 2. Even worse, it seems to be impossible (according to a discussion of the empirical material that I have been working on with Ehrenberg) to interpret the cross-section of 10^{-24} claimed for fast neutron capture as an effect of the slow neutrons. This effect can also be observed at reasonably reproducible scales, if you use deuteron-deuteron interactions as your neutron source. But then it is hard to see where the slow neutrons could come from. Admittedly, so far only the experiments of Bjerge and Westcott have studied this. These were done before the Fermi effect was discov-

[239] H.A. Bethe, 'Theory of Disintegration of Nuclei by Neutrons', *Phys. Rev.* **47**, 747–59 (1935).

ered, and dirt probably played a role.[240] Moreover, we have the fact that (for example) the cross-section reported for fast neutrons in Ag has a value of 10^{-24} because only 1/1000 of the neutrons have a cross-section of 10^{-21}, while the others have no cross section at all. This fact, which I wanted to start this long sentence with, implies that a coating of about 10^{-2} cm Ag could totally absorb the effect of "fast" neutrons even without any paraffin. But we can't find any report of such an effect. The experiment appears never to have been done in this manner, but it is, for instance, certain that neither thick Cd- nor I-coatings significantly reduce the effect in Ag. I will try to persuade Ehrenberg to repeat the experiment with an Ag absorber and Ag indicator, to really be sure, but it seems very unlikely that the situation will change again. If the cross-section for fast neutrons is really 10^{-24}, however, then it can only be explained by long-range forces which would have to be rather strong to affect them. Or perhaps something else is going on here, and we are still too naïve to understand it.

Chadwick has withdrawn his latest panic news from Cambridge regarding the distribution of photoneutron directions,[241] by the way, after I explained to him in a dramatic way what its result would be. He now says that the geometry was so imprecise that you can't explain anything with these experiments. New Wilson chamber measurements seem to confirm the $\sin^2$-law, but there are not yet enough of these to do any statistics.

Concerning the statistics of your paper, I couldn't tell from your notation whether you treat every element as having several isotopes and thus only observe the gross average. I hope you haven't ignored this basic fact. Recently I sent you the reprints of Neutron II[242] and

[240]See E. Fermi, E. Amaldi, B. Pontecorvo, F. Rasetti and E. Segré, 'Azione di sostancze idrogenate sulla radioattività provocata da neutroni', *La Ricerca Scientifica* **2**, 9–10 (1934).

[241]See J. Chadwick and M. Goldhaber, 'Disintegration by slow neutrons', *Nature* **134**, 65 (1934); J. Chadwick and M. Goldhaber, 'Disintegration by slow neutrons', *Proc. Camb. Phil. Soc.* **31**, 612–616 (1935); J. Chadwick and M. Goldhaber, 'The Nuclear Photoelectric Effect', *Proc. Roy. Soc.* **A151**, 479–493 (1935).

[242]H.A. Bethe and R.E. Peierls, 'The Neutrino', *Nature* **133**, 689–90 (1934).

the London conference,[243] and here I attach a list of other people to whom I have sent the reprints. The list is a bit thin here and there, since first I sent half of Neutron I to you and it turned out to be too pessimistic in its estimate of the proportion of non-American physics. You could probably fill in the gap at your convenience. Are you coming back to senile old Europe this summer? Fankuchen is convinced that you will stay at Cornell, since Franck will be there this summer, but I'm not sure this is a fully convincing reason to stay there.
Sincerely,

R. Peierls

P.S. I just learned that Fröhlich is without a job, as the Russians haven't renewed his residence permit for no understandable reason (the same thing happened to Wasser and other foreigners there). He is travelling around looking for something short-term at least. Could you offer him something over there? His address: c/o Notgemeinschaft Deutscher Wissenschaftler im Ausland, Löwenstr. 3/IV, Zürich.

Also, did you know that Ehrenberg's grant here hasn't been renewed? Blackett doesn't seem to be very keen on his staying; he probably needs less independent staff members, or else they didn't get along with each other very well. So he recommended that Ehrenberg take a job at Charkow, and now the situation is that Ehrenberg has committed himself to go to C. if he is paid in valuta. It's still not certain whether this will be granted, and Fröhlich's dismissal makes it even less certain. Even if he gets it, he's not sure whether it is really the best solution. He probably couldn't stand the tougher sides of life there, because of his "incapacity", as he calls it. He favours this solution mostly because the job at Charkow would be rather untroubled for a couple of years, so he could stay there and do experiments. Could you do something for him in the USA, too?

---

[243] *International Conference of Physics, London, 1934*, London: The Physical Society, 1935.

[36] Hans Bethe to Rudolf Peierls

RMS Queen Mary, 1.8.1936

Lieber Peierls,

ein neuer Monat auf einem neuen Schiff – das ist eine vorzügliche Gelegenheit für gute Vorsätze. Immerhin, da Du netterweise den Deckel des Phys.Rev. als einen jedesmaligen Gruss von mir betrachtest, fühle ich mein Gewissen etwas weniger schlagen (schlägt das Gewissen?) und das erleichtert den Anfang.

Also zunächst: schönen Dank für Deine Standpauke. Ich habe sie glaube ich sehr nötig, weil leider die Amerikaner zu höflich sind, mir welche zu halten. Die nichtamerikanischen Physiker in Amerika sind meist Ungarn, wie Teller und Wigner, und da ist auch nicht viel zu erwarten.

Sachlich bin ich nicht ganz Deiner Meinung. In anderen Worten, es steht mit mir schon ebensoschlimm wie mit Bieberbach: ich glaube sogar, was ich sage.

Zunächst einmal im Prinzip: Ich halte es für äusserst nützlich, wenn gerechnet wird. Nur auf diese Weise kann man doch irgendetwas über die Kernkräfte herausbekommen. Aus dem hohlen Bauch kann man doch gewiss nichts darüber aussagen. Und da die Amerikaner Geduld haben, soll man sie doch dazu ausnutzen.

Eine andere Frage ist, wieweit die bisherigen Rechnungen anständig sind. Da gibt es natürlich so'ne und so'ne. Die anständigsten Rechnungen sind von Present, allerdings nur über das H^3. Er nimmt eine recht gute Eigenfunktion (9 Konstanten, glaube ich) und Variationsprinzip. Resultat: selbst mit Wignerkräften kriegt man nur 5 MV Bindungsenergie bei einer Reichweite von 2.10^{-13} cm, und glaube ich 6 MV bei 1.10^{-13}. Vergleiche das mit 4.4 MV, der doppelten Bindungsenergie des Deuterons — soviel würde man ja selbst bei unendlicher Reichweite der Kräfte bekommen — und 8.4 MV, der beobachteten Bindungsenergie. Der Unterschied gegen Thomas ist einfach, dass Thomas nur zeigt, dass im Grenzfall unendlich kleiner Reichweite die Bindungsenergie von H^3 unendlich wird. Aber das geschieht sehr langsam und sagt gar nichts über die Bindungsenergie bei endlicher Reichweite. Andererseits geben natürlich Rechnungen über das α-Teilchen eine untere

Grenze für die Reichweite, und wenn bei dieser unteren Grenze das H^3 zu kleine Bindungsenergie hat, so zeigt das eben die Existenz von Neutron-Neutron Kräften. Thomas Rechnungen zeigen bloss, dass man aus dem H^3 allein nichts schliessen kann (was trotzdem immer wieder versucht wird).

Nun das α-Teilchen: es scheint, dass Inglis eine anständige Rechnung hat, mit Variationsprinzip und etwa 7 verfügbaren Konstanten. Er hat sich zwar einmal dabei ziemlich verrechnet, aber wenn er das das zweitemal nicht wieder getan hat, so gibt seine Rechnung fast identisch dasselbe wie das korrigierte equivalent two-body problem. Korrigiert das heisst, dass Feenberg eine Störungsrechnung gemacht hat, in der er das equivalent two-body problem rechtfertigt und korrigiert. Soviel ich mich erinnere, kommt seine Sache auf eine Näherung von grossen Reichweiten heraus: Als nullte Näherung kann man etwa den Eigenwert betrachten, den man mit der Eigenfunktion $e^{-\mu(r_{12}^2+r_{13}^2+r_{14}^2+r_{23}^2+r_{24}^2+r_{34}^2)}$ bekommt. Die Korrektur dieses Eigenwertes wird in erster Näherung richtig durch die equiv[alent] two-body Methode gegeben, in zweiter Näherung kommt ein positiver Beitrag dazu (dies gibt die korrigierte two-body Methode), der bei $2 \cdot 10^{-13}$cm Reichweite etwa 2 MV beträgt. Die nächste Näherung gibt, wie üblich, wieder eine Verminderung des Eigenwertes.

Ich glaube also, das Feenbergs Methode lange nicht so schlecht ist wie einem danach werden kann. Allerdings ist dies eine Rechtfertigung a posteriori, und Du hast völlig recht, wenn Du sagst, dass ich das alles nicht wissen konnte, als ich meinen Artikel schrieb. Damals wusste ich nur von Presents Rechnungen und war von ihnen ziemlich überrascht. Aber es war damals schon klar, dass man ungeheuer kurze Reichweite braucht, um das H^3 ohne Neutron-Neutron Kräfte zu erklären, und selbst wenn Feenbergs Rechnungen für das α-Teilchen sehr verkehrt gewesen wären: diese ganz kurzen Reichweiten waren jedenfalls unmöglich. Ausserdem war es nach Presents Rechnungen ziemlich wahrscheinlich, dass Feenbergs Methode auch für das α-Teilchen recht gut sein würde.

Natürlich müssen die Rechnungen noch verbessert werden. Das wird auch geschehen. Ich habe einen Dollar zur Beförderung des Fortschritts der Wissenschaft eingesetzt, indem ich mit Feenberg gewettet habe, dass seine equivalent two-body Methode zu wenig Bindungsenergie für das α-Teilchen gibt. Er sagt in seiner letzten Arbeit, zuviel. Zu entscheiden durch eine anständige Variationsrechnung, die natürlich er machen muss.

Immerhin hat es wohl jetzt schon einen Sinn zu sagen, dass die Bindungsenergien von H^3 und He^4 mit den Beobachtungen über H^2, die Neutron-Proton und die Proton-Proton Streuung, innerhalb der Fehlergrenzen, übereinstimmen. Das ist sehr wichtig, denn es zeigt, dass die Kernkräfte sich als Summe von Kräften zwischen Teilchenpaaren darstellen lassen.

Breit hat ein interessantes Resultat aus der Proton-Proton-Streuung abgeleitet: Es scheint, dass für antiparallellen Spin die Kräfte Proton-Proton und Proton-Neutron bis auf die Coulombkraft identisch sind. Was das bedeutet, weiss man natürlich nicht, aber die Übereinstimmung ist überraschend gut. Es ware interessant, herauszufinden, was die Kräfte zwischen gleichartigen Teilchen bei parallelen Spins tun: Die X-Meister, wie Feenberg & Co. versuchen, das Li^6 zu diesem Zweck zu rechnen.

Ich habe vesucht, mit der Hartreemethode (siehe §36 meines Review-Artikels) den Spin der leichten Kerne fest zustellen. Für Li^7 kriegt man ziemlich automatisch einen $^2P_{3/2}$-Zustand als Grundzustand, und ein magnetisches Moment 3.1 was natürlich viel zu gut mit den beobachteten 3.3 übereinstimmt. Der grösste Teil des Moments kommt daher, dass einfach die Neutronenspins antiparallel sind sodass der ganze Spin-anteil des Kernmoments vom Proton herkommt. Zu den 2,85 "Magnetinos" (Autor: Condon) vom Protonenmoment kommt dann noch ein Beitrag vom Bahnmoment.

Viel mehr als ich haben Wigner und Feenberg über die Kerne von He bis O^{16} gerechnet; tiefste Terme, Abstände zwischen Nachbarniveaus etc. Das Quantitative ist wohl noch sehr vorläufig.

Bohrs Papier über die Kerne gefällt mir vortrefflich. Es ist bezeichnend, dass es am 29. Februar erschienen ist. Es ist erstaunlich klar und sagt genau, was nötig ist. (Vergl. dagegen Bohrs letzten Nature letter über die Energieerhaltung.) Man soll aber doch Breit und Wigner nicht unterschätzen. Es ist zwar miserabel geschrieben, aber sie haben doch alles richtig erkannt und ausserdem eine quantitative Formel gegeben, wo Bohr rein qualitativ bleibt.

Bohrs Papier hat mich zu einer Rechnung über die Anzahl der Niveaus veranlasst. Das Modell ist beliebig schlampig: Fermistatistik, Potentialloch, keine Wechselwirkung zwischen den Teilen. Die Anzahl der Energieniveaus ist dann im wesentlichen die Entropie des Fermigases. Berücksichtigt ist, dass nur Zustände mit bestimmtem Drehimpuls für Prozesse wie Neutroneneinfangung in Frage kommen. Das Resultat unterscheidet sich von Oppenheimers (Seattle Meeting der Amer. Phys. Soc.) dadurch, dass es im Rahmen des Modells richtig gerechnet ist; es sieht infolgedessen sehr anders aus. Ich würde gerne eine vernünftige Methode zur statistischen Abschätzung der Matrixelemente für Neutroneneinfangung und γ-Emission haben, meistens kriegt man ja dafür zuviel heraus.

Nebenbei: Du bist doch von Deinen magnetischen Rechnungen her Expert in Fermistatistik mit allen möglichen komplizierten Wechselwirkungen. Weisst Du nicht eine Methode für die Kernprobleme?

Shanklands Experiment hat mich eigentlich nicht sehr aufgeregt, selbst ehe es als falsch nachgewiesen war. Ich habe vorgezogen, es zunächst nicht zu glauben a) weil nicht sein kann, was nicht sein darf, b) weil mir das Experiment nicht so sehr sauber vorkam, c) weil es soviel andere Dinge, auch mit hochenergetischen Lichtquanten, gibt, bei denen Energieerhaltung gilt. Z.B. Paarerzeugung. Was hält Dirac davon? Ich habe wirklich nicht verstanden, warum Dirac so begeistert davon war. Er schien schon vor 1 1/2 Jahren die Nichterhaltung der Energie zu lieben, und fand die damals noch hohe Masse des Be^9 wunderschön. Was ist mit ihm passiert? Die Neuigkeiten über die Disintegration des Cavendish sind sehr sehr interessant. Besonders dass Oliphant fortgeht. Ist das diesmal wirklich um wiederzukommen? Hast Du übrigens inzwischen einen besseren (in beiden Bedeutungen) Eindruck von ihm bekommen? Bei uns gibt es von Personalnachrichten nur, dass Wigner nach Wisconin berufen ist, an die selbe Universität wo Breit ist. Er scheint gehen zu wollen, aber Princeton hat bisher noch nichts unternommen. Also vielleicht wird er doch in P[rinceton] bleiben.

In Cornell kommen allmählich einige Theoretiker. Ich habe einen research assistant, Rose, früher bei Uhlenbeck, sehr gut und schrecklich fleissig, bezahlt von der American Philosphical Society. Er soll offiziell die Rechnungen über die Vielfachstreuung von Elektronen fertigmachen, unoffiziell rechnet er Kernmomente. Ausserdem kommt nächstes Jahr Konopinski als National Research Fellow.

Du erziehst also den Bhabba. Ich dachte er wäre ganz von der Sorte Hulme-Masse, die in Bezug auf Rechnungen tausendmal schlimmer sind als die Amerikaner.

Was macht die family? Placzek erzählte Wunderdinge von Gabys Intelligenz. Und Ronald Frank muss jetzt auch in dem Alter sein, wo man von seiner Intelligenz reden kann.

Ich habe weiterhin keinerlei family. Immerhin habe ich meinen car der Bestimmung jeden amerikanischen cars zugeführt. Er ist dafür sehr praktisch und auch sonst sehr schön. Wie sehr ich amerikanisiert bin, siehst Du daraus dass ich in einem Jahr 7000 Meilen gefahren bin, obwohl ich ursprünglich geschworen hatte, den Nepomuk nur zu Ausflügen in die Umgebung Ithacas zu benutzen.

Zurzeit fahre ich also zu meiner Mutter, und mit ihr nach Seelisberg, Switzerland. Ich glaube meine Mutter schrieb, wir wohnen im Grand Hotel dortselbst. Sie ist momentan sehr aktiv, also sehr gesund. Viele herzliche Grüsse and Genia und Dich selbst Dein

Hans Bethe

Wiedereröffnung des Briefes nach Lektüre Deiniges:
Man soll sich doch wohl über Goldhabers Experiment nicht aufregen (siehe oben Shankland). Das Resultat ist höchstwahrscheinlich falsch; Tuve und Hafstad haben angeblich das Experiment mit besserer Technik nachgemacht und finden den theoretischen Wert. Der Haupteinwand gegen Goldhabers Anordnung scheint mir, dass beim Durchgang durch den Paraffinabsorber viele von den durchgehenden Neutronen verlangsamt werden (kleine Ablenkungen der Doppelstösse) und dadurch grössere Wahrscheinlichkeit haben, gezählt zu werden. Der Effekt ist von derselben Grössenordnung wie die Absorption. Ausserdem ist natürlich die theoretische cross section jetzt gar nicht mehr so gross, nachdem Fermi den Faktor 4 von der chemischen Bindung entdeckt

hat; seine letzten Messungen entsprechen einem virtuellen Niveau bei 120 kV, was den Wirkungsquerschnitt bei 200 kV auf etwa 2/3 des früheren Wertes reduziert.

RMS Queen Mary, 1.8.1936

Dear Peierls,
A new month on a new ship — this is a perfect opportunity for new resolutions. As you always take the cover of *Phys. Rev.* as a greeting from me,[244] I feel my conscience pulse less (does the conscience pulse?) which eases the beginning.

So first: thank you for your philippic. I think I needed it, because the Americans are unfortunately too polite to do that. The non-American physicists in America are mostly Hungarian, like Teller and Wigner,[245] and one can expect much from them, either.

Regarding the facts, I don't really agree with you. In other words, I have become nearly as bad as Bieberbach:[246] I even believe what I say.

First, a statement of principle: I think that it's immensely useful to do calculations. It is the only way to find out anything about nuclear forces. You cannot just "figure something out" about them. The Americans are quite patient, so you should take advantage of that.

Another question is whether the calculations we have done so far are correct. They are so-so. Present's[247] are the best, but only those about H^3. He chooses a quite good eigenfunction (with 9 free parameters, I believe), and uses a variational principle. The result: even if you take powers of the Wigner function you only get 5 MV of binding energy at a range of 2×10^{-13}cm, but I believe in 6 MV at 1×10^{-13}. Compare this

[244] Hans Bethe was editor of the journal *Physical Review*, hence his name appeared on the cover.

[245] Eugene Paul Wigner (1902–1995), studied in Berlin where he obtained his doctorate in 1925. After teaching at Göttingen and Berlin, he moved to Princeton in 1930 where, with the exception of a brief spell at Wisconsin (1936–1938) he remained for the rest of his career.

[246] Ludwig Bieberbach (1886–1992), at the time professor of geometry at Berlin and outspoken Nazi. Founder, in 1936, of the journal *Deutsche Mathematik*.

[247] R.D. Present, 'Must Neutron-Neutron Forces Exist in H^3?', *Phys. Rev.* **50**, 635–42 (1936).

to 4.4 MV, double the binding energy of deuterons — which you would expect to get if the range of powers was infinite — and to 8.4 MV, the observed binding energy. The difference to Thomas[248] is simple; he just shows that at infinitely short ranges the binding energy of H^3 becomes infinite.[249] But this happens very slowly, and says nothing about the binding energy in the case of finite ranges. On the other hand, calculations of the α-particle give a lower limit on the range; if the binding energy of He^3 is too small at this limit, then it only demonstrates the existence of neutron-neutron forces. Thomas' calculations merely show that you cannot conclude anything from the ^3He alone. (It is being attempted over and over again, anyway.)

About the α-particle: it seems as if Inglis did a good calculation using the variational principle and about 7 available free parameters.[250] At one point he miscalculated rather badly, but if he hasn't done this a second time then the result of his calculation should be almost identical to the corrected equivalent two-body problem.

By 'corrected' I mean that Feenberg[251] has done a perturbation calculation in which he justifies and corrects the equivalent two-body

[248]Llewellyn Hilleth Thomas (1903–1992), studied mathematics at Cambridge where he received his Ph.D. in 1927. He joined the faculty of Ohio State University in 1929 where he stayed until 1943. After war work at the Aberdeen Proving Ground in Maryland, he became a member of staff at the Watson Scientific Computing Laboratory at Columbia University where he remained until 1968. Between 1968 and 1876 he was visiting professor at North Carolina State University.

[249]L.H. Thomas, 'The Interaction between a Neutron and a Proton and the Structure of H^3', *Phys. Rev.* **47**, 903–909 (1935).

[250]David Rittenhouse Inglis (1905–1995) studied at Amhurst College and Ann Arbor, then moved to Ohio State, Pittsburgh, Princeton and John Hopkins. He published a number of papers on the alpha particle, e.g. D.R. Inglis, 'Spin-Orbit Coupling in Nuclei', *Phys. Rev.* **50**, 783, (1936); S. Dancoff and D.R.Inglis, 'On the Thomas Presession of Accelerated Axes', *Phys. Rev.* **50**, 784 (1936); D.R. Inglis, 'Spin-Orbit Coupling in the Alpha-Model of Light Nuclei', *Phys. Rev.* **50**, 1175–80 (1939); D.R. Inglis, 'The Alpha Model of Nuclear Structure, and Nucelar Moments', *Phys. Rev.* **60**, 837–51 (1941).

[251]Eugene Feenberg (1906–1977), studied at the University of Texas, Austin and at Harvard where he obtained his Ph.D. in 1933. He later taught at Harvard, Princeton, NYU and, from 1946, at Washington University in St. Louis, where he was appointed Wayman Crow Professor of Physics in 1964.

problem.[252] As far as I can remember, his calculations amount to an approximation valid for long ranges: As the zero-order approximation you take the eigenvalue of the eigenfunction $e^{-\mu(r_{12}^2+r_{13}^2+r_{14}^2+r_{23}^2+r_{24}^2+r_{34}^2)}$. The first-order correction to this eigenvalue can be determined by using the equivalent two-body method; in the second-order approximation a positive value will be added (this gives you of the corrected two-body method), which at a range of 2×10^{-13}cm is about 2 MV. The result of the next approximation is a reduction of the eigenvalue, as usual.

I also think that Feenberg's method is not as bad as it may seem.[253] This is an *a posteriori* justification, however, anyway, and you would be absolutely right to say that I couldn't have known that when I wrote my article. At the time I only knew of Present's calculations, and was quite surprised by their agreement with Feenberg's. But it was already clear at the time that you need very short ranges to explain the H^3 without neutron-neutron powers. Even if Feenberg's calculations for the α-particle had been wrong, such extremely short ranges were impossible anyway. Also, according to Present's calculations, it was very unlikely that Feenberg's method would be helpful for the α-particle.

The calculations obviously have to be improved. This will actually happen. I have put a dollar on the promotion of scientific progress by betting Feenberg that his equivalent two-body method gives too little binding energy for the α-particle. In his last paper, he says there is too much. It will need to be decided by a proper variational calculation, which, of course, he will have to do if he wants to win.

At least it is reasonable to say that the binding energies of H^3 and He^4 correspond to observations of the neutron-proton and the proton-proton diffusion in H^2, within the margin of error. This is very important, as it shows that the powers of the nucleus can be expressed as a sum of powers between particle pairs.

[252] E. Feenberg, 'On Relativity Corrections in the Theory of the Deuteron', *Phys. Rev.* **50**, 674 (1936).

[253] The German play on words does not translate well.

Breit[254] has derived an interesting result from proton-proton scattering:[255] It seems as if in the case of opposite spins the proton-proton and the proton-neutron forces are identical, except for the Coulomb force. We're not sure what to make of this, but the agreement is surprisingly good. It would be interesting to find out what the forces are between similar particles in case of parallel spins: X-masters like Feenberg *et al.* are trying to calculate the properties of Li^6 for this purpose.

I have tried to find the spin of lighter nuclei using the Hartree method (see §36 of my review article).[256] For Li^7 you automatically get the basic $^2P_{3/2}$-condition and a magnetic moment of 3.1, which corresponds with the observed value of 3.3 rather too well. The greater part of the moment is simply a result of the neutron spins being antiparallel, so that the entire spin component in the nucleus moment is determined by the proton. A value for the orbital moment then has to be added to the proton moment 2.85 "Magnetinos" (author: Condon[257]). More than me, Wigner and Feenberg have calculated details concerning the nuclei from He to O^{16}: lowest values, distances between proximate levels, etc. Their quantitative [results] seem to be rather tentative.

I like Bohr's paper on nuclei very much. It is significant that it came out on February the 29th.[258] It is surprisingly clear, and says exactly what is needed. (Contrast this with Bohr's last *Nature* letter about the conservation of energy[259]). But you shouldn't underestimate Breit

---

[254]Gregory Breit (1899–1981) studied at John Hopkins where he received his Ph.D. in 1921, in Leiden and at Harvard, before moving to Washington in 1924, New York University (1929–1934), the University of Wisconsin (1934–47) and Yale (1947–1968).

[255]G. Breit, E.U. Condon and R.D. Present, 'Theory of scattering of protons by protons', *Phys. Rev.* **50**, 825–45 (1936).

[256]See letter [26], note 146.

[257]Edward Uhler Condon (1902–1974), studied at Göttingen, Munich, Berkeley, lectured at Columbia, Princeton and Minnesota before becoming associate director of research at Westinghouse and serving on the National Defense Research Committee during the Second World War.

[258]Niels Bohr had given a lecture to the Danish Academy on 24 January 1936 on neutron capture and nuclear constitution which was summarised in N. Bohr, 'Om Atomkernernes Egenskaber og Opbygning', *Dansk. Vid. Selsk. Virksomh. 6/35–5/36*, 39 (1936).

[259]N. Bohr, 'Conservation Laws in Quantum Theory', *Nature* **138**, 25–26 (1936).

and Wigner, either.[260] [Their paper] is badly written, but they have understood everything correctly and also give a quantitative formula, while Bohr is still just qualitative.

Bohr's paper inspired me to do a calculation about the number of levels. The model is as sloppy as you like: Fermi statistics, hole of potentials, no interaction between the particles. The number of energy levels is essentially equal to the entropy of the Fermi gas. It takes into consideration that only states with a specific angular momentum are relevant for processes like neutron capture. It is different from the work of Oppenheimer (Seattle Meeting of the Amer. Phys. Soc.[261]) in that it has been calculated correctly, within the given model; it therefore looks very different. I would like to have a reasonable method for estimating the matrix elements for neutron capture and γ-emission statistically. The results for these processes are often too high.

By the way: Because of your calculations on magnetism you are an expert on Fermi statistics, with all possible complicated interactions. Don't you know a method for these nuclear problems?

I wasn't too upset about Shankland's experiment,[262] even before it was disproved.[263] I preferred not to believe in it a) because what may not be cannot be, b) because the experiment didn't seem to be carefully done, and c) because there are so many cases where conservation of energy applies, for example high-energy light quanta and pair production. What does Dirac think of that? I really didn't understand why Dirac was so excited about it. He seemed to have loved the non-conservation of energy $1\frac{1}{2}$ years ago, and at the time thought that the high mass of Be^9 was wonderful.[264] What has happened to him?

[260] G. Breit and E. Wigner, 'Capture of Slow Neutrons', *Phys. Rev.* **49**, 519–531 (1936).

[261] 'Minutes of the Seattle Meeting', *Phys. Rev.* **50**, 384–93 (1936).

[262] Robert S. Shankland (1908–1982), received his Ph.D. from the University of Chicago in 1935. He later took up posts at Case Western Reserve University (1940–58) and Argonne National Laboratory (1953–69). Shankland's experiments were published as Robert S. Shankland, 'Scattering of Gamma Rays', *Phys. Rev.* **50**, 571 (1936), Robert S. Shankland, 'An Apparent Failure of the Photon Theory of Scattering', *Phys. Rev.* **49**, 8–13 (1936).

[263] Rudolf Peierls had published a paper about the experiments. R.E. Peierls, 'Interpretation of Shankland's experiment', *Nature* **137**, 904–5 (1936).

[264] P.A.M. Dirac, 'Does Conservation of Energy Hold in Atomic Processes?', *Nature* **137**, 298–99 (1936).

The news about Cavendish's disintegration[265] is very very interesting, especially that Oliphant[266] is leaving. Will he really come back again this time? Have you had a better (in both senses) impression of him in the meantime?

The only job news here is that Wigner has been appointed to the University of Wisconsin, the same university where Breit is. He seems to want to go there, but Princeton hasn't undertaken anything yet. So Wigner will probably stay at P[rinceton] anyway.

At Cornell, some theorists are coming step by step. I have a research assistant, Rose,[267] formerly of Uhlenbeck's, very good, and terribly industrious; he is paid by the American Philosophical Society. Officially he is supposed to finish the calculations about the multiple diffraction of electrons; unofficially he is calculating nucleus moments. Besides that, next year Konopinski[268] will come as a National Research Fellow.

You are educating Bhabba.[269] I thought he was just like Hulme-Massey. Regarding calculations, those two are a thousand times worse than the Americans. How are the family doing? Placzek told me miraculous stories about Gaby's intelligence. And Ronald Frank has to be of an age now that you can talk about his intelligence.

---

[265]In 1935 Chadwick left the Cavendish to accept a Chair at Liverpool. Oliphant replaced him as Assistant Director of Research. Blackett moved to London, Kapitza did not return from Russia, and in 1936 Oliphant left to become Poynting Professor of Physics at Birmingham.

[266]Marcus Laurence Oliphant (1901–2000), studied at Adelaide and Cambridge. In 1937 he took up an appointment as Poynting Professor of Physics at Birmingham, a post he held until his move to his native Australia where he became the first Director of the Research School of Physical Sciences of the Australian National University in 1950. He became the first President of the Australian Academy of Science and later he was appointed State Governor of South Australia.

[267]Morris Edgar Rose (1911–), joined Hans Bethe as a research assistant before moving on to Yale, Oak Ridge and in the 1960s to the University of Virginia.

[268]Emil Jan Konopinski (1911–1990), graduated from the University of Michigan in 1933, later worked at Los Alamos and co-patented the device that triggered the first Hydrogen bomb.

[269]Homi K. Bhabha (1909–1966), studied mechanical engineering, maths and physics at Cambridge where he worked with Dirac. He did research at the Cavendish before joining Raman's institute at Bangalore in 1940. In 1945 he set up the Tata Institute of Fundamental Research in Bombay.

I still have no family. But at least with my car I have been doing what has to be done with any American car. It is very convenient and very nice, too. That I have been driving 7000 miles per year shows just how much I have become Americanised, even though at first I swore that I would use the Nepomuk just for little trips in the vicinity of Ithaca.

Currently I am on my way to my mother and together we will go to Selisberg, Switzerland. I think my mother told me that we will be staying in the Grand Hotel. She is very active at the moment, so very well.[270] Lots of greetings to you and Genia.

Yours

Hans Bethe

A reopening of this letter after having read yours:
You shouldn't be upset about Goldhaber's experiment (see above Shankland). It is very probably wrong; Tuve and Hafstad[271] have, supposedly, repeated the experiment with a better technique, and have found the theoretical value. The main objection to Goldhaber's method seems to be that a lot of the neutrons are slowed down as they pass through the paraffin absorber (small deflections or double impacts). They thereby have a higher probability of being counted. This effect is of the same order of magnitude as the absorption. Of course the theoretical cross section is no longer too big, now that Fermi has discovered a factor of 4 in the chemical bond;[272] his latest measurements correspond to a virtual level at 120 kV, which reduces the cross section at 200 kV to about 2/3 of its former value.

---

[270]Bethe's mother, a talented musician, had suffered from a hearing impairment as a result of contracting influenza. This had lasting physical and psychological effects.

[271]Tuve, Hafstad and Heydenberg had carried out a series of experiments on proton-proton scattering and on neutron-proton scattering. M.A. Tuve, N.P. Heydenburg and L.R. Hafstad, 'The Scattering of Protons by Protons', *Phys. Rev.* **49**, 402 (1936); M.A. Tuve and L.R. Hafstad, 'The Scattering of Neutrons by Protons', *Phys. Rev.* **50**, 490–91 (1936); M.A. Tuve, N.P. Heydenburg and L.R. Hafstad, 'The Scattering of Protons by Protons', *Phys. Rev.* **50**, 239–46 (1936); M.A. Tuve, N.P. Heydenburg and L.R. Hafstad, 'The Scattering of Protons by Protons', *Phys. Rev.* **56**, 1078–91 (1939).

[272]E. Amaldi and E. Fermi, 'On the Absorption and the Diffusion of Slow Neutrons', *Phys. Rev.* **50**, 899–928 (1936).

[37] Rudolf Peierls to Hans Bethe

Cambridge, 6.8.1936

Lieber Bethe,
Vielen Dank für die quantenhafte Ausstrahlung. Ein Brief, würdig des Schiffes auf dem er geschrieben ist! Du bist also über Southampton und vermutlich über London gefahren ohne anzuhalten, heisst das etwas, dass Du auch auf dem Rückweg England schneiden willst?

Im Prinzip bin ich natürlich mit Dir darin einig, dass es dem Fortschritt der Wissenschaft im allgemeinen nützlich ist, wenn viel gerechnet wird, vorausgesetzt dabei ist aber, dass diese Rechnungen nicht in irreführender Weise geschrieben oder auch nur von ihren Verfassern eingeschätzt werden, denn wenn gewisse Feststellungen über die Eigenschaften von Gleichungen als sicher hingestellt werden, ohne es zu sein, so ist dies der Wissenschaft nur schädlich. Die Amerikaner haben zwar Geduld, aber auch die Tendenz, ihre jeweilige Beschäftigung für die letzte Weisheit zu halten, und da sollst Du eben auf sie aufpassen.

Auch was Present betrifft, bin ich noch nicht ganz Deiner Meinung, denn Thomas hat bereits eine Eigenfunktion, die einen wesentlich tieferen Eigenwert gibt als Presents, und seine Methode hat ausserdem die Annehmlichkeit, dass man auch eine obere Grenze bekommen kann. In dieser Beziehung reicht sie also beinahe an die Feenbergsche Methode mit gleichwahrscheinlichen positiven und negativen Fehlern heran. Ich muss andererseits wieder zugeben, dass mir dieses Resultat von Thomas bei Abfassung meines Schimpfbriefes nicht bekannt war.

Der Grund, dass Present trotz 9 Konstanten noch kein so gutes Resultat gibt, ist wohl der, dass auch n Konstanten nichts helfen, wenn sie nicht in der richtigen Weise in die Eigenfunktion eingeführt werden, man kann natürlich leicht 9 Konstanten angeben, die überhaupt nicht zu Bindung führen. Man darf eben nicht vergessen, dass das Problem eine ganz andere Struktur hat als die gewohnten Atom- und Molekülprobleme, und es ist viel schwieriger, eine anständige Funktion zu erraten. Wenn man etwa an das Heliumatom denkt, so kriegt man schliesslich mit jeder Variationsfunktion von einem Parameter, die nicht zu blöd gewählt ist, einen Eigenwert von richtiger Grössenordnung, während das He^3 scheusslich empfindlich gegen die richtige Wahl der Funktion ist (wahrscheinlich mehr als das α-Teilchen). Glaubst Du eigentlich dem Feenberg wirklich seine Rechnung, die von grossen Abständen her entwickelt? Ich habe die Rechnung zwar nicht gesehen, bin aber bereit, einen Shilling zu wetten, dass die ersten vier Näherungen von der gleichen Grössenordnung sein müssen.

Die Rechnung über die Anzahl der Zustände bei unabhängigen Teilchen und Fermistatistik habe ich übrigens auch gemacht, und hatte eigentlich die Absicht, sie zu veröffentlichen. Hast Du darüber schon etwas abgeschickt? Sonst könnte man evtl. poolen. Ich habe zunächst allgemein gerechnet, d.h. angenommen, dass die Anzahl der Protonen- und Neutronenzustände mit Energien unter E zwei beliebige Funktionen $Z_1(E)$ und $Z_2(E)$ sind. Die Anzahl der Zustände pro Energieintervall lässt sich dann allgemein als Funktion eines Parameters, nämlich $dZ_1/dE + dZ_2/dE$ ausdrücken. Wenn man nur Zustände mit gegebenem Spin haben will, so geht noch etwas mehr ein, aber auch nur in die relativ unwichtige Funktionaldeterminante, nicht in den Exponenten.

Ich hatte zunächst einige Zuversicht, dass diese Rechnung etwas mit der Wirksamkeit zu tun hat, denn obwohl die freien Teilchen in einem Kasten sicher ein beliebig schlechtes Modell zur Beschreibung von Kernprozessen sind, sollte das statistische Problem (im wesentlichen die spezifische Wärme) viel weniger vom Modell anhängig sein. Man kann übrigens leicht abschätzen, was kleine Störungen ausmachen; wenn W die Störungsenergie ist, so bekommt man Korrektionsterme in $(W/kT)^2$. Nun ist für eine Anregung von 10 MV die “Temperatur” von der Grössenordnung 2MV, es scheint also auf den ersten Blick, als ob das noch ziemlich viel Freiheit für Vernachlässigungen lässt. (Dies ist jedenfalls insofern befriedigend, als die Kopplung zwischen den verschiedenen Zuständen individueller Teilchen, die nach Bohr vorhanden sein muss, um die Ausstrahlung gross zu machen, nicht mit der Bedingung $W \ll 2\text{MV}$ im Widerspruch ist, denn aus der Anwendbarkeit der Bohrschen Theorie folgt in Strenge nur, dass W grösser sein muss als die Strahlungsbreite der γ-Linien). Bei genauerer Betrachtung wird man aber doch sehr zweifelhaft, denn in den höheren Näherungen des Störungsverfahrens gibt es nicht nur Terme wie $(W_{ii}/kT)^2$ (W_{ii} = Matrixelement der Störung) sondern auch $(W^2)_{ii}/(kT)^2$. Damit dies und alle höheren Terme klein sind, muss die Wechselwirkungsfunktion überall d.h. für alle Werte der Koordinaten, etc. die mit irgendeiner Wahrscheinlichkeit vorkommen, nicht grösser als 2MV sein. Eine solche Bedingung ist doch aber offenbar Unsinn, denn in dem Modell ist ja sicher die starke Korrelation zwischen Nachbarteilchen vernachlässigt, und diese gehört also in die Störung. Andererseits wissen wir, dass diese von einer Energie herrührt, die für kleinen Abstand der Nachbarn den Wert 50 MV (oder so) erreicht.

Ich habe mir den Kopf zerbrochen, ob man nicht einen umgekehrten Grenzfall behandeln kann; das Topfmodell ist ja offenbar gut für schwache Wechselwirkung, genauer gesagt, für starke Korrelation zwischen den Teilchen. Kann man nicht starke Korrelation als umgekehrten Grenzfall behandeln? Wenn die Kräfte mit der kurzen Reichweite nicht so verflucht pathologisch wären, wäre die Antwort hierauf sehr einfach, denn dann würde man sagen: Macht man die Kräfte zwischen Nachbarn immer stärker und erhält dadurch die Korrelation, so muss aus dem "gasartigen" Kern ein Flüssigkeitstropfen und schliesslich ein festes Gebilde werden. Wenn wir für einen Augenblick einmal diese Annahme machen, so ist das Problem natürlich sehr leicht zu behandeln, man muss das Fermigas nur durch ein Bosegas (Statistik der harmonischen Eigenschwingungen) ersetzen. Allerdings unterscheidet sich dieser Fall von dem Fermi-Fall dadurch, dass man nicht nur eine Konstante, nämlich die Dichte der Eigenwerte am oberen Rande der Fermiverteilung, sondern die ganze Frequenzverteilung der Eigenschwingungen braucht. Für einen hinreichend grossen Kern ist es natürlich vernünftig, ein $\nu^2 d\nu$ Gesetz für die Eigenschwingungen anzunehmen, das führt dann natürlich auf T^3 für die spezifische Wärme. Dann wird der Exponent in der Verteilung der Energiewerte des Gesamtsystems proportional zu $U^{3/2}$ (U = Anregungsenergie) statt $U^{1/2}$ wie bei der Fermistatistik. Der Faktor im Exponenten hängt von der "Schallgeschwindigkeit" im Kern ab und man kann daher über die Grössenordnung nicht viel sagen. Sollte es aber später mal möglich sein, die Dichte der angeregten Zustände in zwei verschiedenen Gebieten (etwa für langsame und schnelle Neutronen) empirisch zu bestimmen so kann der Unterschied zwischen $U^{1/2}$ und $U^{3/2}$ schon eine grosse Rolle spielen.

Du wirst natürlich einwenden, dass dieses Modell ganz pathologisch ist, da es wohl kaum im Kern Gleichgewichtslagen für die Teilchen geben kann, eben wegen der kurzen Reichweite. Dagegen gibt es ein Argument: Im flüssigen Helium hat man offenbar einen Fall, wo zwar die Wechselwirkungen ganz andere sind, wo aber auch die Reichweite der Kräfte so kurz (oder die Kräfte so schwach) ist dass die Nullpunktsenergie jedes Atom aus der Gleichgewichtslage herausbefördern kann, trotzdem fällt seine spezifische Wärme bei tiefen Temperaturen stark ab, und obwohl man so einen Fall nicht rechnen kann, sieht man doch aus diesem empirischen Argument, dass es sich weitgehend wie der Fall mit Gleichgewichtslagen benimmt.

Bei diesem Modell entsteht dann die lustige Frage der Rotation. Wenn man nämlich die Gleichgewichtslagen ernst nimmt, so kann der Kern offenbar als Ganzes rotieren. Das würde zu Rotationsniveaus von etwas 1000 volt führen. Obwohl eine Feinstruktur der Linien von dieser Grössenordnung nach Ellis gerade beobachtbar wäre, so ist dies nicht notwendig ein Widerspruch, da die Linien vermutlich keine Feinstruktur zeigen würden, siehe unten.

Die Frage ist nun: Wie ändern sich die Rotationsniveaus, wenn man die Gleichgewichtslagen auflockert, so dass die Teilchen häufig ihre Plätze wechseln können. Dabei glaube ich, dass man sich die Verhältnisse im Kern so vorstellen darf, dass zwar ein Proton sehr häufig mit seinem Nachbarn den Platz wechseln kann, dagegen dass es sehr lange braucht, um von einem Ende des Kerns zum anderen bezw. ganz um diesen herumzukommen. Frage: Sind die Abstände der Rotationsniveaus unter diesen Verhältnissen eher von der Grössenordnung derer des festen Kerns oder der von individuellen Teilchen? Diese Frage habe ich noch nicht in befriedigender Weise beantworten können, ich halte es aber für wahrscheinlich, dass die Rotationsniveaus noch sehr eng liegen müssen. Man kann die Frage z.B. für zwei Teilchen streng diskutieren die eine Wechselwirkung miteinander haben und zusammen in einem ringförmigen Rohr herumlaufen (Letzteres um die Zentrifugalkraft zu vermeiden, deren Resultate man ohnehin abschätzen kann, und um zunächst nur einen Rotationsfreiheitsgrad zu haben.) Das lässt sich dann streng machen, und ist ganz amüsant. Von hieraus ist aber der Sprung auf n Teilchen noch ziemlich schwer.

Ich habe noch einen anderen Grund, warum ich an das Tropfenmodell eher glaube; nämlich die grosse Zahl von Quadrupollinien, die man in den Kernspektren findet. Denkt man nämlich an die einzelnen Teilchen, so sollte es eine Menge von Uebergängen mit Dipolmoment geben, und die sollten vorzugsweise vorkommen. Quadrupollinien sollten nur von metastabilen Zuständen aus kommen, d.h. um so seltener je höher die Anregungsenergie. Anders jedoch im Falle des schwingenden Modells. Hier erscheint nämlich die Symmetrie der Ausstrahlung gar nicht als bedingt durch die Rotationsquantenzahl. Es ist vielmehr so, dass die Rotationszustände so dicht liegen, dass unter den möglichen Endzuständen gerade der herausgegriffen wird, der das richtige Impulsmoment hat, das zu dem Ausstrahlungstypus (Dipol, Quadrupol) gehört. Die Bestimmung des Ausstrahlungstypus dagegen geschieht durch die Symmetrie des Schwingungszustandes. Wenn der nichtschwingende Kern näherungsweise kugelsymmetrisch ist, so lassen sich die Schwingungen in Polarkoordinaten separieren. Jeder erlaubte Uebergang entsteht durch Verminderung einer Schwingungsquantenzahl um 1 und je nachdem ob die zugehörige Schwingung zu einer Kugelfunktion mit $l = 1, 2, 3, ...$ gehört, wird eine Dipol- Quadrupol- oder Oktopolwelle ausgestrahlt. In diesem Modell können aber die Dipolmomente leicht sehr klein werden, denn es handelt sich um eine Schwingung, an der alle Teilchen teilnehmen, und wenn man sich die Ladungs- und Massendichte als über den Kern konstant verteilt vorstellt, so werden sie auch bei der Schwingung einander angenähert proportional bleiben. Die Abweichung von der Proportionalität wird wohl nur von den ungepaarten Protonen und Neutronen herrühren, und daher entsprechend klein sein. Daher muss man erwarten, dass die Dipolmomente, jedenfalls bei den langwelligen Schwingungen, von derselben Grössenordnung wie die Quadrupolmomente werden. Ich stimme natürlich sehr damit überein, dass man nach den glänzenden Bestätigungen, die das Heisenbergsche Prinzip im Falle Shankland und besonders Anderson erfahren hat, es vorziehen soll, das Experiment von Goldhaber provisorisch nicht zu glauben. Es ist aber nicht ganz leicht, herauszureden, jedenfalls tut es der Effekt nicht, den Du erwähnst. Sein Streuer ist nämlich so dünn, und sein Raumwinkel so klein, dass selbst wenn man annimmt, dass der Wirkungsquerschnitt die theoretische Grössenordnung hat, und dass jedes Neutron, welches im Streuer mehr als einmal gestreut worden ist (die einmal gestreuten müssen, um in den Zähler zu kommen, sehr stark nach vorn gehen und verlieren praktisch keine Energie) in dem Zähler mit derselben Wahrscheinlichkeit gezählt wird, wie thermische Neutronen ohne Paraffinmantel, so bekommt man gerade erst einen Effekt der erforderlichen Grössenordnung. Andererseits bedingt aber der dicke Paraffinmantel, der die Zählung schneller Neutronen erleichtert, auch sicher eine starke Absorption*, und Streuung, sodass eine solche Annahme völlig unmöglich erscheint.

Goldhaber selbst hat noch Bedenken wegen Streuung von Tisch und Wänden, und obwohl er dafür Kontrollversuche hat, wird er deswegen das Experiment noch unter verbesserten Umständen wiederholen.

Wie kannst Du übrigens Hulme und Massey in einen Topf tun? Hulme rechnet doch eigentlich nur nützliche Sachen, und wenn ich ihn auch nicht darum beneide, so sind es doch im allgemeinen Sachen, die unbedingt gemacht werden müssen (Ich sehe von seiner letzten Arbeit über Retardierung ab, an der Møller schuld war). Bhabhas Arbeit über Austausch bei der Streuung von Positronen an Eletronen war doch sehr hübsch. Dies nur zur Verteidigung von Cambridge. Ich habe übrigens jetzt einen Doktoranden, der ein bisschen was versteht. Es ist direkt eine Erholung, einen Mann zu haben, der auf Benutzung der Spinorschreibweise besteht (ich habe ihm das übrigens schon ausgeredet) wenn ich meinem bisher besten Doktoranden mit allen Einzelheiten erklären musste, wie man im Falle von Spin- und Bahnmoment den Gesamtdrehimpuls auf Diagonalform bringt.

Sonst habe ich es inzwischen sehr weit gebracht, nämlich zum M.A. (Cambridge) D.Sc. (Mancheser) und Member (nicht etwa Fellow!) of St. John's, mit dem besonderen Privileg, dort bis zu zweimal wöchentlich auf eigene Kosten zu dinieren! Mit Oliphant bin ich von Kopenhagen zurückgefahren (wo es übrigens sehr nett war, besonders wegen Heisenbergs Schauertheorie) und er stellte sich dabei als sehr nett heraus. Leider hat er eine ganz unmögliche Frau.

Gaby hat schon sehr ausgesprochene politische Meinungen. Als sie neulich in der Zeitung das Bild des High Comissioner for Palestine sah und auf Anfrage erklärt bekam, wer das ist, sagte sie "Oh, the High Compromissioner!" Jetzt habe ich aber schon den Schreibkrampf!

Ja übrigens man ist hier sehr böse dass Du Deine Rechnungen über Energie-Reichweitebeziehungen zwar sämtlichen Amerikanern zur Verfügung stellst, aber weder Dein Versprechen hälst, sie herzuschicken, noch sie publizierst!
Viel Spass und viele Grüsse an Deiner Mutter,
Dein

Peierls

(handwritten addition by Genia)
Lieber Bethe, wir haben ein "estate" und alles eigenes von Spargel bis Lavendel. Cambridge ist sehr schön und hier hat man sogar sehr viele nette Frauen und Männer kommen auch zum Tee. Kommst Du auch? Auf Wiedersehen,

Genia

Liebe Frau Bethe, ich bin ein Schwein und zwei Jahre quält mich mein Gewissen (weil ich Ihnen nicht geantwortet habe). Kommen Sie uns mal besuchen?

*Bei Goldhaber ist sehr viel Paraffin auch vor dem Zähler:

Cambridge, 6.8.1936

Dear Bethe,
Thank you for the quantum-like dissemination. A letter worthy of the ship it was written in![273] So you came via Southampton, and presumably via London without stopping; does this mean that you also won't stop in England on your way back either?

I basically agree that for the advancement of science it is generally necessary to do a lot of calculations, provided that these calculations are not written in a misleading way or being described in such a way by the authors. Whenever the conclusions drawn from equations are said to be certain, but really are not, it is detrimental to

[273] Refers to the 'Queen Mary', the ship on which Bethe had been travelling from America to Europe.

science. The Americans might be patient, but they have the tendency to take their results for ultimate truth, and this is where you should look after them.

I don't yet really agree with you concerning Present, since Thomas has already found an eigenfunction with a much lower eigenvalue than Present and his method also has the nice advantage that you can also get an upper limit. In this respect it comes close to Feenberg's method, with equally probable positive and negative errors. On the other hand, I have to confess that I didn't know Thomas' result when I wrote my letter of complaint.

The reason why Present doesn't get as good a result, despite having 9 parameters, might be that even n parameters don't help if they are not applied to the eigenfunction correctly. It is easy to find 9 values which do not lead to a binding. You shouldn't forget that the structure of the problem is totally different from that of a normal atom or molecule problem, and it is much more difficult to guess the proper function correctly. If you think of the helium atom, for example, then any reasonable variational function of one parameter will give you an eigenvalue with the right order of magnitude. For ^{3}He, on the other hand, the eigenvalue is terribly sensitive to one's choice of function (probably even more so than the α-particle). Do you really believe Feenberg's calculation, which expands with long ranges? I haven't seen the calculation, but I would bet a shilling that the first four approximations are of the same order of magnitude.

I have also calculated the number of states for independent particles obeying Fermi statistics, and I was thinking about publishing it. Have you submitted anything about this? If not, you and I could possibly pool our efforts. I first calculated it in a general way; i.e., I said that the number of proton and neutron states with energies less than E were the arbitrary functions $Z_1(E)$ and $Z_2(E)$ respectively. The number of states per energy interval can then be generally expressed as a parametric function of $dZ1/dE + dZ2/dE$. If you just want to know the number of states with a given spin, then a bit more has to be put in. This just affects the relatively unimportant functional determinant, however, not the exponent.

At first I was quite hopeful that this calculation had something to do with the effective potential. Even though free particles in a box are surely a bad and arbitrary model for nuclear processes, this statistical problem (which is basically the specific heat) should be much less model-dependent. You can easily estimate the effect of small perturbations, by the way; if W is the perturbation energy, then you get corrections of order $(W/kT)^2$. Now for an excitation energy of 10 MV the "temperature" is on the order of 2 MV, so at first glance it looks as if this still leaves you with plenty of freedom for disregarding minor details. (This is at least satisfying, in as much as the interconnection between different states of individual particles, which — according to Bohr have to be there to increase the radiation — do not contradict the condition $W << 2$ MV. Bohr's theory only shows that W must be bigger than the width of the radiation's γ-line). But if you take a closer look then this is no longer certain, since the higher-order perturbation approximations contain terms like $(W^2)_{ii}/(kT)^2$ as well as terms like $(W_{ii}/kT)^2$ (where W_{ii} = a matrix element of the perturbation). In order for this and all higher-order terms to be small (i.e., for all values of the coordinates, etc. which are there with any probability), the interaction function must be no bigger than 2 MV at any point. Such a condition is obviously nonsense, as this model of course neglects the strong correlation between nearby particles, so it must belong to a perturbation. On the other hand, we do know that it is related to an energy which can reach 50 MV (or so) for small distances between neighbours.

I have racked my brains to find a contrary boundary case to analyse; the liquid drop model is obviously good when there are weak interactions, or to be more precise when there are no strong correlations between the particles. Couldn't we treat strong correlations as a contrary boundary case? If the short-range forces were not so terribly pathological the answer would be very easy, as then you could say: As you increase the forces between neighbours and get stronger correlations, then the "gas-like" nucleus has to become a liquid drop at first, and finally a solid body.

If we assume this for the moment, the problem is certainly easy to solve. You just have to replace the Fermi gas with a Bose gas (the statistics of harmonic natural oscillations). But this case differs from the

Fermi case in that you need more than just one constant (the density of eigenvalues at the upper boundary of the Fermi distribution). Rather, you need the frequency distribution of all the eigenoscillations. For a sufficiently large nucleus, it is of course reasonable to assume a $\nu^2 d\nu$ law for the eigenoscillations. This leads to a value of T^3 for the specific heat. Then the exponent in the distribution of total energies for the system will be proportional to $U^{3/2}$ (U is the excitation energy) instead of $U^{1/2}$ as is the case for Fermi statistics. The constant factor in this term depends on the "speed of sound" in the nucleus, so you can't say much about its order of magnitude. If it were possible to empirically measure the density of excited states in two different areas (for example, for slow and fast neutrons), then the difference between $U^{1/2}$ and $U^{3/2}$ might turn out to be more important.

You will surely argue that this model is completely pathological, because the short-range interactions won't permit any point of equilibrium in the nucleus for the particles. But there is a counterargument: in liquid helium there is obviously a case where interactions are completely different, but where they act over such short ranges (or are so weak) that the zero-point energy can haul every atom away from this equilibrium position. Nonetheless, its specific heat will decrease sharply at low temperatures; even if you can't do the calculations for such a case, this empirical argument shows that the nucleus will basically behave just as if it were in the equilibrium position.

This model then raises the funny question of rotation. If you take the idea of equilibrium positions seriously, then it would be possible for the nucleus to rotate as a whole. This would lead to rotation levels of about 1000 Volts. Even though according to Ellis[274] it is just barely possible to observe a structure of lines with this order of magnitude, this is not necessarily a contradiction as the lines would presumably show no structure at all (see below).

Now we come to the question: How do the levels of rotation change, if you loosen the equilibrium positions so that the particles could change places often? I believe that you can imagine the conditions in the nucleus

[274]See 'Minutes of the Stanford Meetings', *Phys. Rev.* **45**, 130–139 (1934).

as follows: a proton might often change position with a neighbour, but it takes a very long time to get from one end of the nucleus to the other or around the whole nucleus. Question: Under these conditions, is the order of magnitude of the gaps between rotation levels more likely to be dependent on the solid nucleus, or on the individual particles? I haven't been able to answer this question in a satisfying way, but I think it is likely that the rotation levels still have to be very close together. You can discuss this question more strictly for two interacting particles running around together inside a ring-shaped tube. (The latter to avoid the centrifugal force, whose effects you can estimate anyway, and also to have only one degree of rotational freedom.) This can be done rigorously, and is a quite amusing example, but it is still rather difficult to make the leap from this case to n particles.

I also have another reason why I think the liquid drop model is more likely: the large number of quadrupole lines we are finding in nuclear spectra. If you think of single particles, there would have to be quite a lot of dipole moment transitions, and those should be found even more often. Quadrupole lines should just be built from metastable conditions; i.e., they should become more infrequent as the excitation energy increases.

The case of an oscillating model is different, in that the symmetry of the radiation is not dependent on the quantum number of the rotation. Instead, the rotation conditions are so close together that among all possible final states exactly the one is chosen which has the correct angular momentum which corresponds to the radiation type (dipole, quadrupole). The radiation type is determined by the symmetry of the oscillation conditions. If a nucleus which doesn't oscillate approximates spherical symmetry, the oscillations can be separated into polar coordinates. Any permitted transition results from decreasing the oscillation quantum number by 1, and if the relevant oscillation belongs to a spherical harmonic with $l = 1, 2, 3\ldots$ then a dipole, quadrupole, or octopole wave is radiated respectively. In this model the dipole moments can get very small quite easily, as this is an oscillation that <u>all</u> particles must take part in. If the charge and mass densities are uniform throughout the nucleus, they will also be proportional to each other during the oscillation. Any deviations from proportionality will just result from unpaired protons and neutrons, and will therefore be accordingly small. Because of this you can expect the dipole moments, at least at long wavelengths, to be of the same order of magnitude as the quadrupole moments.

Of course, I totally agree that because of Shankland's and especially Andersen's magnificent validation of Heisenberg's principle we should not believe Goldhaber's experiment for the moment. It's not very easy to make excuses for him, at least not with the effect you mentioned. His scattering apparatus is so narrow, and his solid angle so small, even if you accept his cross-section as a theoretical order of magnitude, and believe that every neutron scattered more than once in the machine (the neutrons scattered only once collect towards the front and lose almost no energy) is being counted with the same probability (like thermal neutrons without a paraffin coating), then you just get an effect with the necessary order of magnitude. On the other hand, the thick paraffin coating makes it easier to count fast neutrons and also results in strong absorption* and scattering, so such an assumption seems absolutely impossible. Even Goldhaber himself is not totally sure, because of scattering from the table and walls. Although he has did checking tests, he will repeat the experiment under improved conditions.

How can you compare Hulme[275] and Massey[276], by the way? Hulme's calculations are basically useful and even if I don't begrudge him doing them, these are mostly things which have to be done in any case (I'm not talking about his last paper on retardation, which was Møller's fault).[277] Bhabha's paper about exchanges in the scattering of positrons on electrons was very nice![278] This is just to defend

[275]Henry Rainsford Hulme (1908–1991), studied at Cambridge where he obtained his Ph.D. in 1932. After work at Cambridge, Leipzig, and the Air Ministry and posts in New Zealand, he accepted a post at the Atomic Weapons Research Establishment at Aldermaston in 1959 where he remained until his retirement in 1973.

[276]Harrie Stewart Wilson Massey (1908–1983), studied at Melbourne and Cambridge where he obtained his Ph.D. in 1932; After lecturing in Belfast he took up an appointment at University College London where he stayed for most of his career.

[277]Christian Møller (1904–1980) studied physics at Copenhagen where he received his doctorate in 1932 and continued to teach until his retirement in 1975.

[278]H.J. Bhabha, 'The Scattering of Positrons by Electrons with Exchange on Dirac's Theory of the Positron', *Proc. Roy. Soc.* **A154**, 195–206 (1936).

Cambridge. By the way, I have a doctoral student now who understands at least a few things.[279] It is kind of a relief to have someone who insists on writing spinors (I have talked him out of this now, by the way); even my best doctoral student previously needed to be taught that in the case of spin and of orbital moment you can express the total angular momentum in diagonal form.

By the way, I have been promoted: namely, to the titles of M.A. (at Cambridge), D. Sc. (at Manchester), and Member (not Fellow!) of St. John's, with the extra privilege of being allowed to dine there twice a week at my own expense! I came back from Copenhagen with Oliphant (where, by the way, it was very nice, especially because of Heisenberg's shocking theory), and he turned out to be very nice. Unfortunately, he has a terrible wife.

Gaby is already thinking very politically. She recently saw a picture of the High Commissioner for Palestine in the newspaper, and after we explained who it was she said: "Oh, the High Compromissioner!" But now, I'm getting writer's cramp!

By the way, some people are very upset that you are making your calculations of the energy-range relations available to all Americans, but neither keeping your promise to send it here nor publishing it! Have fun and best wishes also to your mother,
Yours,

Peierls

(handwritten addition by Genia)
Dear Bethe,
we have an "estate" and everything from asparagus to lavender. Cambridge is very nice, and there are even a lot of nice women and men coming for tea. Are you coming too? Goodbye,

Genia

[279]This refers to Fred Hoyle. See below letter [43].

Dear Miss Bethe,
I know that I am impertinent and I have had a terrible conscience for the past two years (since I haven't answered you). Will you visit sometime?

*In Goldhaber's experiment there is a lot of paraffin also in <u>in front of</u> the counter:

[38] Hans Bethe to Rudolf Peierls

Selisberg, 19.8.1936

Die quantenhafte Ausstrahlung wird kontinuierlich — wenn die Frequenz unseres Briefwechsels sich weiter so vermehrt, konvergiert es gegen einen Häufungspunkt in ein paar Wochen.

Also zuerst meine humble apologies für den nichtgemachten Besuch. Aber so schlimm wie Du Dir das denkst, war es nicht. Ich fuhr nicht über Southampton, sondern verliess die Queen in Cherbourg. Sie machte sich besonders gut vom Tender aus. Auch bei der Rückreise wird wohl diesmal nichts mit einem Besuch in England werden: wegen summer school teaching in Ann Arbor ist es diesmal sehr knapp mit der Zeit in Europa. Ich würde sonst gern das hochherrschaftliche Haus sehen, in dem sich sogar Deine Eltern wohlfühlen, und zum Tee kommen, zu dem auch Männer zugelassen sind: Aber das nächste mal – bis dahin darfst Du vielleicht jeden Tag im St. Johns College auf eigene Kosten dinieren.

Der Hauptgrund dieses Schreibens (Handschreibens, würde Placzek sagen) sind natürlich die Kerne. Ich hätte mir eigentlich denken können, dass Du die Niveaudichte auch ausgerechnet hast. Fermi hat es nebenbei auch getan, und Oppenheimer verkehrt. Ich hätte gern unsere Weisheit gepoolt, aber das geht leider nicht mehr, denn meine Arbeit ist schon vor vier Tagen im Phys. Rev. erschienen (so jedenfalls schrieb man mir). Sie enthält nur das freie-Teilchen-Modell, und dies mit einer nicht sehr eleganten Ableitung – ich habe zu viel gerechnet, weil ich es zu speziell machte.

Das andere Modell — fester Körper – leuchtet mir eigentlich sehr ein. Ich hatte schon auch daran gedacht, ungefähr 3 Tage ehe Dein Brief kam. Dein Argument mit der spezifischen Wärme des He scheint mir ziemlich beweisend; und die Argumente gegen das freie-Teilchen-Modell sind natürlich recht schlimm.

Ich habe versucht, etwas über dies Modell zu rechnen, ohne grossen Erfolg. Zunächst finde ich, dass der Exponent wie $U^{3/4}$ geht (siehe Anlage). Du schriebst $U^{3/2}$. Auch so dürfte der Unterschied gegen $U^{1/2}$ beobachtbar sein. Dann enthält der Exponent einen Faktor $A^{1/4}$ (A = Atomgewicht) gegen $A^{1/2}$ beim freien-Teilchen-Modell; beides unter der üblichen Voraussetzung Kernvolumen proportional Teilchenzahl. D.h. dass die Niveaudichte bei schweren Kernen nicht so rasch steigt, was mir im Hinblick auf die experimentellen Daten nicht unsympatisch ist. Weiter habe ich probiert, die Schallgeschwindigkeit auszurechnen, indem ich sie auf die Kompressibilität zurückführe, die man recht gut aus einer Majorana-Heisenberg-Weizsäcker-Rechnung ableiten kann. Dann kriegt man einen viel zu grossen Niveau-Abstand, nämlich 3MVC(!) bei $A = 100$ und $U = 8MV$. Aber ist natürlich die Annahme, dass die Kompressibilität massgebend ist, verkehrt, M[it] a[nderen] W[orten], die transversalen Schwingungen müssen sich ganz anders verhalten als die longitudinalen, für die die Kompressibilität sicher richtig ist. Das ist kein Wunder, denn schliesslich ist der Kern bestenfalls eine Flüssigkeit und kein fester Körper. Und da ist die Frequenz der Transversalwellen sicher sehr viel kleiner, besonders derjenigen langer Wellenlänge die ja für spezifische Wärme bei niedrigen Temperaturen wesentlich sind. Abschätzen kann man das ja wohl nicht.

Von den empirischen Daten her kann man versuchen, den Elastizitätsmodul für die Transversalschwingungen abzuschätzen: Der Abstand Δ zwischen benachbarten Niveaus ist wohl etwa 100 Volt; da nach der Rechnung $\Delta \approx 30 \cdot 10^6 e^{-5}$ Volts, so muss $S = \log 300000 = 12.6$ sein, d[ann] i[st] $12.6/2.3 = 5.5$ mal so gross wie der Wert den ich aus der Kompressibilität bekam. Da S proportional $E_0^{-3/4}$, so muss E_0 etwa kleiner sein. E_0 ist aber proportional u, also proportional der Wurzel aus dem Elastizitätsmodell. Letzteres muss also für Transversalschwingungen etwa 100 mal kleiner sein als für longitudinale.

Wegen der Wichtigkeit der Transversalschwingungen und des Unterschiedes zwischen Flüssigkeit und Festkörper glaube ich auch nicht sehr an Deine Rotation des Kerns als Ganzes. Er würde sich dadurch doch stark deformieren. Was hälst Du von folgender, sehr klassischer Überlegung zur Bestimmung der Rotationsniveaus: Man bestimme die Form und Energie des rotierenden Kerns als F[unktion] der Winkelgeschwindigkeit ω. Das ist wahrscheinlich einfach eine Frage der Oberflächenspannung. Wenn $E = \frac{1}{2}\Theta\omega^2$, dann ist Θ das "effektive Trägheitsmoment", das sicher viel kleiner ist als das wahre. Der Abstand der Rotationsniveaus wird entsprechend grösser.

Sollen wir diese Dinge über das Tropfenmodell zusammen publizieren? Die Idee ist ja allerdings im wesentlichen von Dir, also ist dieser Vorschlag meinerseits eigentlich eine Unverschämtheit. Ich glaube jedenfalls, es würde dem Fortschritt der Wissenschaft förderlich sein, dies der Oeffentlichkeit zu übergeben.

Da wir dich beim Publizieren sind: Was macht Dein Buch über die Quantentheorie? Ich habe allen Leuten in USA so davon vorgeschwärmt, dass mich jetzt jeder fragt: Wann kommt endlich? So frage ich auch — ich würde es gern meinen Studenten in die Hand drücken.

Mit der Energie-Reichweite-Beziehung war ich natürlich ein Schwein. Ich habe die Daten in Baden-Baden und werde hoffentlich daran denken. Ausserdem werden bei Breit Messungen gemacht, bei kleinen Protonengeschwindigkeiten bis etwa 2MV.

Viele herzliche Grüsse an Dich und Genia

Dein

Hans Bethe

Meine Mutter lässt auch grüssen

(Anhang)

1. Exponent in der Niveaudichte und Energie $\mathfrak{U}$ des System
   Annahme: $\mathfrak{U} = \alpha\tau^n$ ($\tau = kT, \alpha =$ fester Koeffizient)
   Dann ist

$$\underline{S} = \int \frac{c_v}{kT} dT = \int \frac{d\mathfrak{U}}{d\tau}\frac{d\tau}{\tau} = \frac{\alpha n}{n-1}\tau^{n-1}$$

$\tau = (\mathfrak{U}/\alpha)^{1/n}$, also

$$\underline{S} = \frac{n}{n-1}\alpha\left(\frac{\mathfrak{U}}{\alpha}\right)^{\frac{n-1}{n}} = \underline{\frac{n}{n-1}\alpha^{\frac{1}{n}}\mathfrak{U}^{\frac{n-1}{n}}}$$

Beim Schwingungsmodell ist n=4, also $S = \frac{4}{3}\alpha^{1/4}\mathfrak{U}^{3/4}$.
Wenn die Energie pro Volumenteinheit nur von der Dichte der Teilchen abhängt, und diese Dichte bei allen Teilchen konstant ist, so ist α proportional der Anzahl der Teilchen A, also $S \sim A^{1/4}\mathfrak{U}^{3/4}$.

2. Explizite Formel für die Niveaudichte $\rho(\mathfrak{U})$
F ist definiert durch

$$e^{-F(\beta)} = \sum_i e^{-\beta Ei} = \int \rho(\mathcal{U}')d\mathcal{U}'e^{-\beta\mathcal{U}'}\beta = 1/kT$$

Der Integrand hat ein scharfes Maximum bei $\mathcal{U}' = \mathfrak{U}$, wo $\mathfrak{U} = \frac{dF}{d\beta}$ die mittlere Energie des Systems bei einer Temperatur $1/\beta$ ist. Wir setzen $\rho(\mathfrak{U})e^{-\beta\mathfrak{U}} = \lambda(\beta)e^{-F(\beta)}$ wo λ eine langsam veränderliche F. Dann ist genau

$$\int \rho(\mathfrak{U}')d\mathfrak{U}'e^{-\beta\mathfrak{U}'} = \lambda(\beta)\int e^{-F(\beta')+\beta'\mathcal{U}'-\beta\mathfrak{U}'}d\mathcal{U}'$$

wo β' die zu $\mathfrak{U}'$ gehörige Temperatur. Folglich

$$\lambda(\beta)\int e^{F(\beta)-F(\beta')+(\beta'-\beta)\mathfrak{U}'}d\mathfrak{U}' = 1$$

Der Exponent ist

$$(\beta-\beta')\left(\frac{dF}{d\beta}\right)_{\beta'} + \frac{1}{2}(\beta-\beta')^2\frac{d^2F}{d\beta^2} + (\beta-\beta')\mathfrak{U} = \frac{1}{2}(\beta'-\beta)^2\frac{d\mathfrak{U}}{d\beta}$$

da $\mathfrak{U} = \frac{dF}{d\beta}$. Also

$$\frac{1}{\lambda(\beta)} = \int e^{\frac{1}{2}(\beta'-\beta)^2\frac{d\mathfrak{U}}{d\beta}}d\mathcal{U}' = \int e^{\frac{1}{2}(\mathfrak{U}'-\mathfrak{U})^2/\frac{d\mathfrak{U}}{d\beta}}d\mathfrak{U}'\sqrt{2\pi|\frac{d\mathfrak{U}}{d\beta}|}$$

($\frac{\mathfrak{U}}{d\beta}$ ist immer negativ). Demnach Abstand benachbarter Niveaus:

$$\Delta(\mathfrak{U}) = \frac{1}{\rho(\mathfrak{U})} = \frac{1}{\lambda(\mathfrak{U})} e^{F-\beta\mathfrak{U}} = \sqrt{2}\pi |\frac{d\mathfrak{U}}{d\beta}| e^{-s}$$

wo

$$\frac{dS}{d\beta} = -\frac{dF}{d\beta} + \mathfrak{U} + \beta\frac{d\mathfrak{U}}{d\beta} = \beta\frac{d\mathfrak{U}}{d\beta}, \quad \frac{dS}{d\tau} = \frac{1}{\tau}\frac{d\mathfrak{U}}{d\tau}$$

Wenn $\mathfrak{U} = \alpha\tau^n$, so ist

$$\Delta = \sqrt{2\pi n}\alpha^{-\frac{1}{2n}} \mathfrak{U}^{\frac{n+1}{2n}} e^{-\frac{n}{n+1}\alpha^{1/n}\mathfrak{U}^{n-1/n}}$$

3. Schallgeschwindigkeit nach Kompressibilität
Wenn Kernvolumen $= \Omega = \frac{a\pi}{3}R^3$ und Schallgeschwindgkeit $= u$ für Longitudinal- und Transversalwellen, so ist die Anzahl der Schwingungen pro $d\omega$

$$\mathfrak{N}(\omega)d\omega = 3\frac{\Omega}{(2\pi)^3}4\pi\omega^2 d\omega = \frac{2}{\pi}\frac{R^3}{u^3}\omega^3 d\omega$$

die Energie

$$\mathfrak{U} = \frac{2}{\pi}\left(\frac{R}{u\hbar}\right)^3 \int \frac{(\hbar\omega)^3 d(\hbar\omega)}{e^{\hbar\omega/\tau} - 1} = \frac{2}{\pi}\left(\frac{R}{\hbar u}\right)^3 \tau^4 \int_0 x^3 dx (e^{-x} + e^{-2x} + \ldots)$$

bei tiefer Temperatur: $\int = 6(1 + \frac{1}{2^4} + \frac{1}{3^4} + \ldots) = \frac{6\pi^2}{90} = \frac{\pi^2}{15}$;
$\alpha = \frac{2\pi}{15}\left(\frac{R}{\hbar u}\right)^3 A$ wo $R^3 = r_0^3 A$, r_0 unabhängig von A.
Nun ist $u^2 = \frac{\rho^2}{M}\frac{d^2E}{d\rho^2}$, wo M die totale Kernmasse, E die totale Kernenergie und ρ die Dichte der Kernmaterie. Da $\rho \sim R^{-3}$, so ist auch $u^2 = \frac{1}{9M}R^2\frac{d^2E}{dR^2}$.
Die Energie als Funktion von $x = const/R$ ist nach Formel (159, 159s) ein meinem Kernartikel

$$\frac{E*}{AT_o} = \frac{3}{10}x^2 - \frac{B/T_0}{\sqrt{x}}(2x^{-3} - 3x^{-1} + (x^{-1} - 2x^{-3})e^{-x^2} + \sqrt{\pi}\Phi(x))$$

$T_0 = \frac{\hbar^2}{Ma^2}$. Zweimal differenziert gibt das:

$$x^2 \frac{d^2E}{dx^2} \left(R^2 \frac{d^2E}{dR^2} \right)$$

$$= 6\mathrm{AT}_0 \left[\frac{1}{10} x^2 + \frac{B}{\sqrt{\pi}\mathrm{T}_0} x^{-3} (x^2 - 4 + (x^4 + 3x^e + 4) e^{-x^2}) \right]$$

Mit den Werten in (164) etc. ($T_0 = \frac{70}{9.6} = 7.3$ MV, $\frac{B}{T_0} = 9.6$, $x = 2.7$) gibt das

$$R^2 \frac{d^2E}{dx^2} = 6 \cdot 7.3 \cdot 1.65A = 72A \text{ MV}$$

$$u^2 = \frac{8MV}{M_0} (M_0 = \text{Protonenmasse})$$

$$\sqrt{\left(\frac{\hbar u}{r_0}\right)^2} = \sqrt{\frac{\hbar^2}{M_0^2} \cdot 8MV} = 13.3MV$$

$\alpha \approx (0.75 \frac{r_0}{\hbar\omega})^3 A = \frac{A}{E_0^3}$, wo $E_0 = \frac{13.3MV}{0.75} = 18MV$. Folglich

$$S = \frac{4}{3} \alpha^{1/4} \mathfrak{U}^{3/4} = \frac{4}{3} A^{1/4} \left(\frac{\mathfrak{U}}{E_0}\right)^{3/4} = 2.3 \text{ für} A = 100. \mathfrak{U} = 8MV$$

$$\sqrt{2\pi n \alpha^{-\frac{1}{2n}} \mathfrak{U}^{\frac{n+1}{2n}}} = \sqrt{8\pi A^{-1/8} E_0^{3/8} \mathfrak{U}^{5/8}} = 5 \cdot \frac{1}{1.8} \cdot 18^{3/8} \cdot 8^{5/8} = 30MV$$

$$\underline{\Delta = 30 \cdot e^{-2.3} = 3MV}.$$

Selisberg, 19.8.1936

Dear Peierls!
The quantum-like emission will become continuous — if the frequency of our correspondence keeps on increasing, it will converge to an accumulation point in a few of weeks.

So first my humble apologies for the visit that didn't happen.[280] But it wasn't as bad as you believe. I didn't pass Southampton, but left the

[280] Bethe had intended to visit the Peierls family in England but had to change his travel plans going via France to Switzerland instead.

Queen in Cherbourg. It looked especially great from the jollyboat. On the way back, there will be not be time for a visit to England either: because of summer school teaching in Ann Arbor, my time in Europe is especially short this time. I would really like to see the very stately house that even your parents feel comfortable at and have tea, which even men are allowed to drink. Next time — by then, you may be allowed to dine at St. John's College every single day at your own expense.

The main reason for this letter ('hand-letter', as Placzek would say) is the nuclei, of course. I might have known that you calculated the density of levels as well. Fermi has done the same, by the way, and Oppenheimer too, but Oppenheimer did it wrong. I would have liked to combine our wisdom, but unfortunately that's impossible now as my paper was published in the *Phys. Rev.* 4 days ago (that's what I was told).[281] It only contains the free particle model, and this with a derivation that is not very elegant — I did too many calculations, because I did it for too specific a case.

The other model — solid state — is rather convincing. I also thought about it, about three days before I received your letter. Your argument concerning the specific heat of He seems to be rather evident, and the arguments against the free particle model are very faulty.

I have tried to calculate something about this model, without great success. First, I found out that the exponent behaves like $U^{3/4}$ (see appendix). You wrote $U^{3/2}$. Its difference from $U^{1/2}$ can also be described. Then this also includes the term $A^{1/4}$ (A=atomic weight), compared to $A^{1/2}$ in the free-particle model; both provide a condition that the volume of the nucleus is proportional to the number of particles. This means that for heavier nuclei the density of a level doesn't increase so fast, which is not displeasing to me in view of the experimental data. I have also tried to calculate the velocity of sound by tracing it back to the compressibility, which you can easily derive from a Majorana-Heisenberg-Weizsäcker calculation. Then you will get too large a level distance $3MVC$(!) at $A = 100$ and $U = 8MV$. But it is an exception

[281]H.A. Bethe, 'An Attempt to Calculate the Number of Energy Levels of a Heavy Nucleus', *Phys. Rev.* **50**, 332–341 (1936).

that the compressibility is decisive. I[n] o[ther] w[ords], the transverse oscillations must be totally different from the longitudinal, for which the compressibility is likely to be correct. This is not very surprising, since the nucleus is a fluid at best and not a solid body. The frequency of the transverse waves is surely much smaller, especially at the longer wavelengths which are decisive for specific heat at low temperatures. I don't think that you can estimate that.

You can try to estimate the modulus of elasticity for transverse oscillations according to the empirical data: the distance Δ between adjacent levels might be about 100 Volts; as according to the calculation $\Delta \approx 30 \times 10^6 e^{-5}$ Volts, we must have $S = \log 300\,000 = 12.6$, and t[his] i[s] 12.6/2.3=5.5 times larger than the value I got from the compressibility. As S is proportional to $E_0^{-3/4}$, E_0 has to be a little smaller. But E_0 is proportional to u, so is also proportional to the root of the elasticity modulus. The latter thus has to be 100 times smaller for transverse oscillations than for longitudinal.

Since the transverse oscillations and the difference between fluid and solid bodies are so important, I don't really believe your idea that the whole nucleus rotates. It would be severely deformed by that. What do you think of the following, very classical idea for calculating the level of rotation: You have to find the shape and form of the rotating nucleus as a f[unction] of the angular velocity ω. This might simply be a question of surface tension. If $E = \frac{1}{2}\Theta\omega^2$, then Θ is the "effective moment of inertia", which surely is much smaller than the real one. The distance between rotation levels will accordingly be bigger.

Should we publish these ideas about the drop model together? The idea is basically yours, so my proposition is rather impertinent. I believe that it would be conducive to the progress of science to make it public.

As we are talking about publishing: How is your book about the quantum theory going? I have told everybody here in the USA about it in such an enthusiastic way that now everybody is asking: When will it finally appear?[282] So I ask you again — I would like to give it to my students.

---

[282]The book was finally published in 1955. R.E. Peierls, *Quantum Theory of Solids* Oxford: Oxford Univ. Press, 1955.

Regarding the energy-range relation I have behaved like a pig. I have the data in Baden-Baden, and hopefully won't forget about it. There will also be some measurements made by Breit, for small proton velocities up to about 2 MV.
Best wishes to you and Genia
Yours

Hans Bethe

Greetings from my mother

(Appendix)

1. Exponent in the level density and energy $\mathfrak{U}$ of the system Assumption: $\mathfrak{U} = \alpha\tau^n$ ($\tau = kT, \alpha =$ fixed coefficient) Then

$$\underline{S} = \int \frac{c_v}{kT} dT = \int \frac{d\mathfrak{U}}{d\tau}\frac{d\tau}{\tau} = \frac{\alpha n}{n-1}\tau^{n-1}$$

$\tau = (\mathfrak{U}/\alpha)^{1/n}$, so

$$\underline{S} = \frac{n}{n-1}\alpha\left(\frac{\mathfrak{U}}{\alpha}\right)^{\frac{n-1}{n}} = \underline{\frac{n}{n-1}\alpha^{\frac{1}{n}}\mathfrak{U}^{\frac{n-1}{n}}}$$

According to the oscillation model we have $n = 4$, so $S = \frac{4}{3}\alpha^{1/4}\mathfrak{U}^{3/4}$. If the energy per unit volume depends only on the density of particles, and this density is constant for all particles, then α is proportional to the number of particle A, and $S \sim \mathrm{A}^{1/4}\mathfrak{U}^{3/4}$.

2. Explicit formula for the density of levels $\rho(\mathfrak{U})$
F is defined by

$$e^{-F(\beta)} = \sum_i e^{-\beta E_i} = \int \rho(\mathfrak{U}')d\mathfrak{U}'e^{-\beta\mathfrak{U}'}\beta = 1/kT$$

The integrand has a sharp maximum at $\mathfrak{U}' = \mathfrak{U}$, so $\mathfrak{U} = \frac{dF}{d\beta}$ is the average energy of the system at a temperature $1/\beta$. We put $\rho(\mathfrak{U})e^{-\beta\mathfrak{U}} = \lambda(\beta)e^{-F(\beta)}$ where λ is a slowly changing function. Then it is enough to calculate

$$\int \rho(\mathfrak{U}'d\mathfrak{U}e^{-\beta\mathfrak{U}'} = \lambda(\beta)\int e^{-F(\beta')+\beta'\mathfrak{U}'-\beta\mathfrak{U}}d\mathfrak{U}'$$

where β' is the temperature belonging to $\mathfrak{U}$. Therefore

$$\lambda(\beta)\int e^{F(\beta)-F(\beta')+(\beta'-\beta)\mathfrak{U}'}d\mathfrak{U}'=1$$

The exponent is

$$(\beta-\beta')\left(\frac{dF}{d\beta}\right)_{\beta'}+\frac{1}{2}(\beta-\beta')^2\frac{d^2F}{d\beta^2}+(\beta-\beta')\mathfrak{U}=\frac{1}{2}(\beta'-\beta)^2\frac{d\mathfrak{U}}{d\beta}$$

for $\mathfrak{U}=\frac{dF}{d\beta}$. Therefore

$$\frac{1}{\lambda(\beta)}=\int e^{\frac{1}{2}(\beta'-\beta)^2\frac{d\mathfrak{U}}{d\beta}}d\mathfrak{U}'=\int e^{\frac{1}{2}(\mathfrak{U}'-\mathfrak{U})^2/\frac{d\mathfrak{U}}{d\beta}}d\mathfrak{U}'\sqrt{2\pi|\frac{d\mathfrak{U}}{d\beta}|}$$

($\frac{\mathfrak{U}}{d\beta}$ is always negative). Therefore the distance between adjacent levels is:

$$\Delta(\mathfrak{U})=\frac{1}{\rho(\mathfrak{U})}=\frac{1}{\lambda(\mathfrak{U})}e^{F-\beta\mathfrak{U}}=\sqrt{2\pi}|\frac{d\mathfrak{U}}{d\beta}|e^{-s}$$

where

$$\frac{dS}{d\beta}=-\frac{dF}{d\beta}+\mathfrak{U}+\beta\frac{d\mathfrak{U}}{d\beta}=\beta\frac{d\mathfrak{U}}{d\beta},\quad\frac{dS}{d\tau}=\frac{1}{\tau}\frac{d\mathfrak{U}}{d\tau}$$

If $\mathfrak{U}=\alpha\tau^n$, then we have

$$\Delta=\sqrt{2\pi n}\alpha^{-\frac{1}{2n}}\mathcal{U}^{\frac{n+1}{2n}}e^{-\frac{n}{n+1}\alpha^{1/n}\mathfrak{U}^{n-1/n}}$$

3. Velocity of sound using the compressibility.
If the nucleus volume $=\Omega=\frac{a\pi}{3}R^3$, and the velocity of sound $=$ u for both longitudinal and transverse waves, then the number of oscillations per $d\omega$ is

$$\mathfrak{N}(\omega)d\omega=3\frac{\Omega}{(2\pi)^3}4\pi\omega^2d\omega=\frac{2}{\pi}\frac{R^3}{u^3}\omega^3d\omega.$$

The energy

$$\mathfrak{U}=\frac{2}{\pi}\left(\frac{R}{u\hbar}\right)^3\int\frac{(\hbar\omega)^3d(\hbar\omega)}{e^{\hbar\omega/\tau}-1}$$

$$=\frac{2}{\pi}\left(\frac{R}{\hbar u}\right)^3\tau^4\int_0 x^3dx(e^{-x}+e^{-2x}+\cdots),$$

or at low temperatures $\int = 6(1 + \frac{1}{2^4} + \frac{1}{3^4} + \cdots) = \frac{6\pi^2}{90} = \frac{\pi^2}{15}$; $\alpha = \frac{2\pi}{15}\left(\frac{R}{\hbar u}\right)^3 A$ where $R^3 = r_0^3$A, and r_0 is independent of A. Now we have $u^2 = \frac{\rho^2}{M}\frac{d^2E}{d\rho^2}$, where M is the total nucleus mass, E the total nuclear power, and ρ the density of the nuclear matter. As $\rho \sim R^{-3}$, we find $u^2 = \frac{1}{9M}R^2\frac{d^2E}{dR^2}$.

The energy of a nuclear particle as a function of $x = const/R$ is, according to formulas (159, 159s)[283]

$$\frac{E*}{AT_o} = \frac{3}{10}x^2 - \frac{B/T_0}{\sqrt{x}}(2x^{-3} - 3x^{-1} + (x^{-1} - 2x^{-3})e^{-x^2} + \sqrt{\pi}\Phi(x)).$$

$T_0 = \frac{\hbar^2}{Ma^2}$. The second derivative is:

$$x^2\frac{d^2E}{dx^2}\left(R^2\frac{d^2E}{dR^2}\right)$$

$$= 6\text{AT}_0\left[\frac{1}{10}x^2 + \frac{B}{\sqrt{\pi}\text{T}_0}x^{-3}(x^2 - 4 + (x^4 + 3x^e + 4)e^{-x^2})\right]$$

With the values given in formula (164), etc. ($T_0 = \frac{70}{9.6} = 7.3$MV, $\frac{B}{T_0} = 9.6$, x=2.7) results in

$$R^2\frac{d^2E}{dx^2} = 6 \cdot 7.3 \cdot 1.65A = 72AMV$$

$$u^2 = \frac{8MV}{M_0}(M_0 = \text{proton mass})$$

$$\sqrt{\left(\frac{\hbar u}{r_0}\right)^2} = \sqrt{\frac{\hbar^2}{M_0^2}} \cdot 8MV = 13.3MV$$

$\alpha \approx (0.75\frac{r_0}{\hbar\omega})^3 A = \frac{A}{E_0^3}$, where $E_0 = \frac{13.3MV}{0.75} = 18MV$. Therefore

$$S = \frac{4}{3}\alpha^{1/4}\mathfrak{U}^{3/4} = \frac{4}{3}A^{1/4}\left(\frac{\mathfrak{U}}{E_0}\right)^{3/4} = 2.3 \text{ for } A = 100.\ \mathfrak{U} = 8MV$$

$$\sqrt{2\pi n}\alpha^{-\frac{1}{2n}}\mathfrak{U}^{\frac{n+1}{2n}} = \sqrt{8\pi}A^{-1/8}E_0^{3/8}\mathcal{U}^{5/8} = 5 \cdot \frac{1}{1.8} \cdot 18^{3/8} \cdot 8^{5/8} = 30MV$$

$$\underline{\Delta} = 30 \cdot e^{-2.3} = \underline{3MV}$$

[283] Letter [26], note 146.

[39] Rudolf Peierls to Hans Bethe

Cambridge, 25.8.1936

Lieber Bethe,
Unsere Korrespondenz nimmt in der Tat ein atemberaubendes Tempo an!
Die Rechnungen, die Du schreibst, sind im grossen und ganzen dieselben, die ich gemacht habe. (Der Exponent 3/2 in der Entropie war natürlich ein misprint, denn die Rechnung ist so einfach, dass selbst ich dabei den Exponenten nicht falsch machen kann). Der Faktor in der Entropie ist dagegen bei Dir falsch, denn ich möchte eigentlich dafür sein, dass $\sum_1^\infty n^{-4} = \pi^4/90$, nicht, wie Du schreibst $\pi^2/90$ ist. Also sogar der grosse Bethe macht Rechenfehler? Teller hatte übrigens grossen Erfolg in Kopenhagen mit der Bemerkung, dass eine Tabelle in Deinem Handbuchartikel zwei numerische Rechenfehler enthält und dass Gamow dies entdeckt hat. Ist das Amerika oder das Auto?

Deine Abschätzung der Schallgeschwindigkeit aus der Kompressibilität scheint mir aus zwei Gründen inkonsequent: 1. weil, wie Du selbst sagst, die Transversalwellen vermutlich viel kleinere Frequenzen haben, und 2. weil es nicht erlaubt ist, in diesem Modell mit der Heisenbergschen Methode zu rechnen, die doch fast unabhängige Teilchen vorausgesetzt, während andererseits wir doch jetzt auf dem Standpunkt stehen, dass das Modell mit den beinahe unabhängigen Teilchen nicht einmal die richtige spezifische Wärme gibt! Es scheint mir daher notwendig, den Faktor α in dem Gesetz $E = \alpha(kT)^4$ offen zu lassen. Uebrigens soll man wohl auch den Faktor 3 von der Polarisation auf jeden Fall durch einen Faktor 2 ersetzen, da eben praktisch nur die transversalen Schwingungen etwas ausmachen. Ich würde versuchen, umgekehrt das α so zu bestimmen, dass die Sache vernünftig wird. Dazu muss die Entropie offenbar von der Grössenordnung 10 werden (im Gegensatz zum Fall der Fermistatistik, wo wegen der vielen Nebenbedingungen wie Konstanz der Neutronenanzahl und Protonenanzahl und des Impulsmoments, der Faktor vor dem e^s viel kleiner wird). Hieraus folgt:

$$\alpha \sim 6(MV)^{-3}$$

Andererseits hat α die folgende Beziehung zur Grundgeschwindigkeit des Kerns:

$$\epsilon = \gamma \left(\frac{\pi^5}{45\alpha} \right)^{1/3}$$

Hierin ist ϵ die kleinste Anregungsenergie, γ ist ein numerischer Faktor, der für eine elastische Kugel sehr nahe an 1 ist, und es ist vorausgesetzt, dass die Kugel nahezu inkompressibel ist (nur bei Transversalwellen). Folglich müsste dann ϵ von der Grössenordnung 1 MV sein, was wohl zu hoch.

Jedenfalls weiss man von den radioaktiven Kernen, dass dort die Anregungsenergie höchstens 100 000 Volt ist, auch dazu muss man schon alle die vorhandenen Niveaus von 4.10^4 Volt auf Feinstruktur oder auf Rotation schieben, was ich nur ungern tun würde. Nun kann man sich allerdings darauf ausreden, dass man ja wenig über die Resonanzniveaus von so schweren Kernen weiss, und es wäre ja denkbar, dass für die radioaktiven Kerne wirklich der Exponent beträchtlich grösser, d.h. die virtuellen Niveaus bereits selbst für thermische Neutronen praktisch kontinuierlich sind. Aber dazu müsste die Schallgeschwindigkeit mit wachsendem Atomgewicht ziemlich rasch abfallen, was zu ad hoc gemacht ist.

Nun scheint es mir auch nicht sicher, dass es richtig ist, die Debyesche Formel so Ernst zu nehmen. Denn wir wissen ja, dass der Kern nicht ein Kristall mit Gleichgewichtslagen ist, sondern eher eine Flüssigkeit wie flüssiges He. Je mehr ich mir darüber den Kopf zerbreche, umsomehr scheint es mir, dass dieses ein neuer Aggregatzustand mit definierten Eigenschaften sein muss. Denn flüssiges He ist offenbar makroskopisch eine Flüssigkeit, jedenfalls bei nicht zu tiefen Temperaturen, andererseits muss bei sehr tiefen Temperaturen seine Entropie verschwinden. Hieraus folgt unmittelbar, dass bereits bei einer endlichen Temperatur sehr kurze Transversalwellen nicht mehr die Frequenz Null haben können (obwohl lange das offenbar noch tun), denn wenn man z.B. annimmt, dass Wellen von der Länge L noch keine Frequenz haben, so hat ein Tropfen He von der linearen Dimension $D(n/L)^3$ Freiheitsgrade ohne potentielle Energie und daher eine abschätzbare Entropie (obwohl es mir noch nicht gelungen ist, diese Abschätzung vernünftig durchzuführen, man muss das natürlich quanteln, und dabei gibt es unangenehme Unklarheiten). Aus einer — bisher noch nicht vorhandenen — anständigen thermischen Untersuchung von He bei extrem tiefen Temperaturen könnte man also eine Abschätzung für L bekommen. Ferner erwarte ich auch Aufschluss von Untersuchungen an Ultraschallwellen in flüssigem He, die jetzt hier gerade angefangen werden.

Es ist jedenfalls wohl sicher, dass sehr kurze Transversalwellen eine endliche Frequenz haben werden, während lange das nicht haben. Das kann man sich auf zwei Weisen einfach vorstellen: Entweder nimmt man an, dass die Substanz sich für lange Wellen wie eine Flüssigkeit, für kurze wie ein Kristall benimmt, wobei die kritische Wellenlänge dann von der Temperatur abhängen müsste. Oder aber, was mir seit heute früh sympathischer scheint, man nimmt an, dass die Frequenz in diesem komischen Aggregatzustand proportional zum Quadrat des Wellenvektors ist.

(Eine dritte Möglichkeit wäre, dass die Elastizitätsgrenze von der Wellenlänge abhängt, und bei langen Wellen bereits bei der Amplitude Null liegt. Dies ist natürlich schlampig ausgedrückt, denn wenn die Gleichungen nicht mehr linear sind, so darf man überhaupt nicht mehr von einzelnen Wellen reden.)
Nimmt man die Proportionalität zwischen ω und k^2 an, so wird die Entropie $U^{3/5}$ statt $U^{3/4}$, was natürlich den Faktor nicht mehr stark ändert, aber die Anregungsenergie des Kerns wird viel kleiner.

Für das flüssige He ergibt sich, dass U prop $T^{5/2}$ wird, was allerdings ein bisschen zu schwach scheint. Es liegt aber immerhin innerhalb der Messgenauigkeit, es scheint nur sicher zu sein, dass die spezifische Wärme rascher als proportional mit T verschwindet.

Es müsste eigentlich möglich sein, diese Frage theoretisch zu entscheiden, aber jedenfalls etwas mehr herauszubekommen darüber, wie ein Modell mit Gleichgewichtslagen aber mit Nullpunktsenergie > Potentialschwelle sich benehmen sollte. Hierüber verstehe ich jedoch vorläufig noch immer Bahnhof.

Mit Deinem Vorschlag, die Rotationsniveaus klassisch abzuschätzen, bin ich eigentlich gar nicht einverstanden, und zwar aus folgendem Grunde: was man klassisch untersuchen kann, sind im wesentlichen zwei Dinge: Erstens gibt es eine Zentrifugalkraft, unter der der Kern deformiert wird. Diese Zentrifugalkraft ist aber erstens sehr klein, weil die Winkelgeschwindigkeit bei einem schweren Kern sehr klein ist, und zweitens bewirkt sie eine Erhöhung des Drehmoments (d.h. Erhöhung der Rotationsenergie bei fester Winkelgeschwindigkeit) und daher eine Verkleinerung der Rotationsenergie bei festem Drehimpuls. Zweitens kann klassisch z.B. im Fall einer Flüssigkeit ein Teil des Kerns unabhängig vom Rest rotieren, und hierzu würde natürlich ein kleineres Trägheitsmoment gehören. Damit beweist man aber nur, dass es auch Rotationszustände mit höherer Energie gibt — woran man natürlich sowieso nicht zweifelt aber klassisch gibt es jedenfalls sicher auch die Rotationsform, wo der Kern als Ganzes rotiert, bis auf eine von der Zentrifugalkraft herrührenden Deformation.

Dagegen gibt es einen quantenmechanischen Effekt, der im Prinzip die Rotationsenergie sehr wohl erhöhen kann, der aber klassisch nicht zu behandeln ist. Man kann sich von der Existenz am Besten etwa so überzeugen: Denken wir uns n Teilchen mit beliebiger Wechselwirkung, die aber, statt umeinander herumlaufen zu können wie im wirklichen Kern, an einen Kreisring gebunden sind, so dass man nur ein n-dimensionales Problem hat (so wird man die Eulerschen Winkel los) und bezeichnen wir die Lagen der Teilchen durch N Winkel. Die Energie hängt dann von deren Differenzen ab (natürlich mod 2π) und die Randbedingung lautet, dass die Wellenfunction ungeändert bleiben soll, wenn man irgendeinen dieser Winkel um 2 vermehrt. Man kann nun leicht das gesamte Impulsmoment das natürlich eine Integrationskonstante ist, abseparieren, indem man in der Eigenfunktion den Faktor $e^{2\pi i(\phi_1+\phi_2+\ldots\phi_n)m/N}$ abspaltet. Der Rest der Wellenfunktion enthält dann noch die $N-1$ Differenzen, und genügt dann einer Wellengleichung, die bis auf den Term $m^2\hbar^2/2NMR^2$ in der Energie von m unabhängig ist. Dagegen hat sich nun die Randbedingung geändert und verlangt jetzt, dass bei Änderung eines Winkels um 2π die Eigenfunktion sich mit $e^{2\pi im/N}$ multiplizieren soll. Diese Veränderung der Randbedingung bewirkt sicher eine Erhöhung der Energie, wieviel hängt natürlich von den Kräften ab. Für $N = 2$ ist die Sache natürlich trivial, man kann dort die Erhöhung leicht für die Fälle anschreiben, wo das W.K.B. erlaubt ist und zwar entweder mit überall reeller oder streckenweise komplexer Wellenzahl. Im ersten Fall wird die Energieerhöhung $\frac{1}{2}dE/dn$

(n = Quantenzahl der Relativbewegung) d.h. die Rotationsenergie wird gleich der inneren Änderungsenergie. Im anderen Grenzfall dagegen, wenn es ein grosses Gebiet mit V.E. gibt, so ist die Erhöhung proportional $\exp(-\int pdq)$ über diesen Bereich also sehr klein. Der Übergang zwischen diesen beiden Grenzfällen wird hier gegeben durch die Bedingung dass das eine Teilchen relativ zum anderen gerade eine merkliche Wahrscheinlichkeit bekommt, einmal ganz herumzulaufen. Es fragt sich man, wie dies auf N Teilchen zu veallgemeinern ist, ob die Bedingung dort lautet, dass ein Teilchen relativ zu allen andern gerade eine merkliche Wahrscheinlichkeit haben soll, einmal um den ganzen Ring herumzulaufen (was natürlich nicht der Fall ist) oder ob die Wahrscheinlichkeit nur beträchtlich sein soll, dass die Teilchen ihre Reihenfolge wechseln (was sicher der Fall ist). Diese Frage muss entscheidbar sein, aber ich sehe noch keinen Weg dazu.

Was die Frage der Publikation betrifft, so habe ich den Eindruck, dass wir im Augenblick auf ziemlich verschiedenen Pfaden wandeln. Ausserdem aber scheint mir die ganze Sache noch nicht publikationsreif, bevor die diversen ungelösten Fragen (elastische Eigenschaften der "Heliumphase", Grösse der Rotationsquanten, etc.) etwas weiter geklärt sind. Ich würde daher vorschlagen, dass man dies abwartet; wenn es dann soweit ist, können wir ja immer noch eine Pfütze (ich meine pool) machen. Oder bestehst Du auf einer Publikation in amerikanischem Tempo?

Ich würde übrigens auch auf jeden Fall warten, bis ich von Bohr gehört habe, ob er das alles schon weiss, bezw. machen will, denn ich habe keine Lust, in unfertige Bohrsche Arbeiten hineinzupfuschen. Mein Buch hatte ich seit Oktober nicht mehr angerührt. Dein gutes Zureden wie das von Mott haben aber jetzt den Erfolg gehabt, dass ich es wieder ausgegraben habe, und dass mir die Stelle wegen deren ich schon die Hoffnung verlor, eigentlich gar nicht mehr so schlecht vorkommt. Vielleicht wird doch noch etwas daraus.

Thomas hat mir jetzt eine Methode zur Behandlung von H^3 erzählt, die wirklich wunderhübsch ist und für ein gegebenes Potentialfeld mit beliebiger Genauigkeit durchgeführt werden kann (jede weitere Näherung dauert etwa zwei Tage, Konvergenz wie eine geometrische Reihe mit einem Faktor von ca. $\frac{1}{2}$).

Resultate damit hat er noch keine. Sein früheres Resultat, wonach Presents Methode sehr schlecht ist, hat sich übrigens als Rechenfehler herausgestellt. Er glaubt aber immer noch an die Richtigkeit seiner Behauptung. Bloch betreibt Gamowsche Physik, die dann in der Physical Rev. in der Gamowsprache veröffentlicht wird (wieso lässt Du als Redakteur das eigentlich zu?
Nu aber Schluss! Beste Grüsse
Dein

Peierls

[handwritten addition]
Hier sind übrigens viele Leute "settled". Born hat die Darwinsche Professur in Edinburgh, Berg eine Stelle bei Kodak in London, Ehrenberg scheint bei HMV sehr zufrieden und hat eine kleines Auto.

[handwritten addition by Genia Peierls]
Ach wie schade dass Du nicht zum Tee kommst. I am sooo sorry! Aber, I hope, nächstes Jahr. Ja übrigens bist Du noch nicht berühmt genug in U.S.A um mich (das heißt Rudi) einzuladen? Wenn nicht, hast Du zwei Jahre verloren. Hast Du den Gamow[284] gesehen? Family blüht und gedeiht und Gaby ist so hübsch geworden (von wo kommt das) dass sie wird für Reklame photographiert. Ronny ist ein judisches Affe, aber für Junge das macht nichts. Ich habe eine amerikanische Party gehabt und 20 Amerikaner haben gefressen wie 200 Krokodile! Warum warst Du so verhungert? Viele Grüsse auch an Deine Mutter.

Genia

[284]George Gamow (1904–1968), studied at Odessa and Leningrad, Göttingen, Copenhagen and Cambridge. After further work in Russia, he defected in 1933 and emigrated to the US in 1934 where he became Professor of Physics at George Washington University in 1934. In 1956 he took up a position at the University of Colorado.

Cambridge, 25.8.1936

Dear Bethe,
Our correspondence has now reached a really breathtaking speed![285]

The calculations you sent me are more or less the same as those I have done. (The exponent 3/2 in the entropy must have been a misprint, as the calculation is so easy that even I couldn't make an error with the exponent). Your factor in the entropy is wrong, because I would say that $\sum_1^\infty n^{-4} = \pi^4/90$ and not, as you wrote, $\pi^2/90$. Even the great Bethe makes calculation errors? Teller in Copenhagen has been very, very successful with his comment that there are two numerical calculation errors in one of the tables in your review article,[286] and that Gamow discovered them. Is that due to America or the car?

Your estimate of the sound velocity from the compressibility seems to be inconsequential for two reasons: 1. because you say that the transverse waves are supposed to have much smaller frequencies, and 2. because this model cannot be calculated with the Heisenberg method which requires a nearly infinite number of particles. We now claim that even a model with a nearly infinite number of particles does not give the correct specific heat!

Therefore it would seem to be necessary to leave the factor α open in the law $E = \alpha(kT)^4$. You should also substitute the polarisation factor of 3 for a factor of 2 in any case, as only the transverse oscillations are actually important.

On the contrary, I would try to define α in such a way that the whole thing is still reasonable. For that you obviously need an order of magnitude for the entropy of around 10 (in contrast to Fermi statistics, where the factor in front of the e^s is much smaller because of numerous secondary conditions like having a constant number of neutrons and protons, and constant angular momentum). It follows that

$$\alpha \sim 6(MV)^{-3}.$$

[285]Reference to Bethe's comment in letter [38].

[286]See letter [26], note 146.

On the other hand α has the following relationship to the basic velocity of the nucleus:

$$\epsilon = \gamma \left(\frac{\pi^5}{45\alpha} \right)^{1/3} .$$

Here ϵ is the smallest excitation energy, γ is a numerical factor, which in the case of an elastic ball is very close to 1, and requires that the ball be almost incompressible (to transverse waves). ϵ therefore has to be on the order of 1MV, which seems to be too high. As we know, the excitation energy of radioactive nuclei is at most 100 000 Volts, and even for that you have assign all the existing levels above 4×10^4 Volts to hyperfine structure or rotation, which I wouldn't like to do. For now you have the excuse that we don't know much about the resonance levels of heavy nuclei, and that the exponent could be considerably higher for radioactive nuclei; i.e., the virtual levels might actually be continuous for thermal neutrons. But for that the sound velocity has to fall off rapidly as the atomic weight increases, which has been done ad hoc.

I'm not sure if it's right to take the Debye formula too seriously, because we know that the nucleus is not a crystal in equilibrium with the positrons but more like a fluid such as liquid He. The more I think about that, the more it seems to me that there has to be a new condition of aggregation with defined characteristics. Liquid He obviously seems to be a macroscopic fluid, at least at temperatures which are not too low, otherwise its entropy would have to disappear in the case of very low temperatures. The result is that at finite temperatures very short transverse waves can not have a frequency of zero (even though they still do that for a long time), because if you say for example that waves of length L have zero frequency, then in its linear dimensions a drop of He has $D(n/L)^3$ degrees of freedom without potential energy and therefore an appreciable entropy (even though we haven't been able to estimate it in a reasonable way, you have to quantise that and it may become quite unclear). A permanent thermal analysis of He at extremely low temperatures — which hasn't been done yet — could lead to an estimate of L. I expect to learn something new from the analysis of ultrasonic waves in liquid He, which has just been started here.[287]

[287] In 1933 the Mond Laboratory had opened in Cambridge and in 1934, largely spurred by Kapitza's efforts, the first liquid helium had been produced there in 1934.

It seems to be certain that very short transverse waves will have a finite frequency, while long waves will not. You can see this in two different ways: Either you suppose that the substance will behave like a fluid for long waves and like a crystal for short waves, in which case the critical wave length would depend on the temperature. Or, as appears to me more pleasing since this morning, you suppose that the frequency in this strange aggregate state is proportional to the square of the wave vector.

(A third possibility would be that the elasticity limit depends on the wavelength, and in the case of long waves would have zero amplitude. This isn't very precise, to be sure. If the equations are not linear anymore, you can't talk about single waves at all.)

If you take ω and k^2 to be proportional, then the entropy will be $U^{3/5}$ instead of $U^{3/4}$ which doesn't change the factor very much. But the excitation energy of the nucleus would be much smaller.

For liquid He the result is that U is prop[ortional to] $T^{5/2}$, which seems to be a bit weak. But at least it is still within the precision of measurements, it seems to be clear that the specific heat disappears at a rate which is steeper than simple proportionality with T.

It must be possible to decide these questions in a theoretical way, or at least to find out more about how a model with equilibrium conditions but zero-point energy > the threshold potential should behave. But I still don't understand that at all.

I don't agree with your proposal to estimate the rotation levels in a classical way for the following reasons: You can basically analyse two things in a classical way: first, there is a centrifugal force by which the nucleus is deformed. But this centrifugal force is very small, as the angular velocity for a heavy nucleus is very small, yet still causes an increase in the moment of inertia (i.e., an increase of the energy of rotation at fixed angular velocity) and therefore a reduction of the energy of rotation at a fixed angular momentum. Second, you can rotate part of the nucleus, in case of a liquid, independently of the other parts

---

After Kapitza's detention in Russia, John Cockroft and Ernest Rutherford supervised work at the Cambridge facilities and managed to make some significant contributions to low temperature physics until the war interrupted work.

in a classical way. For that you need a smaller moment of inertia. Doing this, you can only prove that there are conditions of rotation with higher energy — which surely nobody doubts — but according to the classical theory there is a kind of rotation where the nucleus rotates as a whole, except for a deformation caused by the centrifugal force.

On the other hand, there is a quantum mechanical effect which can raise the energy of rotation in principle, but which can't be explained in a classical way. Its existence can easily be proven like this: Think of n particles with an arbitrary reciprocal interaction, which are bound to an annulus instead of going all over the place as they do in real nuclei, so that you have just one n-dimensional problem (with that you get rid of Euler's angle) and describe the positions of the particles by N angles. The energy results from the angular differences (mod 2π of course), and the boundary condition says that the wavefunction has to be unchanged if you increase any of these angles by 2π. Then you can easily separate the whole angular momentum, which is an integration constant, if you separate the factor $e^{2\pi i(\phi_1+\phi_2+\cdots\phi_n)m/N}$ from the eigenfunction. The remainder of the wavefunction consists of the $N-1$ differences, and is enough for a wave equation which is energetically independent of m except for the value $m^2\hbar^2/2NMR^2$. Instead, the boundary condition has changed and requires that the eigenfunction has to be multiplied by $e^{2\pi im/N}$ whenever the angle is changed by 2π. The change in the boundary condition surely causes an increase of energy, but how much it changes depends on the forces. If $N = 2$ the thing is quite easy, you can easily apply the increase in cases where the W.K.B.[288] is allowed, either with generally real or partly complex wave numbers. In the first case the increase of energy will be $\frac{1}{2}dE/dn$ (n = quantum number of relative motions); i.e., the rotation energy equals the internal kinetic energy. In the other border case, if there is a wide area with $V > E$ the increase is proportional to $\exp(-\int pdq)$ around this area, so is very small. The transition between these two border cases is given by the condition that, compared to the other particles, one particle will have a distinct probability of going around the whole annulus once. You could ask how this could be generalised to N particles: whether you have the

[288]Wentzel-Kramers-Brillouin Approximation.

condition that, compared to the other particles, one particle has to have a distinct probability of going around the annulus once (which is surely not the case); or whether the probability just has to be high enough that the particles will change their order (which is definitely the case). This question has to be solved, but at the moment I can't see how.

Regarding the publication, I have the impression that we are doing rather different things at the moment. The whole thing also doesn't seem ready for publication, and won't be before these diverse and unsolved questions (elastic characteristics of the "helium period", quantization of the rotation, etc.) are made a bit clearer. Therefore, I would suggest waiting a bit longer; when the time comes, we can make a puddle (a pool I mean). Or do you insist on publishing at American speeds?

I would also wait until we know whether Bohr already knows all this or is planning to do so, since I don't want to botch up his unfinished works. I haven't worked on my book since October. But your and Mott's support have convinced me to go back to it. The part which nearly made me abandon all hope now doesn't seem to be too bad. Something will probably come out eventually.

Thomas has told me about a method of treating H_3 which is really very nice, and works on any given potential to arbitrary precision. (every subsequent approximation takes about two days, converging like a geometrical series with a factor of about $\frac{1}{2}$) I have no results yet. His former result, which makes Present's method look very bad, turned out to be a calculation error. But he still believes that his proposition is correct.[289]

Bloch is doing Gamow physics, which has been published in the Physical Rev. in Gamow's language (how can you allow that as an editor?)[290]

I have to come to an end! Best wishes

Yours

Peierls

[289]See letter [36].

[290]Felix Bloch had been working on magnetic scattering of neutrons and on γ-ray emission. He had published several papers one of which was with Gamow. F. Bloch and G. Gamow, 'On the probability of gamma-ray emission', *Phys. Rev.* **50**, 260 (1936); F. Bloch, 'On the continuous gamma-radiation accompanying the beta-decay', *Phys. Rev.* **50** (1936).

[handwritten addition]
A lot of people are "settled" here. Born[291] got the Darwin professorship in Edinburgh, Berg a job with Kodak in London, and Ehrenberg seems to be very happy with HMV and has a small car.

(handwritten addition by Genia Peierls)
What a pity that you don't come for tea. I am sooo sorry! But you will, I hope, next year. Aren't you famous enough to invite me (that means Rudi) to the U.S.A.? If not, you have lost two years. Have you seen Gamow? The family is thriving and prospering, and Gaby has become so pretty (where does that come from) that she is being photographed for advertisements. Ronny is a Jewish monkey, but for a boy it doesn't matter. I had an American party with 20 Americans, who ate like 200 crocodiles! Why did you have to starve? Many greetings to your mother as well.

Genia

[40] Hans Bethe to Rudolf Peierls

Seelisberg, 27.8.1936

Lieber Peierls,
Fortsetzung über den Fortschritt der Wissenschaft:

1. Ich habe versucht, die Kernniveaus mit dem Tropfenmodell abzuschätzen unter der Annahme, dass nur die Oberflächenspannung für die Transversalschwingungen eine Rolle spielt. Die Oberflächenenergie nimmt man natürlich proportional zur Oberfläche an. Dann spielt bei einer angegebenen Oberfläche nur die Verschiebung der Teilchen senkrecht zur Oberfläche eine Rolle.

[291] Max Born (1882–1970) had left his Chair at Göttingen in 1933, and after short spells at Cambridge and Bangalore he became Tait professor in Edinburgh in succession to Charles Galton Darwin.

Wenn die Oberfläche in der xy-Ebene liegt und u die Verschiebung ist, so ist die relative Änderung der Oberfläche

$$\delta\sigma = \frac{\int dxdy \left(\sqrt{1 + \left(\frac{\partial u_z}{\partial x}\right)^2 + \left(\frac{\partial u_z}{\partial y}\right)^2} - 1\right)}{\int dxdy}$$

$$= \frac{1}{2} \frac{\int d\sigma \left[\left(\frac{\partial u_n}{\partial x}\right)^2 + \left(\frac{\partial u_n}{\partial y}\right)^2\right]}{\int d\sigma}$$

Wenn nun $u = \underset{\sim}{a} \sin \underset{\sim}{k} \cdot \underset{\sim}{r}$ so ist $\delta\sigma = \frac{1}{2} a_n^2 k_\parallel^2 \overline{\cos^2 k \cdot r}$, wo $k_\parallel$ die Komponente von k in der Oberfläche. Bei einer Kugel muss man natürlich über die Oberfläche mitteln. Nehmen wir $\underset{\sim}{k}$ parallel zu z, $\underset{\sim}{n}$ parallel zu x, so ist in Polarloordinaten

$$a_n^2 k_\parallel^2 = a^2 k^2 \sin^4 \vartheta \cos^2 \varphi$$

gemittelt $= a^2k^2/10$.

Also die Änderung der Oberflächenenergie$= \frac{1}{20} S_o a^2 k^2 \overline{\cos^2 kr}$ wo S_o die Oberflächenenergie ohne Schwingung ist. Wenn M die Gesamtmasse des Kerns ist, so ist dann sie Schallgeschwindigkeit $u = \sqrt{S_o/10M}$.

S_o kann man aus den empirischen Kernenergien abschätzen (à la Weizsäcker); man bekommt dann mit den neuen Kernradien (siehe unten) $S_o = 10A^{2/3}M$, wo $A =$ Atomgewicht. Wenn man dann die Formel für die Totalenergie des Kerns schreibt:

$$\mathfrak{U} = \frac{4\pi}{45} \left(\frac{R}{\hbar u}\right)^3 \tau^4 = \frac{\tau^4}{E_o'^3} A$$

so ist

$$E_o' = \left(\frac{45}{4\pi}\right)^{1/3} \cdot \sqrt{\frac{\hbar^2}{M_o r_o^2 A} \frac{S_0}{10}} = \left(\frac{45}{4\pi}\right)^{1/3} \sqrt{\frac{\hbar^2}{M_o r_o^2} \cdot \frac{10MV}{10}} A^{-1/6}$$

$$= E_o A^{-1/6}$$

(M_o = Protonenmasse; $r_o = 2 \cdot 10^{-13}$cm entsprechend den Bemerkungen unter 2) Also $\frac{\hbar^2}{M_o r_o^2} \approx 10MV$, $E_o = 4.8MV$

$\mathfrak{U} = \frac{\tau^4}{E_o^3} A^{3/2}$; $\alpha = \frac{A^{3/2}}{E_o^3}$ (siehe vorheriger Brief)

$$\tau = (\mathfrak{U}/\alpha)^{1/4} = \mathfrak{U}^{1/4} E_o^{3/4} A^{-3/8}$$

$$S = \frac{4}{3}\frac{\mathfrak{U}}{\tau} = \frac{4}{3}\left(\frac{\mathfrak{U}}{E_o}\right)^{3/4} A^{3/8}$$

$$\Delta = \sqrt{8\pi}\sqrt{\mathfrak{U}\tau}e^{-5} = \sqrt{8\pi}\mathfrak{U}^{5/8} E_o^{3/8} A^{-\frac{3}{16}} e^{-\frac{4}{3}\cdot 5.61 \cdot 1.67^{3/4}} = \text{ca.}200V_0$$

Das sieht eigentlich recht vernünftig aus!

2. Kernradius: Man hat die Radien der radioaktiven Kerne bisher immer mit der Einkörpertheorie ausgerechnet und etwa $9 \cdot 10^{-13}$cm gefunden. Jetzt muss man natürlich auch hier die Vielkörpertheorie anwenden. Das heisst, dass es schon unwahrscheinlich ist, dass das α-Teilchen überhaupt gebildet wird, und dass daher der Potentialberg viel niedriger werden muss. Die reziproke Lebensdauer ist $\lambda = \gamma\alpha e^{\frac{-G}{\hbar}}$, wo γ_α die "α-Breite" des Kernzustandes wäre, wenn es keine Barriere gäbe. e^{-G} ist die Durchlässigkeit des Potentialbergs. Die Hauptfrage ist also die nach der Grösse von γ_α. Die einzige Evidenz, die man über Niveaubreiten hat, ist die von Neutronenexperimenten. Dort findet man etwa 0.005 Volt. Die Neutronenbreite enthält einen Faktor. $v \sim \frac{1}{\lambda\!\!\!^-}$, wo $\lambda\!\!\!^-$ die Wellenlänge ist. Als "normale" Neutronenbreite kann man etwa die Breite bezeichnen, die man bei einer Wellenlänge $\lambda\!\!\!^-$ von der Grössenordnung des Kernradius enthält, d.h. bei einer Neutronenenergie von vielleicht 500 000 Volt. Dann wird die Breite von der Grössenordnung $\sqrt{500000/5} \cdot 0.0005 \sim$ 1 Volt. Über α-Breiten weiss man natürlich nichts, ich habe einfach auch $\gamma_\alpha = 1$ Volt angenommen. Dann kann man G, und daraus die Radien der radioaktiven Kerne berechnen. Die meisten liegen zwischen 12.5 und $13.5 \cdot 10^{-13}$ cm. Nimmt man $R = r_o A^{1/3}$ an, so wird r_o etwas mehr als $2 \cdot 10^{-13}$cm. Konservativerweise, und der Einfachheit halber, habe ich $r_o = 2 \cdot 10^{-13}$cm genommen.

> Wenn man das in die Formel für die Niveaudichte mit <u>freien</u> Teilchen einsetzt, kriegt man viel zu kleine Abstände der Niveaus; Bruchteile eines Volt schon bei $A = 100$ und $\mathfrak{U}$=8 MV.

Es erhebt sich die technische Frage der Abfassung unseres Meisterwerkes, die womöglich vor meiner Abreise nach USA beendet sein sollte. Wollen wir die Schreibarbeit verteilen? Vorschläge erbeten!

Seit Sonntag ist Plazcek (im Dublettzustand) hier. Wir haben gestern eine Tour gemacht, wobei er hauptsächlich seinen Puls fühlte.
Viele schöne Grüsse
Dein

Hans Bethe

Cambridge, 27.8.36

Dear Peierls,
I proceed with the progress of science:

1. I have tried to evaluate the nuclear levels using the liquid drop model, assuming that only surface tension is important for the transverse oscillations. The surface energy has to be proportional to the surface area, of course. For a given surface, the shifting of particles <u>perpendicular</u> to the surface is important. If the surface lies in the xy-plane, and u is the shift, then the relative change in the surface area will be

$$\delta\sigma = \frac{\int dxdy \left(\sqrt{1 + \left(\frac{\partial u_z}{\partial x}\right)^2 + \left(\frac{\partial u_z}{\partial y}\right)^2} - 1\right)}{\int dxdy}$$

$$= \frac{1}{2}\frac{\int d\sigma \left[\left(\frac{\partial u_n}{\partial x}\right)^2 + \left(\frac{\partial u_n}{\partial y}\right)^2\right]}{\int d\sigma}$$

   If $u = \underset{\sim}{a} \sin \underset{\sim}{k} \cdot \underset{\sim}{r}$ then $\delta\sigma = \frac{1}{2} a_n^2 k_\parallel^2 \overline{\cos^2 k \cdot r}$, where $k_\parallel$ is the component of k parallel to the surface. In the case of a ball, you have to average over the surface. If $\underset{\sim}{k}$ is parallel to z then $\underset{\sim}{n}$ is parallel to x, so in polar coordinates the result is

$$a_n^2 k_\parallel^2 = a^2 k^2 \sin^4 \vartheta \cos^2 \varphi$$

averaged $= a^2k^2/10$.

Therefore the change in the surface energy $= \frac{1}{20} S_0 a^2 k^2 \overline{\cos^2 kr}$ where S_0 is the surface energy without any oscillations. If M is the total mass of the nucleus, the speed of sound is $u = \sqrt{S_0/10M}$ S_0 can be estimated by considering the empirical nuclear energy (à la Weizsäcker);[292] with a new nuclear radius (see below) you get $S_0 = 10A^{2/3}M$, if A = atomic weight. If you write down the formula for the total energy:

$$\mathfrak{U} = \frac{4\pi}{45}\left(\frac{R}{\hbar u}\right)^3 \tau^4 = \frac{\tau^4}{E_0'^3} A$$

then you get

$$E_0' = \left(\frac{45}{4\pi}\right)^{1/3} \cdot \sqrt{\frac{\hbar^2}{M_0 r_0^2 A}\frac{S_0}{10}} = \left(\frac{45}{4\pi}\right)^{1/3} \sqrt{\frac{\hbar^2}{M_0 r_0^2} \cdot \frac{10MV}{10}} A^{-1/6}$$

$$= E_0 A^{-1/6}$$

(M_0 = proton mass; $r_0 = 2\cdot 10^{-13}$cm corresponds to the comments under item 2 below). We therefore have $\frac{\hbar^2}{M_0 r_0^2} \approx 10MV$, $E_0 = 4.8MV$

$\mathfrak{U} = \frac{\tau^4}{E_0^3} A^{3/2}$; $\alpha = \frac{A^{3/2}}{E_0^3}$ (see the previous letter)[293]

$$\tau = \mathfrak{U}/\alpha^{1/4} = \mathfrak{U}^{1/4} E_0^{3/4} A^{-3/8}$$

[292]In 1935, von Weizsäcker had formulated a semi-empirical approach explaining the difference between the mass of a nucleus and the mass of its constituent protons and neutrons, the binding energy, as being the result of energies associated with the interactions of the nucleus. Bethe and R. Bracher simplified Weiszäcker's formulation in 1936 and E. Wigner extended to formula in 1937.

[293]Letter [38].

$$S = \frac{4\mathfrak{U}}{3}\frac{}{\tau} = \frac{4}{3}\left(\frac{\mathfrak{U}}{E_0}\right)^{3/4} A^{3/8}$$

$$\Delta = \sqrt{8\pi}\sqrt{\mathfrak{U}\tau}e^{-5} = \sqrt{8\pi}\mathfrak{U}^{5/8}E_0^{3/8}A^{-\frac{3}{16}}e^{-\frac{4}{3}\cdot 5.61\cdot 1.67^{3/4}} = \text{ca.}200V_0$$

This looks quite reasonable!

2. Radius of the nucleus: Until now we have always calculated the radius of radioactive nuclei with the one-body theory, finding a value of about 9×10^{-13} cm. Now we have to apply the many-body theory as well. This means that it becomes unlikely that the α-particle will be built at all, and that the potential accumulation has to be much lower. The reciprocal lifetime would be $\lambda = \gamma\alpha e^{\frac{-G}{\hbar}}$, if γ_α were equal to an "α-width" under the conditions in the nucleus, and if there were no [potential] barrier. The factor e^{-G} gives the permeability of the potential barrier. The main question is what value to use for γ_α. The only evidence concerning level widths comes from neutron experiments, where you find about 0.005 Volt. The width of the neutrons is of the order $v \sim \frac{1}{\lambda}$, where λ is their wavelength. The "normal" size of neutrons using their wavelength λ is what you get from order of magnitude calculations on the radius of the nucleus; i.e., for a neutron energy of about 500 000 Volts. This gives a size on the order of magnitude of $\sqrt{500000/5} \cdot 0.0005 \sim 1$ Volt. We know nothing about the widths of α-levels, of course. I just used $\gamma_\alpha = 1$ Volt. Then you can calculate G, and from this the radii of the radioactive nuclei. Most of them come out in between 12.5 and 13.5×10^{-13} cm. If you assume that $R = r_0 A^{1/3}$, then you find that r_0 is slightly above 2×10^{-13}cm. I have calculated it in a more conservative way because it's easier to do so, so I take $r_0 = 2 \times 10^{-13}$cm.

   If you put that into the formula for the density of levels with free particles, you will get a level spacing that is much too small; only fractions of a volt in the case of $A = 100$ and $\mathfrak{U}$=8 MV.

This raises technical questions regarding the writing of our masterwork, which should probably be finished before my departure to the USA. Shall we share the work? Proposals gratefully received!

Plazcek (in a doublet condition) has been here since Sunday. Yesterday we went on a tour, and he was always taking his pulse.
Many greetings.
Yours,

Hans Bethe

[41] Rudolf Peierls to Hans Bethe

Cambridge, 30.8.36

Lieber Hans,
Ich bin ja ganz einig, aber, aber, muss ich sagen. Oder in Variation des neuesten Kopenhagener Schlagers: Erst kommt das Denken, dann das Integral!

Oder mit noch anderen Worten: Wir sind in der Tendenz natürlich sehr einer Meinung, nicht aber in den Ansätzen, was zunächst die Kernradien betrifft, so habe ich so etwas ähnliches auch gerne immer haben wollen, war aber immer zu faul, es mir zu überlegen. Du hast natürlich völlig recht damit, dass die Kernradien in Wirklichkeit grösser sind als aus der Gamowschen Rechnung folgt, und natürlich wird Dir auch gern geglaubt, dass sie dann ein bisschen grösser als 10^{-12} werden. Nicht geglaubt wird dagegen die Behauptung, dass die α-Breiten von derselben Grösse wir die Neutron-Breiten werden, denn ich kann auf Wunsch ebenso gute Argumente produzieren, dass sie grösser wie dass sie kleiner sein sollten (Man könnte natürlich nach dem Rezept des "equivalent two-body-problems" hieraus schliessen, dass die Gleichsetzung eine ausgezeichnete Näherung sein muss. Es ist natürlich richtig, dass der Radius von dieser Breite ziemlich schwach abhängt, für einen Fehler in γ ist der zugehörige Fehler in r:

$$\frac{\delta r}{r} = \frac{1}{2}\frac{\lambda\!\!\!^{-}_0}{r}\frac{\delta\gamma}{\gamma}$$

wobei $\lambda\!\!\!^{-}$ die Wellenlänge eines α-Teilchens mit der Energie $4Ze^2/r$ ist. Numerisch also ungefähr $\frac{\delta r}{r} \sim 0.06\frac{\delta\gamma}{\gamma}$.

Damit Du also berechtigt bist, den Kernradius mit einer Genauigkeit von 5% hinzuschreiben, wie Du das tust, müsste γ bis auf einen Faktor 2 bekannt sein, was es offenbar nicht ist. Um r bis auf 20% zu bekommen, (und weniger hat offenbar keinen Sinn) kann ich Dir nur einen Fehler von einem Faktor 15 in der Breite erlauben. Du musst mich aber noch davon überzeugen, dass die Breite nicht ebensogut 12 statt 1 Volt sein kann.

Es hilft auch nichts, wenn man einen Vergleich aller α-Strahle[n] zu Hilfe nimmt. Denn wenn man in allen γ um denselben Faktor falsch macht, das bei allen denselben Fehler in r und man bekommt noch immer vernünftige Verhältnisse der Radien. (Dagegen stimmte eben das Gamowsche Bild so schön). Ich bin übrigens auch noch nicht so restlos davon überzeugt, dass die Neutronenbreite prop. v geht, d.h. dass die anderen Faktoren, die ausser dem v eingehen, für langsame Neutronen konstant werden. Es klingt natürlich plausibel.

Ich möchte auch gegen die Behauptung protestieren, dass man aus diesen Radien nur das Modell mit freien Teilchen widerlegen kann. Erstens ist nämlich ein Faktor 100 in der Niveaudichte noch nicht so sehr schlimm, denn er bedeutet nur, dass die Entropie um 5, d.h. um 25% falsch ist.

Nun wird zwar Dein Wert für den Radius schon nicht um 25% falsch sein, wohl aber die Annahme eines rechteckigen Potentialtropfens. Wenn Du stattdessen annimmst, dass die Breite des Potentialtropfens mit wachsender Höhe zunimmt, so misst die Zerfallskonstante vermutlich die Breite am oberen Rand, (d.h. am Rand der Fermi-Verteilung) während in die Entropie ein Mittelwert eingeht.

Selbst wenn sich aber aus diesem Modell eine zu grosse Niveaudichte ergibt, so ist das nicht notwendig ein Widerspruch, denn es bedeutet nur dass ein Teil der Zustände nicht stark genug mit dem Kontinuum kombiniert, um als virtuelle Niveaus beobachtbar zu werden. Gerade im Fall fast freier Teilchen wäre das ganz vernünftig. Nehmen wir an, dass die Kopplung der Teilchen durch einen Faktor s gemessen wird, so wird die Wahrscheinlichkeit der Anregung eines Niveaus bei dem n Teilchen nicht mehr im Grundzustand sind, den Faktor s^n enthalten. Durch geeignete Wahl von s kann man dann einen beliebigen Bruchteil der Niveaus ausschalten, denn für $s \to 0$ kommt man offenbar auf Deine alte Theorie zurück. Man braucht s nicht sehr von 1 verschieden anzunehmen, um den Schaden zu beheben.

Ich glaube natürlich nicht an das freie-Teilchen-Modell, aber ich glaube nicht, dass man es so widerlegen kann. Nun zum Tropfenmodell. Ein Teil meiner Bedenken stehen schon in meinem letzten Brief, der sich offenbar mit Deinem gekreuzt hat. Mein Haupteinwand gegen Deine Rechnung ist folgende: Nimmt man das Tropfenmodell mit Oberflächenspannung Ernst, so hat es offenbar eine von Null verschiedene Nullpunktsentropie, denn es gibt eine Menge von Deformationen, die weder das Volumen, noch die Oberfläche ändern, und daher zur Frequenz Null gehören. Beispiel: man denke sich die Kugel in eine Reihe konzentrischer Kugelschalen beliebiger Dicke eingeteilt und verdrehe jeder dieser Schalen in beliebiger Weise gegen ihre Nachbarn. Nun wirst Du natürlich sagen, dass es in der Quantentheorie anders ist und zwar aus den verschiedenen Gründen. Erstens muss in der Quantentheorie diese Rotation der Kugelschalen mit einer minimalen Rotationsgeschwindigkeit vor sich gehen, und dadurch entsteht eine Zentrifugalkraft, die dann die Oberfläche doch deformiert. Dies ist zweifellos richtig, aber die hierzugehörende Frequenz ist offenbar viel kleiner als Du sie abschätzt, denn wenn man die Oberflächenspannung gegen unendlich gehen lässt, so geht die auf diese Art von Rotation wirkende Kraft keineswegs gegen unendlich sondern sie verwandelt sich nur in eine Rotation ohne Deformation. Andererseits kannst Du sagen, dass die Bose- oder Fermistatistik schon einen Teil der Entartungen aufhebt, sodass im Gegensatz zum klassischen Fall durch Angabe der äusseren Form der Zustand des Systems bereits mehr oder weniger eindeutig angegeben ist. So etwas ist möglicherweise richtig. Ich habe das aber noch nicht zu Ende denken können, aber wenn es richtig ist, so ist offenbar die Hydrodynamik nicht mehr anwendbar, und der Zusammenhang zwischen Schallgeschwindigkeit und Entropie geht verloren. Mir scheint, es gibt zwei mögliche Standpunkte:

A: Der Kern ist eine Flüssigkeit bis auf die Oberflächenenergie. Mit anderen Worten, wenn man sich einen makroskopischen Tropfen aus Kernmaterie bauen würde, so würde er wirklich eine endliche Nullpunktsentropie haben. (Genauer gesagt, die Temperatur, bei der die Entropie zu verschwinden beginnt, nimmt mit wachsenden Dimensionen ab. In diesem Fall muss man sich überlegen, welches die Kräfte sind, die dann den Grundzustand einfach machen und welches diejenigen Bewegungen sind, die zur kleinsten Anregungsenergie gehören. Dass es auch Bewegungen mit grösserer Anregungsenergie gibt, nützt gar nichts, wir wissen ja ohnehin schon dass es Longitudinalwellen mit sehr hoher Frequenz gibt. Daher erscheint mir der Ansatz einer ebenen Welle mit festem Wellenvektor ebenso falsch als wenn man eine Welle mit beliebiger Polarisation ansetzen würde (also keine Transversalität voraussetzen) und dann über die Polarisationsrichtung mitteln [würde]. So bekommt man natürlich selbst in einer idealen Flüssigkeit 1/3 der Longitudinalgeschwindigkeit als mittlere Schallgeschwindigkeit, was offenbar Unsinn ist.

Der zweite mögliche Standpunkt ist B: dass die Entropie bereits im makroskopischen Fall gegen Null geht. Dann bekommt man den in meinem letzten Brief vorgeschlagenen neuen Aggregatzustand. Mir ist diese Idee deswegen sympathischer, weil sie offenbar bei Helium realisiert ist und wahrscheinlich doch die Bindungskräfte dieselbe qualitative Charakteristik haben. In diesem Falle haben dann die Transversalschwingungen eine endliche definierte Frequenz, und die Berechnung der Entropie bleibt, ungefähr jedenfalls, zulässig. Aber da man weiss, dass die transversalen Schwingungen jedenfalls für unendlich lange Wellen nicht existieren, so muss man dann feststellen, was dort passiert, und das können eben wohl die drei Möglichkeiten sein, dass die Schallgeschwindigkeit mit zunehmender Wellenlänge entweder kontinuierlich oder diskontinuierlich gegen Null geht, oder dass die elastische Grenze für wachsende Wellenlänge gegen Null geht. Je mehr ich es mir überlege, scheint eigentlich der letzte Fall, jedenfalls für flüssiges Helium der wahrscheinlichste. Aber natürlich ist er der unangenehmste zum Rechnen. Immerhin würde er wohl heissen, dass im Kern, wo ja lange Wellen ohnehin nicht in Frage kommen, vermutlich die Näherung vom festen Körper her ganz anständig sein sollte.

Es tut mir leid, dass ich das Meisterwerk durch so viel Pedanterie aufhalte, aber ich glaube, man sollte sich das alles anständig überlegen. Vielleicht publizierst Du mal zunächst die Sache mit den Kernradien, die ja hiermit nicht direct was zu tun hat? Grüsse Placzek. Hat das Dublett (JJ) oder Russell-Saunders-Kopplung? Ich möchte von ihm

gerne wissen, wie weit die im Werden begriffene Arbeit von Bohr auch solche Modellfragen behandelt. Ausserdem ist er herzlich eingeladen, uns hier zu besuchen; ich empfehle die Route über Cambridge als sehr praktischen Weg von Seelisberg nach Kopenhagen! Wann schiffst Du Dich wieder ein?
Grüsse

[Rudi]

Cambridge, 30.8.1936

Dear Hans,
I totally agree, but I have to say "but". Or as a variation of the latest hit from Copenhagen: First comes the thinking, and then comes the integral![294] Or in other words: We are basically in agreement with each other, but we take another approach. Concerning the nuclear radii, I have always wanted something like that but was too lazy to work it out. You are absolutely right that the nuclear radii are in fact much bigger than what you get with Gamow's calculation, and you are right that they are thus a little bigger than 10^{-12}. But I don't believe that the α particles will be of the same size as the neutrons. If you wish, I could produce arguments that they are bigger which are just as good as those saying that they are smaller. (Of course one could deduce that, following the principle of the "equivalent two-body problems" equalization is an excellent approximation.) It is absolutely right that given the radii this size is not too important. For a given error in γ the corresponding error in r is:

$$\frac{\delta r}{r} = \frac{1}{2}\frac{\lambda\!\!\!^{-}_0}{r}\frac{\delta\gamma}{\gamma}$$

where λ is the wavelength of an α-particle with energy $4Ze^2/r$. Numerically, this is about $\frac{\delta r}{r} \sim 0.06\frac{\delta\gamma}{\gamma}$.

To give the radius of the nucleus to an precision of 5% as you do, γ would have to be known to within a factor of 2, which is obviously not

[294] Adaptation of the German saying: 'Erst kommt das Fressen, dann kommt die Moral.'

the case. To get an error of 20% on r, (less seems to be senseless) I can only allow you a factor of 15 in error on the width. But you have to persuade me that the width could just as well be 12 Volts as 1 Volt.

A comparison between different α-rays doesn't help either. If you make an error of the same factor in all the γ, and make the same error in r, you will still get reasonable proportions for the radii. (In contrast, Gamow's results were very nice). I am not totally convinced, by the way, that neutron widths are prop. to v; i.e., that all other factors except for v are constant for slow neutrons. This seems plausible, at least on the surface.

I also want to oppose the opinion that given the radii, you can only disprove the principle of free particles. First, a factor of 100 in the density of levels is not too bad since it just means that the entropy is wrong by a factor of 5, or 25%.

Your value of the radius won't be wrong by a factor of 25%, but your proposition of a rectangular potential well will be wrong. If you assume instead that the width of the potential well increases along with its depth, then the disintegration constant probably measures the width of the upper boundary (i.e., the edge of the Fermi distribution), and you can use an average value to calculate the entropy.

Even if the level density is too high going by this principle, this is not necessarily a contradiction, since it just means that some of the conditions some of the conditions do not correspond to observable virtual levels. In the case of weakly bound particles, this would be especially reasonable. If we presume that the number of two-particle combinations is given by the factor s, then the probability of any level being stimulated will depend on the factor s^n, where n particles are not in the ground state. By the correct choice of s you can cut out an arbitrary fraction of the levels; letting s→0 leads back to your old theory. s doesn't have to be very different from 1 to solve the problem.

I don't believe in the free-particle model, of course, but I also don't believe that you can disprove it this way.

Now, about the liquid drop model. I already mentioned some of my concerns in my last letter, which obviously crossed yours in the mail. My main objection to your calculation is this: if you take the model of a liquid drop with surface tension seriously, you will obviously get a zero-point entropy which is different from zero. This is because there are a lot of deformations which change neither the volume nor the surface area, and therefore belong to the frequency zero. Example: Think of a ball made from a series of concentric spherical shells of arbitrary thickness, and twist each of these shells in an arbitrary fashion relative to the next. Now you will say that in quantum theory it is different, for different reasons. First, in quantum theory the rotation of the spherical shells has to occur with a minimal angular velocity, which results in a centrifugal force which will deform the surface. This is definitely right, but this frequency is obviously much smaller than you estimate. As the surface tension goes to infinity, the forces that influence this kind of rotation don't go to infinity at all; rather, the motion changes to a rotation without deformation. On the other hand, you could say that Bose or Fermi statistics annihilate some of the deformations so that in contrast to the classical approach, the state of the system is more or less precisely indicated by the outer form. This might be correct. I couldn't finish my analysis of this problem, but if this is correct then hydrodynamics cannot be applied anymore, and the relationship between sound speed and entropy is lost. I think there are two different positions one can take:

A: The nucleus behaves like a liquid except for the surface energy. In other words, if a macroscopic drop were built of nuclear matter then it would actually have a finite zero-point entropy. (To be more precise, the temperature at which the entropy starts to disappear would decrease as its dimension increased). In this case you <u>have to</u> think about which forces make the ground state simple, and which motions belong to the <u>smallest</u> excitation energy. It is no help that there are <u>also</u> motions with higher excitation energies, as we already know that there are longitudinal waves with very high frequencies. The approach of a plane wave with fixed wave vector therefore seems to be equally wrong, as if you were to take a wave with arbitrary polarisation (that means no given transversality) and average over the direction of the polarisation. Then even in a perfect fluid you will get an average sound speed equal to 1/3 of the longitudinal velocity, which is obviously nonsense.

The second position B is this: The entropy approaches zero in the macroscopic case. Then you will get the state of aggregation that I proposed in my last letter. I prefer this idea, because it is obviously realised in helium and the binding forces are likely to have the same qualitative characteristics. In this case every transverse oscillation has a finite, defined frequency, and this way of calculating the entropy is basically appropriate. But we know that there are no transverse oscillations for infinitely long waves. There are then three possibilities for what might happen: as the wavelength increases the speed of sound approaches zero, either continuously or discontinuously; or as the wavelength increases the elasticity of the boundary approaches zero. The more I think about it, the more the last case seems to be the most likely, at least in the case of liquid helium. But it is certainly the most difficult to calculate. At least it would mean that within the nucleus, where there are no long waves anyway, the solid body approximation will be highly accurate.

I am sorry for all this pedantry and the delay of our masterwork, but I believe that we really have to think about all this properly. You could probably publish that thing with the radii of the nuclei first, which is not directly related to all this? Greetings to Placzek. Does his doublet have a (JJ) or Russell-Saunders interface? I would like to ask him whether Bohr's impending paper also deals with these models. I also wanted to invite him to come here; I recommend the route through Cambridge as a very practical way to get from Seelisberg to Copenhagen!
When do you re-embark?
Greetings,

Peierls

[42] Hans Bethe to Rudolf Peierls

Seelisberg, 4.9.[19]36

Lieber Peierls,
Bohr würde sagen: Wir sind uns einiger, als Du denkst. So Ernst habe ich die Integrale vor dem Denken nicht genommen, und ich bin ganz Deiner Meinung, dass man zur Zeit das flüssige Helium noch nicht genügend versteht, um die Kerne anständig zu behandeln. Die Absicht,

bei dem vorgeschlagenen Meisterwerk war mehr destruktiv als konstruktiv: Ich fühle mich nämlich etwas schuldig wegen der Publikation des Fermi-Thomas-Modells. Die Amerikaner werden das glauben, genau wie sie vorher die Hartree-Näherung für Kerne zu wörtlich geglaubt haben. Und da wollte ich ihnen mitteilen, dass ich selbst nicht daran glaube. Aber wenn Du meinst, dass das zu destruktiv für ein Meisterwerk ist — und ausserdem sehe ich ein, dass ich für eine destruktive Arbeit etwas viel integriere. Das ist Amerika.

Das Einzige, was ich ernst meine, sind die Kernradien. Die sind doch wirklich besser als Du schriebst. Nach Deiner Formel ist

$$\frac{\delta R}{R} = \frac{\lambda\!\!\!^{-}_0}{2R}\frac{\delta\gamma}{\gamma}$$

$\lambda\!\!\!^{-}$ ist die Wellenlange eines α-Teilchen der Energie $V_0 = \frac{2Ze^2}{R}$
Mit $R = 12.5 \cdot 10^{-13} = 4.5\frac{e^2}{mc^2}$ ist $V_0 = \frac{2 \cdot 90}{4.5}mc^2 = 20MV$, also $\lambda\!\!\!^{-}_0 = \frac{\hbar}{\sqrt{2MV_0}} = \frac{4.5 \cdot 10^{-13}}{\sqrt{4 \cdot 20}} = 0.5 \cdot 10^{-13}$cm. Demnach $\frac{\delta R}{R} = \frac{0.5}{2 \cdot 2.5}\frac{\delta\gamma}{\gamma} = 0.020\frac{\delta\gamma}{\gamma}$ (nicht $0.06\frac{\delta\gamma}{\gamma}$).

Eine Änderung von R um 5% entspricht also einer Änderung von $\log\gamma$ um einen Faktor von 12. Deshalb kann man wohl sagen, dass R auf 10% genau ist, denn Du wirst mir wohl zugeben, dass γ wahrscheinlich zwischen $\frac{1}{100}$ und 100 Volt liegt.

Bezüglich der Neutronenbreite bin ich doch von dem Faktor v sehr überzeugt: Das Matrixelement, das die Neutronenbreite bestimmt, enthält die Eigenfunktion des compound-Kerns und des Neutrons. Für den compound Kern ist es sicher ganz gleich, ob er mit $8MV$, mit 10 oder auch mit $15MV$ angeregt ist, weil auf jeden Fall die Anregungsenergie minimal ist gegen die totale Bindungsenergie. Die Neutronenfunktion enthält einen Normierungsfaktor, der eben das v-Gesetz gibt, und eine Ortsabhängigkeit. Letztere kann aber sicher nichts ausmachen, solange die Neutronenwellenlänge grösser als der Kernradius ist. Es sieht mir jetzt eigentlich plausible aus, dass die kritische Wellenlänge nicht der Kernradius ist, sondern etwa die Reichweite der Kernkräfte — oder, was dasselbe ist, der Abstand benachbarter Teilchen im Kern. Das würde die Abschätzung des γ um etwa einen Faktor von 5 erhöhen und R um ein paar % verkleinern.

Mit der Bemerkung, dass alle α-Strahlen dasselbe R geben, wollte ich nicht die Richtigkeit meines $\gamma's$ beweisen. Immerhin ist es nicht ganz trivial: Dadurch dass R soviel grösser wird, wird der Potentialberg niedriger. Die abgekürzte Gamowsche Formel für den Exponenten im Durchlässigkeitskoeffizienten, $G = \frac{\pi\, Ze^2}{\hbar v} - \frac{2e}{\hbar}\sqrt{2zZMR}$ gilt ja nur wenn $E \ll \frac{zZe^2}{R}$. Wenn die Energie vegleichbar ist mit der Höhe des Potentialberges, kann G nicht mehr geschrieben werden als ein Term, der nur von der Energie abhängt minus einem zweiten Term, der nur vom Radius abhängt, sondern es kommt noch etwas dazu, was von beiden Grössen abhängt. Es ist deshalb nicht trivial, dass das Geiger-Nuttall-Gesetz noch genügend genau erfüllt ist.

Die neuen Kernradien sind mir hauptsächlich noch deshalb sympathisch, weil sie die absolute Ausbeute der Deuteronen-Reaktionen in schweren Elementen verständlich machen. Mit den alten Radien kam die Ausbeute immer zu klein heraus — mit der Bohrschen Theorie sowohl wie mit der uralten Gamowschen. Über diese Sachen will ich einen letter to the editor (also an mich selbst) schreiben, es ist ja ganz unabhängig von den Fragen der Niveaudichte.

Die Geschichte mit der Oberflächenspannung hat Teller, der drei Tage hier war, viel gescheiter verstanden als ich sie meinte: Er sagte, dass die Oberflächenspannung ein ganz gutes Mass für die Scherungskräfte sein könnte, da man ja bei der Scherung zwei benachbarte Schichten von der günstigeren Position in eine weniger günstige bringt, also einen Teil einer freien Oberfläche erzeugt. Allerdings macht das meine Integrale nicht besser. (Und wahrscheinlich ist es für die Transversalschwingung etwa dieselbe Schallgeschwindigkeit wie für die longitudinalen.) Dein Einwand gegen die Oberflächenenergie, dass die Entropie bei einer mit wachsender Grösse abnehmenden Temperatur verschwindet, ist natürlich richtig. Ich bin auch ganz davon überzeugt, dass es noch andere Kräfte gibt, die sich der Scherung widersetzen. Es wäre aber an sich denkbar, dass diese Kräfte nur bei sehr grossen Tropfen wesentlich werden, und dass bei kleinen Tropfen, wie wir sie in Kernen haben, die Oberflächenspannung den Hauptbeitrag liefert.

Mit Teller sprach ich auch über die Frage der Rotation des ganzen Kerns. Er hatte sich das schon früher überlegt, und es schien uns herauszukommen, dass auf jeden Fall die Rotationsniveaus von derselben Grössenordnung sind wie bei einzelnen Teilchen, und zwar wegen der Symmetrie der Rotationseigenfunktion. Bei n festgebundenen Teilchen mit Bosestatistik in einem Ring gibt es nur jedes n-te Rotationsniveau, und das hebt das grösste Trägheitsmoment auf. Teller wird nach Cambridge kommen — er ist in London — und kann es Dir viel besser erzählen als ich.

Placzek, der gestern vom Dublett- in den Singlett-Zustand übergegangen ist, sagte, dass man sich um Bohr nicht kümmern dürfte — sonst müsste man einfach die Kernphysik im augenblicklichen Zustand abbrechen. Die Arbeit Bohr & Kalckar wird nach Placzeks Meinung nie erscheinen. Bis zu quantitativen Dingen ist man überhaupt nicht vorgedrungen; anscheinend ist die Hauptsorge eine möglichst komplizierte Ableitung der Dispersionsformel, und es scheint gelungen zu sein, die Breit-Wignersche Ableitung an Komplikation zu übertreffen. Von dieser Seite ist also kaum etwas zu erwarten.

Trotzdem hast Du mich ziemlich überzeugt, dass man das flüssige Helium erst anständiger überlegen soll, bevor man etwas sagt. Nebenbei, die meisten Probleme der Transversalwellen gelten wohl auch für gewöhnliche Flüssigkeiten in derselben Weise. Placzek* hat einmal abgeschätzt, dass Transversalwellen mit einer Frequenz grösser als $\rho u^2/\eta$ sich im wesentlichen wie longitudinale verhalten sollen, wo $\eta =$ Viskosität, $\rho =$ Dichte und $u =$ Schallgeschwindigkeit (hochfrequenter) Wellen. Viel nutzt das nicht, weil die Temperaturabhängigkeit von η wieder problematisch ist.

Sollen wir also die Sache vertagen?

Viele schöne Grüsse and Dich und Genia

Dein

Bethe

*Es ist mir bisher nicht gelungen, Placzek davon zu überzeugen, dass der nächste Weg von Seelisberg nach Kopenhagen über Cambridge führt.

Seelisberg, 4.9.[19]36

Dear Peierls,
Bohr would say that we are more in agreement with one another than you think. I hadn't taken these integrals seriously before I started thinking, and I totally agree that we don't know enough about liquid helium to deal with the nuclei adequately. The proposed masterwork was intended to be more destructive than constructive; I feel a bit guilty about the publication of the Fermi-Thomas model. The Americans will believe in it, just as they have taken the Hartree nuclear approximation too literally. I therefore wanted to tell them that I don't believe in it. But if you think that this is too destructive for a masterwork — I will also accept that I am integrating too much for a destructive paper. That's America for you.

The only result I think is important is the nuclear radii, and those are much better than you say. According to your formula

$$\frac{\delta R}{R} = \frac{\lambda\!\!\!^{-}_0}{2R}\frac{\delta\gamma}{\gamma}$$

$\lambda\!\!\!^{-}$ is the wavelength of an α-particle of energy $V_0 = \frac{2Ze^2}{R}$. With $R = 12.5 \times 10^{-13} = 4.5\frac{e^2}{mc^2}$, we get $V_0 = \frac{2 \cdot 90}{4.5}mc^2 = 20MV$, and thus $\lambda\!\!\!^{-}_0 = \frac{\hbar}{\sqrt{2MV_0}} = \frac{4.5 \times 10^{-13}}{\sqrt{4 \cdot 20}} = 0.5 \times 10^{-13}$cm. Therefore $\frac{\delta R}{R} = \frac{0.5}{2 \cdot 2.5}\frac{\delta\gamma}{\gamma} = 0.020\frac{\delta\gamma}{\gamma}$ (not $0.06\frac{\delta\gamma}{\gamma}$).

A change of 5% in R thus corresponds to a factor of 12 change in $\log\gamma$. Therefore we could presumably agree that R is in fact known reasonably exactly to 10%, as you should agree that γ is likely to lie between $\frac{1}{100}$ and 100 Volts.

Concerning the width of the neutrons, I am very much convinced of the factor v; the important matrix element is composed of the compound nucleus and neutron eigenfunctions. As for the compound nucleus, it makes no difference whether it is stimulated with 8, 10, or 15 MV, because the excitation energy is minimal compared to the total binding energy in any case. The neutron eigenfunction contains a normalisation factor which gives the v-law and its dependence on position. The latter is not important, as long the neutron wavelength is bigger than the radius of the nucleus. The range of the nuclear force is important, for example — or equivalently, the distance between adjacent particles in the nucleus. This would raise our estimate of γ by a factor of 5, and diminish R by some %.

When I noted that all α-rays give similar results for R, I wasn't trying to prove the accuracy of my γ's. This is not a trivial point after all: since R is getting so much bigger, the magnitude of the potential is much smaller. A shorter version of the Gamow formula for the exponent of the permeability coefficient, $G = \frac{\pi\ Ze^2}{\hbar v} - \frac{2e}{\hbar}\sqrt{2zZMR}$, is only valid if $E \ll \frac{zZe^2}{R}$. If the energy is comparable to the height of the potential, then G cannot be expressed as a first term dependent only on the energy minus a second term dependent only on the radius. You have to add something dependent on both values as well. It is therefore not a trivial result that the Geiger-Nuttall law is satisfied sufficiently precisely.

I also like the new nuclear radii, as they explain the role of Deuteron reactions in producing heavy elements. Under Bohr's theory and Gamov's very old theory, the yield was always too small using the old radii. I want to report on these things in a letter to the editor (that means to myself); they are fully independent from the question of level densities.

Teller, who was here for three days, has understood the problem of surface tension much better than I. He said that the surface tension could be a good instrument for measuring the shear forces, as the presence of a shear brings two adjacent layers from a good position to a worse one. This means producing a partially free surface, but this doesn't make my integrals any better. (And it is likely that both transverse and longitudinal oscillations have the same velocity.) Your objection to surface energy on the grounds that the entropy goes to zero as the temperature diminishes must be right. I am also convinced that there must be other forces which oppose the shear. It is conceivable that there are other important forces in the case of big drops, and that in small drops such as nuclei the surface tension would be the most important factor.

I also discussed the question of a rotation of the whole nucleus with Teller. He had thought about it before, and we both think that the rotation levels of an isolated particle in the nucleus must be of the same order of magnitude as the rotation levels in any case, because of the symmetry of the rotation eigenfunction. With Bose statistics, in the case of n particles fixed in a ring only every nth rotation level is allowed, and this offsets the larger moment of inertia. Teller will come to Cambridge — he is in London now — and he can explain it to you much better than I.

Placzek, who has changed the doublet condition for the singlet condition, said that you don't have to worry about Bohr — otherwise we would have to leave nuclear physics in its present state. According to Placzek, the work of Bohr & Kalckar will never appear in print.[295] They haven't addressed any quantitative questions at all; it seems as if their main purpose is to find the most complicated derivation of the dispersion formula, and they seem to have surpassed even the Breit-Wigner derivation in complexity. We can expect nothing important from them.

Anyway, you have convinced me to think more carefully about liquid helium before saying anything. By the way, most of the problems concerning transverse waves can also be applied to normal fluids in the same manner. Placzek * once estimated that transverse waves with a frequency greater than $\rho u^2/\eta$ should basically behave like longitudinal ones, where η = viscosity, ρ = density and u = the sound speed of the (high-frequency) waves. This doesn't help a lot, as the dependence of η on temperature makes things difficult again.
Shall we once more postpone the whole thing?
Best greetings to you and Genia.
Yours,

Bethe

*I haven't yet convinced Placzek that the best route from Seelisberg to Copenhagen goes via Cambridge.

[295] This presumably refers to a paper eventually published in 1937. N. Bohr and F. Kalckar, 'On the Transmutation of Atomic Nuclei by Impact of Material Particles. I. General Theoretical Remarks', *Mat.-Fys. Medd. Dan. Vidensk. Selsk.* **14**, 10 (1937).

[43] Rudolf Peierls to Hans Bethe

Cambridge, 13.2.[19]37
(carbon copy)

Lieber Bethe,
Diesmal bin ich derjenige welcher, und noch dazu ohne jeden Vorwand mit dem Titelblatt, denn auf dem Umschlag der Camb. Phil. Soc. steht nur der Name von meinem Freund A.H. Wilson, und Du bist berechtigt, das nicht als Gruss anzuerkennen. Als mildernden Umstand muss ich immerhin plädieren, dass ich immer hoffte, jeweils etwas besseres über die angeregten Kerne und flüssiges He herauszubekommen, aber dann wird es immer Essig (flüssiger).

Was ich bisher weiss, geht nicht weit über das hinaus, worüber wir uns schon im Sommer einigten. So lohnt sich vielleicht jetzt, die Situation zusammenzufassen:

Wenn man ausser der Energie noch r Integrale der Bewegungsgleichungen $F_1 \ldots F_r$ konstant hält, (Protonen- und Neutronenzahl, Spin), so wird die Anzahl der Niveaus pro $dEdF_1 \ldots dF_r$:

$$\sqrt{(2\pi)^{r+1} D_{r+1}} e^s \tag{1}$$

wo s die Entropie ist und $D_r + 1$ eine Determinante von

$$d_{i,k} = \frac{\delta^2 S}{\delta F_i \delta F_k} \qquad i, k = 0, 1, \ldots r, \qquad F_0 \equiv E$$

Dies dürfte Dir keine Neuigkeit sein. Man kann so die Formeln Deiner Arbeit ein bisschen hübscher und symmetrischer ableiten, aber das ist natürlich egal.

Im Falle von $r = 0$ bekommt man natürlich anstatt der Determinante einfach d^2S/dE^2, und der Korrektionsfaktor um den sich der Ausdruck (1) hiervon unterscheidet, geht natürlich wie eine Potenz der Teilchenzahl A. Der Fehler, den man begeht, wenn man diese Korrektur weglässt, ist also ein Term der Entropie der Grössenordnung $\log A$.

Dies bedeutet nun aber, dass eine makroskopische Analogie diesen Faktor nie ergeben kann. (Der Unterschied ist natürlich der zwischen mikrokanonischer und kanonischer Verteilung bezüglich $F_1 \ldots F_r$).

Ferner zeigt dieses Argument aber auch dass es ungerechtfertigt ist, diese Feinheiten mitzunehmen, ohne Oberflächeneffekte in Betracht zu ziehen, denn wenn Oberflächenterme vernachlässigbar sind, so sind es die logarithmischen Terme erst recht.

Solche Oberflächeneffekte sind natürlich in Deiner Formel für das Fermigas vernachlässigt, insofern nämlich als Du die Anzahl der Niveaus für das Einzelteilchen proportional zum Volumen annimmst. Das könnte man im Prinzip zwar leicht richtigstellen, sobald man das Potentialfeld kennen würde. Da man es aber nicht mit hinreichender Genauigkeit kennt (bezw. da man es vermutlich gar nicht mit derartiger Genauigkeit definieren kann) so hilft das nicht viel.

Hieraus würde ich schliessen, dass es ohne ein vollständiges mikroskopisches Modell unmöglich ist, die Oberflächeneffekte zu berücksichtigen, und einem daher auch die logarithmischen Terme, d.h. die $F's$ in (1) schnuppe sein können. Dabei begeht man natürlich im Absolutwert der Dichte einen Fehler von mehreren Zehnerpotenzen, und muss sich mit einer ganz groben Annäherung begnügen.

Wenn man das tut, so hat man

$$N = \sqrt{2\pi \frac{\delta^2 S}{\delta E^2}} e^S$$

und braucht nur noch mit der Funktion $S(E)$ oder $E(T)$ zu spielen. Da die Temperaturen tief sind, so ist es eine gute Annäherung, diese Funktion durch eine Potenz darzustellen

$$E = A.B.T^n$$

Dies entspricht der Annahme völliger Entartung in Deinem Fall und bedeutet, obwohl der Fehler im Exponenten so klein ist, vermutlich auch wieder eine beträchtliche Unsicherheit im Resultat. Ferner ist der Faktor A schon so geschrieben, dass bei Vernachlässigung von Oberflächeneffekten B unabhängig von A wird. Wenn man das annimmt, so bleibt nur noch n und der mehr oder weniger universelle Faktor B verfügbar.

Für n ist es wohl vernünftig, anzunehmen, dass es mit dem entsprechenden Exponenten für flüssiges He identisch ist, und man könnte denken, den letzteren experimentell zu bestimmen. Im Mond hier sind derartige Experimente geplant, sie sind aber nicht leicht, da man wesentlich unter 1° heruntergehen muss, um das asymptotische Gesetz zu finden. Die Leidener Messungen der spezifischen Wärme gehen etwa bis auf 1° und wenn man dort C/T^3 gegen T aufträgt, so bekommt man eine gerade Linie gerade durch den Nullpunkt. Nimmt man diese ernst, so bekommt man $n = 5$, was wohl Blödsinn ist, denn dann sollte das flüssige He schwerer zu Schwingungen anregbar sein als ein fester Körper! Es ist natürlich leicht möglich, dass die Kurve, statt durch den Nullpunkt zu gehen, schliesslich mal umbiegt und horizontal wird, wie das für $n = 4$ erforderlich wäre, aber um weniger als 4 zu bekommen, müsste sie schliesslich gegen unendlich gehen, und das ist ein bisschen viel verlangt. Infolgedessen scheint mir vorläufig $n = 4$ der plausibelste.

Die beste Abschätzung für B scheint mir zu sein, dass für die mittlere Anregungsenergie der radioaktiven Elemente und für $A = 200$, S von der Grössenordnung 1 werden muss. Nimmt man für diese Anregungsenergie 0.2MV, so erhält man für $A = 100$ und $E = 8$MV einen Niveauabstand von $100V$. Das stimmt natürlich viel zu schön, ich glaube man muss mindestens zwei Zehnerpotenzen in jeder Richtung offenlassen aber jedenfalls ist die Sache im Rahmen dieser groben Ueberlegungen in Ordnung. Besser kann man es, glaube ich, nicht bekommen, ohne ein anständiges Modell für die Bewegung des Kerns zu haben und das wäre natürlich das eigentlich interessante, aber dafür habe ich nichts herausgebracht. Das einzige, was sicher ist, ist dieses: Die schlampige Ueberlegung, die ich damals mit Longitudinal- und Transversalwellen machte, lässt sich nicht präzisieren, denn wenn die Substanz kein Kristall ist, so lässt sich ihre Bewegung halt überhaupt nicht mehr streng in Wellen zerlegen, und für die langen Wellen für die das noch geht benimmt sie sich halt wie eine ideale Flüssigkeit.

Der einzige Schritt vorwärts, der aber auch nicht viel zu helfen scheint, liegt in der folgenden Feststellung: Man kann kaum hoffen, die angeregten Zustände besser behandeln zu können als den Grundzustand. Für diesen kommt man aber natürlich nur mit einem Variationsprinzip weiter. Es ist daher ein gewisser Trost, dass es auch für die Zustandssumme ein Variationsprinzip gibt. Es gilt nämlich folgendes: es sei H eine Hamiltonfunktion, und ich bilde ihre Diagonalwerte H_{ii}

für irgendein beliebig gewähltes Orthogonalsystem. Wenn ich dann so tue als ob dies die richtigen Eigenwerte wären, und daher die Funktion bilde

$$F(T) = \sum_i e^{-H_{ii}/kT}$$

so ist diese immer kleiner (d.h. die zugehörige freie Energie immer grösser) als die wahre Zustandssumme. Wenn man ein vollständiges Orthogonalsystem mit freien Parametern nimmt (also etwa die Eigenfunktion eines lösbaren Problems mit Parametern) (im Potential), so kann man vermutlich durch Variation der Parameter der wirklichen freien Energie recht nahe kommen. Ich habe aber bisher für diesen Fall noch kein geeignetes Funktionensystem finden können. Aber vielleicht fällt Dir hierzu etwas Kluges ein.

Das ist natürlich alles herzlich dünn, und ich weiss nicht, ob es sich lohnt, darüber überhaupt etwas zu veröffentlichen, aber vielleicht lohnt es sich immerhin einen Brief an die Phys.Rev. im Anschluss an Deine Arbeit. Die Frage wird noch dadurch kompliziert, dass eine schüchterne Anfrage bei Bohr zu dem Ergebnis geführt hat, dass er sehr bittet, mit einer evtl. Publikation über dieses Thema zu warten, da er sehr ähnliche Dinge macht und er würde mir das Manuskript "nächste Woche" schicken. Da das aber schon sechs Monate her ist, und ich ausserdem davon überzeugt bin, dass er sich für diese Punkte nicht interessiert hat, so braucht man sich, glaube ich, nicht mehr gebunden zu fühlen — vorausgesetzt natürlich, dass es sich überhaupt lohnt, etwas zu schreiben. Was findest Du?

Die Schwierigkeit ist natürlich wieder, dass der Kern, so wie flüssiges He, so wie ein Supraleiter, eben eine Art von Ordnung hat, und man nicht weiterkommt, ohne ein Bild davon zu haben, worin diese Ordnung besteht, und wie sie zu beschreiben ist. Natürlich soll man das nicht so machen wie Herr Slater.

Anlässlich der oben erwähnten Variationsmethode für die Zustandssumme habe ich mir übrigens folgendes überlegt: man kann natürlich für die Wechselwirkung der Metallelektronen etwas ähnliches probieren. Dazu muss man natürlich in nullter Näherung die Zustandssumme ausrechnen, die sich ergibt, wenn man die Diagonalelemente der Gesamtenergie (einschliesslich Wechselwirkung) für die

Eigenfunctionen freier Elektronen ohne Wechselwirkung ausrechnet. Dabei bekommt man etwas sehr lustiges. Bei $T = 0$ kommt natürlich die übliche Verteilung der Elektronen heraus, bis auf einen konstanten Zusatzterm in der Energie. Die spezifische Wärme enthält aber, ausser dem üblichen Term mit T, dessen Koeffizient etwas modifiziert wird, noch einen Term mit $T. \log T$, der natürlich bei tiefen Temperaturen überwiegt. Was das physikalisch heisst, ist mir noch nicht ganz klar. Vermutlich gar nichts, denn obwohl die so berechnete Zustandssumme sicher kleiner ist als die wirkliche, hat die spezifische Wärme, die sich ja durch Differentiation ergibt, keine Extremaleigenschaft, und dieser komische Term kann also leicht bei besserer Annäherung verschwinden. Immerhin zeigt dieses Argument doch, dass die spezifische Wärme in Wirklichkeit nicht proportional zu T ist, und ich bemühe mich ein vernünftiges Funktionssystem mit Parametern zu finden, mit dem man weiter variieren kann.

Sonst habe ich mir den Kopf hauptsächlich über β-Theorie zerbrochen. Ich habe jetzt einen ganz vernünftigen Doktoranden den ich alle möglichen Ausdrücke vom Uhlenbeck-Konopinski-Typus habe ausrechnen lassen. Es gibt natürlich, bereits mit den ersten Ableitungen, ziemlich viel, und insbesondere kann man noch neue Ausdrücke gegenüber den in Deinem Artikel angegebenen finden, indem man zwei Deiner Ausdrücke linear kombiniert. Das ist weniger pedantisch als es auf den ersten Blick aussieht, denn wenn man z.B. die vier Wellenfunktionen anders paart, etwa: $(\psi^*_{\text{Neutron}} \cdot \psi_{\text{Neutrino}})(\psi^*_{\text{Proton}} \cdot \psi_{\text{Positron}})$ so erhält man einen Ausdruck, der sich als Linearkombination von Ausdrücken der von Dir oder U-K angegebenen Form schreiben lässt. Solche Ausdrücke braucht man insbesondere, um in zweiter Näherung eine Majoranakraft zu bekommen.

Die neue Freiheit, die man damit bekommt, kann man insbesondere dazu benützen, um das β-Spektrum am unteren Ende ($E \ll mc^2$) beliebig zu verändern. In ähnlicher Weise kann man auch das obere Ende beliebig anpassen, wenn die Neutrinomasse von Null verschieden ist, was man ja annehmen muss, wenn man die Messungen von Lyman glaubt.

Dieser Doktorand, Hoyle, hat sogar alle Ausdrücke zweiter Ordnung ausgeixt, einschliesslich Linearkombinationen, und kunstvoll derartige Ausdrücke konstruiert, die sich von U-K innerhalb der Messgenauigkeit nicht unterscheiden. Aber sie erscheinen mir etwas zu kunstvoll*, um glaubhaft zu sein.

Wir zerbrechen uns jetzt den Kopf darüber, wie man die ganze Theorie logisch in Ordnung bringen kann, denn wie Du wohl weisst, ist ja die U-K Theorie logisch nicht in Ordnung, da die Hamiltonfunktion sich nicht mit den Vertauschungsrelationen verträgt. Das muss immer passieren, wenn die Hamiltonfunktion, deren Zweck doch nur darin besteht, die zeitlichen Ableitungen von Grössen zu bestimmen, selbst wieder zeitliche Ableitungen enthält. Der plausibelste Ausweg schien mir zunächst der zu sein, dass man dem Neutrino eine Wellengleichung zweiter Ordnung [...][296] Dann ist nämlich die zeitliche Ableitung der Wellenfunktion eine Grösse die zu jeder Zeit noch beliebig vorgegeben werden kann, und die daher auch ungestraft in der Hamiltonfunktion vorkommen darf. Das würde physikalisch bedeuten, dass es einen Spin 3/2 hat, und dass sein Ort ebenso unbeobachtbar ist, wie z.B. der eines Lichtquants. Beides würde mich nicht stören. Aber die Sache scheint an den Zuständen negativer Energie zu scheitern, denn um diese nach dem Diracschen Rezept wegzubringen, braucht man einen positiv definierten Ausdruck für die Dichte, den es nicht gibt, und um das Pauli-Weisskopfsche Rezept anzuwenden, braucht man eine positive Energiedichte, die es im Falle von halbzahligem Spin gleichfalls nicht gibt. Im Augenblick bemühen wir uns, die Sache auf konservativere Weise in Ordnung zu bringen.

Ein anderer ganz anständiger Student arbeitet über das Oppenheimer-Philips Problem. Ich glaube gar nicht, dass dieses Problem als Zweikörperproblem behandelt werden kann, denn wahrscheinlich spielen dort doch Resonanzen eine Rolle. Das Oppenheimersche Argument, dass die kontinuierliche Anregungsfunktion Resonanzen ausschliesst, ist natürlich Unsinn, sobald es sich nicht um reine Capture handelt, sondern das Proton wegläuft. Dann kann sich das Proton eben immer gerade eine solche kinetische Energie aussuchen, dass der Rest in einem Resonanzniveau zurückbleibt. Aber unabhängig davon hat es Interesse, das Zweikörperproblem zu studieren, da das einer der wenigen Fälle ist, in denen man W.K.B. in mehr als einer Dimension anzuwenden wünscht. Streng geht das anscheinend nicht, aber

[296]Missing in manuscript.

man kann offenbar die Abschätzung machen: Offenbar ist ψ von der Form $Ae^{-S/h}$, wo S der Hamilton-Jakobigleichung genügt, aber im allgemeinen komplex ist. Mit $S = a + ib$ folgt dann, dass grad a auf grad b orthogonal ist und dass $(\text{grad}\, a)^2$ grösser ist als $2m(V - E)$. Um nun den Wert von S an einem Punkt x zu bekommen, kann ich den Gradienten von einem klassischen Umkehrpunkt, an dem ψ von der Grössenordnung 1 und daher S klein ist, bis zu x integrieren. Dabei müsste ich als Integrationsweg natürlich eine Fallinie von a wählen. Wenn ich aber das Integral von grad a über alle möglichen Wege betrachte, und mir den kleinsten Wert heraussuche (was mit Hilfe von Variationsrechung leicht geht) so bekomme ich wieder etwas zu kleines. Es folgt also, dass $a(x)$ sicher grösser ist als der kleinste Wert des Integrals von $2m(V - E)$ über beliebige Wege vom Umkehrpunkt bis zu x. Im Falle von Oppenheimer-Philips stellt sich heraus, dass die von O.-Ph. berechnete Zertrümmerungswahrscheinlichkeit bereits grösser ist als die so gewonnene obere Grenze, was natürlich an ihrer adiabatischen Näherung liegt. Wir versuchen jetzt, die Methode zu verbessern, um eine nähere Abschätzung zu bekommen.

Nebenbei spiele ich noch ein bisschen mit Metallen, Halbleitern und tiefen Temperaturen, wo es alle möglichen spassigen Dinge gibt.

Auch sonst ist es in Cambridge noch recht lebendig. Im Mond arbeiten zehn Experimentatoren, teilweise recht hübsche Sache. Der neueste "thrill" ist, dass die Wärmeleitfähigkeit von flüssigem Helium II ca. 2000 Watt/cm.grad ist also noch etwa zehnmal grösser als Keesom behauptet. Das ist doch schon eine erschütternde Grössenordnung! Ausserdem werden wir hier wahrscheinlich bald den Rekord für tiefe Temperaturen haben was natürlich zu nichts führt ausser Reklame.

Das neue Hochspannungslaboratorium ist gebaut und sieht sehr imposant aus (von aussen Kino, von innen Kathedrale) und die erste Stufe mit 1,2 MV funktioniert bereits. Auch ein Cyclotron ist im Bau. Dee und Genossen messen die Winkelverteilung für Proton-Neutron-Streuung sehr sauber, zunächst bei 2 MV. Die Resultate stimmen sehr schlecht mit Gans & Harkins. Apropos, wie gefällt Dir die Theorie von Fisk & Morse die so schön mit Gans & Harkins stimmt, wenn man nur annimmt, dass die Neutronen Rn+Be alle eine Energie von 25 MV haben?

In den letzten beiden Termen hatten wir sogar ein Kolloquium über Kerntheorie, (zu dem Fowler meistens zu kommen vergass) und zu Fowlers grossem Bedauern fangen die meisten Doktoranden an, sich nicht mehr an so blödsinnige Probleme setzen zu lassen, mit denen er sie gewöhnlich beschäftigt.

Ich habe jetzt auch mehr mit Studenten zu tun, Kursvorlesung über Kerne, College supervision (d.h. je eine Wochenstunde Privatunterhaltung mit undergraduates, was hier obligatorisch und sehr nützlich ist) und für all das kriegt man extra gezahlt. Das einzige, was in diesem Winter ein bisschen fehlt, ist Diskussionen mit "Erwachsenen", denn Dirac ist durch seine Heirat zu sehr abgelenkt, Darwin hat als Master über Christs zu viel anderes zu tun, Bhabha und Hulme sind fort. Jetzt sind allerdings meine Studenten schon bald so weit, dass man sich mit ihnen unterhalten kann, aber das dauert noch eine Weile.

Sonst gibt es nette Leute hier natürlich wie immer im Ueberfluss und man hat viel zu wenig Zeit, um alle zu sehen. Wenn nur die politische Lage nicht wäre. Aber wegen der hast Du doch wahrscheinlich Recht, dass Du nach Amerika gegangen bist, denn sehr lange kann das ja hier nicht mehr gut gehen, und dann nützt einem die ganze Prosperity nichts.

Apropos Prosperity, weisst Du in den USA eine Stelle für einen ganz ordentlichen, aber nicht genialem, sehr geduldigen und auf dem Gebiet der Fluoreszenz allwissenden Experimentator namens Hirnschlaff? Er hat sich hier mit einigen Leuten verkracht was wohl zu 25% seine Schuld war und wird durch ca. monatlich verlängerte Stipendien am Leben gehalten. Er würde vermutlich jede Stelle annehmen, und wenn es ein bisschen Aussicht hat, würde er hier wohl genug Geld zusammenkratzen um auf eine Fahndungsreise hinüberzufahren, auch ohne eine Stelle in der Tasche zu haben.

Kommst Du diesen Sommer in das senile Europa, solange es noch nicht in die Luft gegangen ist?

Mit besten Grüssen

Dein

R. Peierls

*Wir haben eben einen sehr glaubhaften Ausdruck dieser Art gefunden, d.h. eine Summe von zwei Gliedern, mit Koeffizienten 1.

Cambridge, 13.2.[19]37

Dear Bethe,
This time I am the one [who has not written], and lack even the pretext of the title page; only the name of my friend A.H. Wilson appears on the cover of the Camb. Phil. Soc., and you are well within your rights not to accept this as a greeting. My only excuse is that every time I thought I was going to discover something better about excited nuclei and liquid He, it always turned to vinegar (liquid vinegar). I don't know much more right now than what we could agree upon this summer. It might be worthwhile to sum up the situation. If keep the spin and the number of protons and neutrons constant as well as the energy within the r integrals in the equations of motion $(F_1 \ldots F_r)$, then the number of levels in an interval $dEdF_1 \ldots dF_r$ is

$$\sqrt{(2\pi)^{r+1} D_{r+1}} e^s, \tag{1}$$

where s is the entropy and D_{r+1} is the determinant of

$$d_{i,k} = \frac{\delta^2 S}{\delta F_i \delta F_k}. \qquad (i, k = 0, 1, \ldots r, \qquad F_0 \equiv E)$$

This shouldn't be new to you. You can derive the formulas in your work a bit more nicely and more symmetrically, but of course this isn't important.

In the $r = 0$ case you will just get d^2S/dE^2 instead of the determinant, and the correction factor which gives the difference between this and proposition (1) will be just a power of the particle number A. The error you make by leaving out this correction is an entropy term on the order of $\log A$.

This means, however, that a macroscopic analogy can never give you this factor. (The difference is obviously that between the microcanonical and canonical distributions involving $F_1 \cdots F_r$). This argument also shows that it is incorrect to consider these nuances without taking into account the surface effects. If you could neglect the surface terms, then you could certainly also neglect the logarithmic terms.

Such surface effects are neglected in your formula for a Fermi gas, as you take the number of levels for the particles to be proportional to the volume. In principle, this could be easily corrected if you knew the potential field. As you don't know it to sufficient accuracy (or rather, as you presumably can't define it to such an accuracy), this isn't very helpful.

From this I would conclude that without a fully microscopic model, it is impossible to treat the surface effects correctly. You therefore don't have to worry about the logarithmic terms (i.e., the F's in (1)) at all. The absolute value of the density is incorrect by several orders of magnitude, however, so you have to be content with a very crude approximation. If you do that, then you get

$$N = \sqrt{2\pi \frac{\delta^2 S}{\delta E^2}} e^S$$

and you only have to play with the function $S(E)$ or $E(T)$. As the temperatures are low, it is reasonable to approximate this function as a power law

$$E = ABT^n.$$

This corresponds to your assumption of total deformation, and seems to lead to a quite considerable uncertainty on the result even though the error in the exponent is very small. Furthermore, the factor A is expressed in such a way that if you neglect the surface effects, B will be independent of A. If you agree to that, you are left with n and the more or less universal factor B.

It seems to be reasonable to presume that n is equal to the corresponding exponent of liquid He, and you might think that you could find out the latter by experiment. Such experiments have been planned at the Mond,[297] but they are not easy to carry out since you have to go to temperatures much lower than 1° to find the asymptotic behaviour of the law. The Leiden measurements of specific temperature go to about 1°, and if you plot C/T^3 against T you get a straight line going right through the zero point. If you take this seriously you get $n = 5$, which seems to be nonsense, since if true liquid He should be more difficult to make oscillate than a solid body! It could easily be the case that instead

[297] See letter [39], note 287.

of going through the zero point, the curve turns around and becomes horizontal as would be necessary for $n = 4$. To get a power less than 4, however, it had to approach infinity and this would seem a bit too much. Therefore $n = 4$ seems to be the most reasonable choice at the moment.

Our best estimate for B seems to come from the average excitation energy of radioactive elements, which tells us that for $A = 200$ S has to be of order 1. If you take the excitation energy to be 0.2 MV, then you get a distance of 100 V between levels for $A = 100$ and $E = 8$ MV. This result is much too nice; I think you should have a leeway of at least two factors of ten in every direction, but in the context of this crude analysis it seems to be all right. Without an adequate model for the motion of the nucleus, I guess one can't get a better result. This model is the most interesting part, but I couldn't figure out anything at all about it. The only thing I am really sure of is this: my sloppy analysis of the longitudinal and the transverse waves can't be made any more precise. If the substance is not a crystal, then the wave motions can't be strictly confined to ratios. In the case of longer wavelengths, for which this is still possible, it behaves like a perfect fluid.

The only step forward, which doesn't seem to be of any great help either, is based on the following statement: there is no hope that the excitation conditions can be more easily dealt with than those of the ground state. This problem can only be approached with a variational principle. It may be of some slight comfort to know that there is also a variational principle for the partition function. The following is correct: for a given Hamiltonian H I can find the diagonal values H_{ii} for any chosen orthogonal system. I just act as if these were the correct eigenvalues and set up the function

$$F(T) = \sum_i e^{-H_{ii}/kT}.$$

The result is always smaller (i.e., the corresponding free energy is always larger) than the true partition function. If you take a fully orthogonal system with some free parameters (for example, the eigenfunction of a solvable problem with free parameters) (describing the potential), then you might come quite close to the real free energy just by varying the parameters. But I still haven't found an appropriate system of functions for this approach. You probably have a better idea.

This is all quite unclear and I don't know whether it is worth publishing, but you could also consider submitting it as a letter to the Phys. Rev. which would follow your work. The question is now even more complicated, as Bohr answered my shy request and asked me to delay its publication. He said that he was working on similar things, and would send me the manuscript "next week".[298] As this was now six months ago, and I am convinced that he is no longer interested in these points, we don't have to feel bound by this anymore — provided that it is worth writing about at all. What do you think?

The difficulty is that the nucleus has a kind of order to it, just like liquid He or a superconductor, and if you don't know more about this order and how to describe it then you can't make much progress. But you shouldn't follow Mr. Slater's example, that's for sure.

Regarding the variational method for the partition function mentioned above, I have had a thought: you could try something similar for the interaction of electrons in a metal. You would first have to calculate the partition function in the zeroth order approximation, which is what you get if you use the diagonal elements of the total energy (including interaction) to define the eigenfunction of free electrons <u>without</u> interaction. Then something very strange happens. In case of $T = 0$ you get a normal electron distribution of course, except for an additional constant energy term. In addition to the normal term with T, whose coefficient has been slightly modified, the specific heat has another term proportional to $T.\log T$ which is of course dominant at low temperatures. I don't really know how this should be interpreted physically. I presume nothing. Even though the partition function calculated this way must be smaller than the real one, the specific heat (which results from differentiation) has no extremal value. This strange term could easily disappear if a better approximation were used. This argument shows that in fact the specific heat is not proportional to T, and I am trying to find a reasonable system of parametric functions that you can keep working on varying.

[298] See letter Niels Bohr to Rudolf Peierls, 17.10.1936, in Lee, *Selected Correspondence*, 1, chapter 3.

Besides this, I have mostly been thinking about the β-theory. I now have a quite reasonable doctoral student,[299] whom I have asked to calculate all possible terms of the Uhlenbeck-Konopinski type. There are already quite a few just with the first derivatives. There are also more terms than those you gave in your article, which result from combining two of your terms in a linear way. This is less pedantic than it sounds. If you combine the four wave functions in some other manner, for example $(\psi^*_{\text{neutron}} \cdot \psi_{\text{neutrino}})(\psi^*_{\text{proton}} \cdot \psi_{\text{positron}})$, then you get a new term which can be expressed as a linear combination of those terms described by you and U-K. You need such terms to get a Majorana force in the second approximation.

You can use this new degree of freedom to change the β-spectrum at the lower end ($E \ll mc^2$) in just the way you want. In a similar way you can also adjust the upper end, if the neutrino mass is not equal to zero. This will be necessary if you believe Lyman's measurements.[300]

Hoyle, my doctoral student, has even ixed all the second-order terms including their linear combinations. He has also skilfully constructed some terms which agree with U-K to within the accuracy of measurement. But they seem a bit too skilful* to be believable.

We are now thinking about how to arrange the whole theory in a logical order, since as you know the U-K theory is logically incorrect. This is because its Hamiltonian is inconsistent with its permutation relation. This will always happen if the Hamiltonian, which is needed to define the time derivative of any value, itself contains time derivatives. The most plausible alternative seemed to be a second-order wave equation for the neutrino, $[\cdots]$[301] because then an arbitrary value can be assigned to the time derivative of the wave equation at any time, which will be absolutely correct for the Hamiltonian. In terms of physics this would mean that it has spin 3/2, and that its position is unobservable

[299]Fred Hoyle (1915–2001) studied at Cambridge where, initially under the supervision of Peierls and later by Maurice Pryce, he completed work for a Ph.D. (which Pryce persuaded him not to submit because of the his disapproval of the degree which was new to Cambridge.

[300]Elisabeth Reed Lyman, initially at Berkeley, then University of Illinois, Urbana.

[301]Missing in manuscript.

just like the light quanta. Neither result bothers me. But the theory seems to fail because of the negative energy conditions. To carry it off according to Dirac's method, you need a density term which is defined to be positive. This doesn't exist. Similarly, to apply the Pauli-Weisskopf method you need a positive energy density, which also doesn't exist for particles with half-integer spin. At the moment we are trying to solve the problem in a more conservative way.

Another very smart student is working on the Oppenheimer-Philips problem.[302] I don't think that this can be treated as a two-body problem, as it seems that resonances play a decisive role. Oppenheimer's argument that a continuous excitation function excludes resonances is surely nonsense, if the proton escapes instead of experiencing a pure capture. In this case the proton can always decide on its kinetic energy in such a way that the remainder is held back in the resonance. Nevertheless, it is interesting to study the two-body problem as it is one of the few situations in which you might want to apply W.K.B. in more than one dimension. Obviously you can't do it properly, but you can still make this estimate: ψ is of the form $Ae^{-S/h}$, where S satisfies the Hamilton-Jakobi equation but is still a complex number. With $S = a + ib$ it follows that grad a is orthogonal to grad b, and that $(\text{grad}\, a)^2$ is bigger than $2m(V - E)$. To find the value of S at point x, I can integrate the gradient between a classical turning point and x when ψ is of order 1 and S is therefore small. As for the direction of integration, I of course had to go in the sense of decreasing a. But if I observe the integral of grad a over all possible paths and choose the smallest value (which is easy with the help of a variational calculation), then I again get something too small. It follows that $a(x)$ is surely larger than the smallest possible value of the integral of $2m(V - E)$ over all possible paths between the turning point and x. If you take the Oppenheimer-Philips result then you will find that their probability of demolition is already larger than the calculated upper limit, which has to do with their adiabatic approximation. We are now trying to improve on their method to get a closer approximation.

[302] Bethe published a paper on the Oppenheimer-Philips problem the following year. H. Bethe, 'The Oppenheimer–Phillips Process', *Phys. Rev.* **53**, 39–50 (1938).

Besides this, I'm also playing with metals and semiconductors at low temperatures, where you can find a lot of funny things.

In Cambridge there is still a lot going on. There are now 10 experimenters at Mond, some of them working on quite nice things. The latest "thrill" is that the heat conductivity of liquid helium II is about 2000 Watt/cm.grad, which is about ten times bigger than Keesom[303] claimed. This is a shocking order of magnitude! We will also establish the record for low temperatures soon, which of course is nothing but advertising.

The new high-voltage laboratory is finished, and it looks impressive (like a cinema from the outside, and like a cathedral from the inside). The first level with 1.2 MV is already working. They are also building a cyclotron. Dee and his colleagues are now precisely measuring the angular distribution of proton-neutron scattering at 2 MV (at first). The results correspond poorly with those of Gans & Harkins.[304] By the way, what do you think about the theory of Fisk & Morse[305] which agrees well with Gans & Harkins, as long as you assume that the neutrons Rn+Be all have an energy of 25 MV?

During the last two terms we have even had a colloquium on nuclear theory (which Fowler mostly forgot to come to). Fowler is very unhappy that most of our doctoral students don't want to study the stupid problems that he normally wants them to deal with.

I now have more to do with students, my lectures about nuclei and College supervision (which means one hour per week of private classes with undergraduates, which is obligatory and very helpful), and this all means extra pay. The only thing I'm missing a little bit this winter is the chance to have discussions with "adults"; Dirac is distracted by his marriage,[306] Darwin as the Master of Christs has too much to do, and

[303]Willem H. Keesom (1876–1956), student of Heike Kamerlingh Onnes; in 1926 Keesom developed a method to solidify helium.

[304]William D. Harkins, David M. Gans, 'Inelastic Collisions with Changes of Mass and the Problem of Nuclear Disintegration with Capture or Non-Capture of a Neutron or Another Nuclear Projectile', *Phys. Rev.* **46**, 397–404 (1934).

[305]J.B. Fisk, Philip M. Morse, 'The Elastic Scattering of Neutrons by Protons', *Phys. Rev.* **51**, 54–55 (1937).

[306]Paul Dirac had married Eugène Wigner's younger sister Margit Balasz (Manci) earlier in the year.

Bhabha and Hulme have gone away. It won't be too long before I can talk with my students in this way, but it will still take a little while.

There are always a lot of nice people here, and one has too little time to see all of them. If only there weren't this political situation. With this in mind, you were probably right to go to America. The situation won't be ok here for much longer, and then all this prosperity won't do us any good.

Apropos of prosperity, do you know of a job in the USA for a quite good and very patient, but not ingenious experimental physicist named Hirnschlaff who is an expert in the field of fluorescence? He has fallen out with some people here, which is 25% his fault. He is only surviving by monthly grants. He would probably accept any job, and given the opportunity he would scrape together the necessary funds for an expedition to the USA even if he wasn't sure to get the job.

Are you planning to come to senile Europe, as long it hasn't been blown up? With best wishes,
Yours,

R. Peierls

*We just found a proper term of this kind, i.e., a sum of two elements with coefficient 1.

[44] Hans Bethe to Rudolf Peierls

[Cornell], [ohne Datum]

Lieber Peierls,
Vielen Dank für Dein Manuskript. Ich finde es wunderschön. Wir, d.h. besonders Weisskopf, hatten Ähnliches versucht, aber wir waren nicht ganz durchgekommen und es war auch sonst wesentlich weniger schön, da W. versuchte, gleich mit den wahren komplexen Eigenwerten zu rechnen statt mit den zu einem bestimmten k gehörenden. Ich habe es schon immer abscheulich gefunden, dass man Störungstheorie anwendet, wo man weiß, dass sie nicht gilt. Ausserdem finde ich es sehr hübsch, dass die Potentialstreuung so ohne jeden Zwang herauskommt.

Was habt Ihr über den Fall größere Breiten herausbekommen? Es hat vielleicht nicht viel Sinn, in diesem Fall überhaupt von der Dispersionsformel auszugehen. Konopinski und ich zerbrechen uns den Kopf über die "Breite ohne Potentialberg", d.h. im Wesentlichen Dein Φ_n, bei niedrigen Energien (siehe die Kontroverse über die Radien). Weisskopf und ich interessieren uns hauptsächlich für den Photoeffekt. Übrigens behauptet Placzek, Du hättest ihm gesagt, die Arbeit von Kalckar, Oppenheimer und Serber über den Photoeffekt wäre in Ordnung. Das kann doch wohl nicht sein (oder Dein) Ernst sein. Sehr traurig fand ich Bohr's Note in Nature über dieses Thema, sowohl die experimentellen Resultate sind nicht, wie er sagt, als auch die "Erklärung" scheint mir reactionär und Unsinn. Er nimmt ja jetzt das Kristallmodell ernster als je irgendjemand das Hartree Modell!

Dieser Brief ist aber nicht gemeint, um zu kritisieren, sondern um zu lernen, was Du im Sommer tust. Ich werde etwa am 18. Juni nach Europa fahren und etwa am 18. September zurück. Können wir uns dann irgendwann sehen: wie wäre es z.B. mit einer gemeinsamen Unternehmung von 1–14 Tagen in der Schweiz? Ich würde aber auch gern nach England kommen — nur ist es wohl unwahrscheinlich, dass noch viele Leute da sind, wenn ich komme. Besonders Mott würde ich gerne sehen, ich werde ihm direct schreiben. Schreib mir auch bitte möglichst gleich, was Euch am besten passen würde.
Wie geht's Genia und der Nachkommenschaft?
Viele herzliche Grüsse,
Dein

Hans

[Cornell], [date unspecified]

Dear Peierls,
Thank you very much for your manuscript. It is beautiful. We, have tried similar things, especially Weisskopf,[307] but couldn't really follow

[307]Viktor F. Weisskopf (1908–2002), studied in Vienna and Göttingen, where he received his Ph.D. in 1931. After research at Leipzig, Vienna, Copenhagen and Zurich he took up a position at Rochester; he joined the Manhattan Project in 1943; in 1945 he joined the faculty at MIT where he remained until his retirement; from 1961 to 1966 he also served as director-general of CERN.

through. Besides, our effort was definitely not as good as yours because Weisskopf tried to do the calculation with true complex eigenvalues instead of eigenvalues belonging to a defined k. I have always thought that it is terrible to apply perturbation theory when you know that it can't be applied. I also find it very nice that the potential scattering comes out just like that.

What did you find out about the case of larger widths? It is not very useful to begin with the dispersion formula. Konopinski and I are racking our brains about the "width without potential barrier", which basically means your Φ_n, in the case of lower energies (see the controversy over the radii). Weisskopf and I are mostly interested in the photoelectric effect. By the way, Placzek says you told him that the work of Kalckar, Oppenheimer and Serber[308] on the photoelectric effect was correct. This can't be true, or at least you can't mean that seriously. I thought that Bohr's note in Nature about this problem was very sad. The results of the experiments are not what he claims them to be, and his "explanation" seems to be reactionary and nonsense. He now takes the crystal model even more seriously than anyone ever took the Hartree model!

This letter was not meant as a criticism, however, but to get to know your plans for this summer. I will travel to Europe around the 18th of June, and come back around the 18th of September. Could we meet at some point? What would you think, for example, of spending 1–14 days together on a trip to Switzerland? I would also like to come to England, but it is likely that not many people will still be there when I come. I would especially like to see Mott, and I will write to him directly. Please tell me as soon as possible what would work best for you.

How are Genia and the offspring?

Many greetings,

Yours

Hans

[308]Robert Serber (1909–1997), obtained his Ph.D. in 1934 from the University of Wisconsin; he then worked with Oppenheimer at Berkeley and later became Associate Professor at Wisconsin and Professor at Berkeley before taking up a Professorship at Columbia University in 1951 where he stayed for the remainder of his academic career.

[45] Hans Bethe to Rudolf Peierls

[Cornell], [ohne Datum]

Lieber Peierls,
Das wäre ein herrlicher Plan, Autofahrt durch Frankreich! Ganz nach meinem amerikanischem Herzen. 10 Tage ist sehr schön. Mein Schiff geht wahrscheinlich zwischen 14. und 21. September. Alright?

Ob ich vorher noch mehr Zeit habe, ist sehr fraglich. Ich fahre am 20. Juli hier weg, bin also wohl am 25. in Cherbourg. Meine Eltern werden wohl auf den größten Teil der vorhandenen Teil reflektieren. (Ausser natürlich, wenn man mich nicht nach Deutschland hereinläßt. Aber der Mann an der Deutschen Botschaft in Washington meinte, es wäre alles bei mir in Ordnung).

Über die Physik ein andernmal. Zum Teil sind Eure Entdeckungen mit unseren identisch, zum Teil verstehe ich sie aber nicht. Grüsse an Genia und Dich

Hans

[Cornell], [date unspecified]

Dear Peierls,
That would be a great idea, a road trip through France! Just as I like it, the American way. 10 days is very nice. My ship will probably depart between the 14th and 21st of September. All right?

It is very doubtful whether I will have any more time beforehand. I will leave here on the 20th of July, so it is likely that I will be in Cherbourg around the 25th. My parents will probably claim most of my time. (Except if I am not allowed to pass the German border, of course. However, the man at the German embassy in Washington told me that in my case everything was all right.)

About physics another time. Partly your results are identical to ours; partly I don't really understand them. Greetings to you and Genia.

Hans

[46] Hans Bethe to Rudolf Peierls

[Cornell], [ohne Datum]

Liebe Peierls'
Also meine Mutter will es nicht Erstens hat sie es nicht gern, wenn ich wegfahre, ehe ich muss. Zweitens ist sie krank und meint wenn ich später zu ihr komme, ist sie eher wieder gesund. Drittens hat sie noch einen mysteryösen Grund, "den sie mir nicht schreiben kann". (Als ob der Deutsche Zensor solche Sätze gerne hätte). The upshot of it all is, dass ich direct zu Euch komme, wenn ich in Europa ankomme. Das ist vermutlich am 25. Juli mit der Bremen. Die geht zwar nach Southampton, ich werd' aber wohl auch von da zu Euch finden. Ich telegraphiere noch meine genaue Ankunftszeit.

Bis dahin bin ich in Ann Arbor. (Physics Department).
Viele schöne Grüsse und auf Wiedersehen!

Hans Bethe

[Cornell], [date unspecified]

Dear Peierls!
My mother doesn't want it.[309] First she doesn't like it at all when I go abroad before I have to. Second, she is ill and thinks she will recover sooner, if I come later. Third, she has a mysterious reason that "she can't write about". (As if the German censor would like such a sentence). The upshot of all this is that I will come to you straight away, once I have arrived in Europe. This will presumably be on the 25th of July with the Bremen. I am arriving at Southampton, but from there I guess I will find my way to you. I will telegraph my exact time of arrival later.

Until then, I will be in Ann Arbor (Physics Department).
Many greetings, and see you then!

Hans Bethe

---

[309] Refers to Hans Bethe's plans for the summer vacations.

[47] Hans Bethe to Rudolf Peierls

[Seelisberg], 22.8.[19]38

Liebe Peierlses,
Ich habe also Eure Eltern (bzw. Schwiegereltern) gesehen. Nicht sehr erfreulich (wie überhaupt die Reise nach Deutschland). Besonders die Mutter ist furchtbar nervös, der Vater nimmt es sehr viel ruhiger. Aber er ist noch nicht entschlossen und findet den Gedanken schrecklich, seinen Kindern "zur Last zu fallen" (oder seinem Bruder). Dabei müssen die Zustände in Berlin noch unerträglicher sein als wir draussen wissen. Die Bitte, Ihr solltet keine Briefe schreiben, ging z.B. zurück auf die persönliche Warnung ob die berechtigt war, weiss ich nicht; es scheint mir, dass einer den anderen verrückt macht in Berlin. Inzwischen sei das nicht mehr aktuell und Ihr könntet ruhig wieder Briefe schreiben.

Dann gabs grosse Aufregung über eine Büchersendung an Euch. Bücher über 20M Wert dürfen nicht mehr ins Ausland geschenkt werden; rechts und links tuns zwar die Nachbarn noch, aber bei Eurem Papa hat sich die Golddiskontbank darüber aufgeregt. Er nahm sich darauf einen Anwalt (!) und schließlich, nach 4 Wochen, gab sich die Golddiskontbank schliesslich zufrieden. Aber immerhin wars eine grosse Aufregung für ihn.

Dann gibts so hübsche Sachen wie Verbot, den Bank-Safe zu öffnen ausser in Anwesenheit eines Beamten der Zollfahndungsstelle; besondere Genehmigung für jede Sendung ins Ausland, die meist mehrere Monate dauert und dann doch abgelehnt wird; und neuerdings Vorschriften der Steuerbehörde über die Höhe der zulässigen Abhebungen vom Bankkonto (Dies allerdings erst ganz vereinzelt, aber so fängts immer an). Hingehen kann man in Deutschland fast nirgens mehr, denn überall beklagen sich die Hotelgäste, wenn man überhaupt aufgenommen wird. Solche Dinge wie das Waldpark-Sanatorium werden (in diesem Fall: sind) verkauft. Also, es wird so systematisch betrieben, dass man seine Freude daran haben könnte, wenn es einen nichts anginge.

Nun sollte man meinen, dass die Entscheidung über Dableiben oder Fortgehen niemand mehr schwer fallen könnte. Und trotzdem gibts da noch immer Zweifel. Angeblich nicht der Verzicht auf Luxus und

Wohlleben, sondern nur der Gedanke, Almosen zu empfangen. Ist auch schwer für jemand, der immer anderen gegeben hat, aber schliesslich ... Und dann hat Dein Vater die Idee, er dürfte das nicht tun, weil er dadurch seiner arischen Frau die schwerverdiente und grosse Pension nimmt. Sie ist übrigens sehr viel mehr entschlossen als er, macht ihn aber durch ihre Nervosität ganz verrückt.

Ich hab versucht, was ich konnte, aber genützt hats glaube ich nicht. Sie sehen nicht ein, dass heutzutage alle normalen Gesichtspunkte bedeutugslos geworden sind. Inzwischen ist also das Affidavit von Amerika gekommen und das hat vielleicht eine gewisse Überzeugungskraft. Aber es warten schon Tausende auf ihr Visum, und die Quoten für 1938 und 39 (!) sind schon vergeben, ausser einem kleinen Rest der für dringende Fälle aufbewahrt wird. Das wissen übrigens Deine Eltern noch nicht. Aber schon die Idee, bis nach Neujahr warten zu müssen, erschien unerträglich. Konsequent oder inkonsequent — na ja. Und unter diesen Umständen dachten sie daran, erst nach England zu gehen und dann später nach Amerika. Ich finde das keine sehr glückliche Idee; alle Komplikationen kommen zweimal, das amerikanische Visum wird von England aus schwieriger zu kriegen sein als von Berlin; der grössere Umzug England–Amerika muss in Pfund bezahlt werden usw. Ich sehe auch nicht ein, warum die plötzliche Ungeduld, nachdem man noch nicht einmal fest entschlossen ist.

Ich habe aber jedenfalls den Auftrag, Euch zu bitten, auch Eurerseits ein Affidavit zu schicken für alle Fälle. Dies soll nur eine Formalität sein und Euch finanziell zu nichts verpflichten — das besorgen die Amerikaner. Es soll auf Universitätsbriefbogen geschrieben sein und sagen, dass Ihr gerne hättet, wenn Eure Eltern nach England kämen, und dass Ihr willens und finanziell imstande seid, sie dort zu ernähren. Auf Englisch natürlich. Im übrigen war man (d.h. nur mütterlicherseits) darüber böse, dass Ihr nie mehr zugeredet hättet, nach England zu kommen (nach dem einen Versuch vor vier Jahren) und man war erstaunt zu hören, dass Ihr sehr für Auswanderung seid usw. Ich hoffe, ich habe getreulich alles ausgerichtet.

Andererseits ist meine Mutter entschlossen, nach Amerika zu gehen. "Ich lasse ja hier nichts zurück", meinte sie. Und ein Zimmer in meiner Nähe in Amerika würde sie glücklicher machen als ein Haus in

Baden-Baden. Wir waren eigentlich beide voneinander überrascht: Ich dachte, ich müsste sie überreden, und sie dachte, ich wäre gegen eine Übersiedlung. Für Mütter sind Visa leichter zu bekommen, wir denken also an einen Umzug nächstes Frühjahr. Mein Vater war sehr überrascht von dem Entschluss und hatte Bedenken, aber meine Stiefmutter möchte am liebsten selbst. Jetzt sind wir also mit Pass in der Schweiz, und es ist wie eine Offenbarung für meine Mutter. Es regent zwar in Strömen, aber Adresse: Seelisberg (Uri), Hotel Sonnenberg. Wir werden bis zu meiner Abreise hier bleiben, und die ist am 16. September mit der Britannic ab Le Havre. Schwierigkeiten habe ich zwar weder beim Rein- noch beim Rausfahren gehabt, aber man sollte nicht übermütig werden.

Meine Geschwister sind sehr gross geworden, und sehr ungezogen. Etwas von Eurer Erziehung täte ihnen gut. Meine Schwester ist lange nicht so gescheit wie Gaby, spricht keine so vollendeten Sätze und kann auch weder rechnen noch schreiben. Dafür kann sie, im Sinne des heutigen Reiches, besser rennen als alle ihre Freunde und besser schwimmen als ihr grosser Bruder (wozu nicht viel gehört). Sie liebt mich sehr und ist sehr zärtlich. Klaus liebt mich auch sehr, aber seine Zärtlichkeit äussert sich hauptsächlich in unvermuteten Rippenstössen.

Seit vorgestern, d.h. seit wir hier sind, denke ich wieder an Physik. Ich dachte, ich könnte beweisen, dass die Unendlichkeiten in der Quantenelektrodynamik nicht eine Folge unzweckmässiger Integration sind, sondern in den Grundformeln drinstecken. Mit Hilfe eines Variationsprinzips. Leider geht das nur für einen tiefsten Eigenwert eines Teilchens. Bei der Nichtlöchertheorie ist aber $E = mc^2$ nicht der tiefste Eigenwert, und bei der Löchertheorie muss man die Differenz zwischen zwei Eigenwerten (nur der "See" und See + Elektron) nehmen, wofür Variationsprinzipe im allgemeinen nicht gelten.

Dank für die Woche in Cornwall. Ich wünschte ich könnte jetzt nochmal hin, denn ich glaube, ich könnte jetzt die quaint old villages and cute ancient churches wieder noch mehr geniessen. Und auch das surfboard riding vielleicht würde ich das schliesslich doch lernen. Aber vor allem Dank für die Aussprache, physikalischer und insbesondere persönlicher Natur. Es hat mir sehr gut getan, und ich wünschte, es wäre weniger weit von Ithaca nach Birmingham.

Natürlich habe ich Eure Adresse doch verloren, ich weiss zwar noch die Nummer 38, aber nicht mehr die Strasse (Calthorpe?). Aber jetzt seid Ihr sowieso beim britischen Esel. Grüsst ihn von mir; kann er mehr als 9 stones 3 tragen?
Schönste Grüsse

Hans

[Seelisberg], 22.8.[19]38

Dear Peierlses,
So, I have seen your parents (or your parents-in-law, respectively). It was not at all pleasurable (just like my whole journey to Germany). Both of them are terribly nervous, especially the mother, but the father is taking it much better. Still, he hasn't decided yet and doesn't at all like the idea "of being a burden" to his children (or his brother). The situation in Berlin must be even more unbearable than it seems from "outside". Their pleas that you should not send any more letters were actually a personal warning, for example. Whether it was justified or not, I don't know; it seems to me as if they all drive each other mad in Berlin. This warning is no longer up to date; you are allowed to send letters again.

Then there was big trouble concerning the package of books for you. It is no longer permitted to send books abroad, if they are worth more than 20M; the neighbours on the left and right are still doing it, but in the case of your Dad the *Golddiskontbank* tried to enforce this rule. He therefore engaged a lawyer (!), and 4 weeks later the *Golddiskontbank* finally relented. But it was still very annoying.

Then there are other nice things, such as: it is forbidden to open your safe-deposit box unless a customs official is present; you need special permission for anything you send abroad, which often takes several months and will finally be refused anyway; and then there are new regulations from the tax authorities regarding the amount you are allowed to withdraw from your account (up till now this has only been applied in some cases, but that's always the way it starts). You can't go anywhere in Germany anymore because the hotel guests complain constantly, assuming they agree to take you at all. Places like the *Waldpark Sanato-*

rium are being (or in this case: have already been) sold out. Well, all this is being done in such a systematic way that it would almost be a pleasure to watch if it had nothing to do with you.

Now you might think that it's not very hard for anyone to decide whether to stay or go. One still can't make up one's mind, even so. They say that it's not because they would have to go without luxury and good living, but rather because they couldn't stand to accept charity. This is especially hard on someone who has always supported others, but in the end ⋯ Your father also thinks that he can't afford to do it, as then he would deprive his Aryan wife of the highly deserved and huge pension. She is much more determined, by the way, but her nervousness is driving him crazy.

I have done what I could, but I'm afraid it hasn't helped. They won't accept that these days, reasonable arguments are no longer of any relevance. In the meantime they have received the affidavit from America, and this might be more convincing. But thousands are waiting for their visas, and the quotas for 1938 and 39 (!) have already been filled (except for a few which are held in reserve for urgent cases). Your parents don't know this yet, by the way. But even the possibility of having to wait until the new year seems unbearable. Consistent or or not consistent — well. In this situation they have thought about going to England first, and later to America. I don't think that this is a good idea. Then you get all the difficulties twice, and American visas are even more difficult to get from England than from Berlin; the move from England to America is more expensive and has to be paid in pounds, etc. I also don't understand why they are suddenly so impatient, while still not being fully determined to do it.

I now have the duty of asking you to send another affidavit just in case. This should be just a formality, and commit you to nothing financially — that's the Americans' job. You should write it on a university letterhead and say that you would like your parents to come to England and that you would be willing and able to support them financially. In English, of course. Besides that, they (i.e., just your mother) were unhappy that you haven't really tried to convince them to come to England (since that attempt four years ago), surprised to hear that you favoured emigration, etc. I hope I have reported everything accurately.

On the other side, my mother is determined to go to America. "I won't be leaving anything behind," she said, and also that she'd be more happy with a room next to me in America than a house in Baden-Baden. We surprised each other; I thought I would have to convince her, and she thought I was against emigration. Visas for mothers are easier to get, so we're thinking about a removal next spring. My father was very surprised and rather concerned by this decision, but my stepmother would like to do the same. At the moment we are in Switzerland with our passports, and it is like a revelation to my mother. It is raining hard, but here's the address: Seelisberg (Uri), Hotel Sonnenberg. We will stay here until my departure, which is with the Britannic, from Le Havre on September 16th. I haven't had any difficulties getting in and out, but I shouldn't get overconfident.

My siblings have grown a lot and are very naughty. Some of your education wouldn't be bad for them. My sister is far from being as smart as Gaby; she doesn't speak in such perfect sentences, and can't do arithmatic or write. Instead, in the spirit of the new Reich, she can run faster than all her friends and swims better than her big brother (which isn't saying much). She loves me very much and very tenderly. Klaus also loves me a lot, but basically shows his tenderness by hitting me in the ribs.

For the past two days, meaning since we arrived here, I have been thinking about physics again. With the help of a variational principle, I thought I could prove that the infinities in quantum electrodynamics do not result from inefficient integration but are just part of the basic formulas. Unfortunately, this only works for the lowest eigenvalue of a particle. In the case of a no-hole theory $E = mc^2$ is not the lowest eigenvalue, and in the case of a hole theory you have to take the difference between two eigenvalues (just the "sea", and the "sea" + electrons). In the latter case you normally can't apply a variational principle.

Thank you for the week in Cornwall. I wished I could go there again, and I think this time I would enjoy the quaint old villages and cute ancient churches more. And then there's the surfboarding — which I would probably learn in the end. But I also thank you for talking with me — about physics, and especially about personal things. This has been very good for me, and I wish it weren't so far from Ithaca to Birmingham.

I have lost your address, of course. I still remember number 38, but not the street (Calthorpe?). But now you are with the British donkey anyway. Say hallo for me; can it carry more than 9 stones 3?
Best greetings,

Hans

[48] Rudolf Peierls to Hans Bethe

Birmingham, 24.8.[19]38
(carbon copy)

Lieber Bethe,
Ich komme eben von der Tagung in Cambridge zurück, die sehr interessant war. Insbesondere gab Blackett einen sehr guten Bericht über die Höhenstrahlung, wonach vieles sehr schön in Ordnung zu kommen scheint. Die Versuche von Auger und anderen, wonach die schrägen Strahlen anscheinend rascher als die senkrechten absorbiert werden, (wenn man Höhenabhängigkeit und Winkelabhängigkeit der Koinzidenzien vergleicht) klären sich dadurch auf, dass das Yukon unstabil ist, und daher die Intensität einfach eine Funktion der Zeit ist. Das stimmt alles auch quantitativ. Die Theorie ist von Heisenberg und Euler.

Ich erzählte Oliphant und Cockroft, was Du über das Cyclotron und die Möglichkeit von Fokussierung durch Abweichungen von der Zylindersymmetrie erzählt hast, und sie waren beide sehr interessiert, und wollten gerne wissen, wieweit die verallgemeinerten Rechnungen schon durchgeführt sind, und wie hoch danach die Grenze (im Prinzip) liegen soll. Kannst Du mir das noch gelegentlich schreiben?

Eine weitere Neuigkeit ist, dass Jones, der mein Nachfolger im Mond-Laboratorium war, ein Readership im Imperial College bekommen hat, und Cockroft daher wieder jemand sucht. Ich erwähnte Placzek und das schien ihn sehr zu begeistern, es wurde daher beschlossen an ihn zu telegraphieren, ob er noch frei ist, da ich nicht wusste, wie sich seine Verhandlungen mit Cornell inzwischen entwickelt haben. Ich wollte Dir dies jedenfalls zur Kenntnis geben. Wenn er sich bezüglich Cornell noch

nicht definitiv geäussert hat, und wenn dieser neue Vorschlag seine Stellungnahme zu Cornell irgendwie beeinflusst, so wird er sich wohl gleich mit Dir direkt in Verbindung setzen. Ich werde Dich aber jedenfalls über die weitere Entwicklung in Cambridge auf dem Laufenden halten. Schreib mir doch bitte auch wann, und mit welchem Schiff Du fährst.
Mit herzlichen Grüssen von uns allen, auch an Deine Mutter,
Dein

[Rudi]

Birmingham, 24.8.[19]38
(carbon copy)

Dear Bethe,
I just got back from the conference in Cambridge,[310] which was very interesting. Blackett gave an especially good report on cosmic radiation, according to which many things seem to be coming together. The experiments of Auger and others,[311] which find that inclined rays are absorbed more quickly than vertical rays (e.g., by measuring the rate of coincident rays as a function of height and angle) can be explained by assuming that the yukon is unstable, so that their intensity is just a function of time. This work is also quantitatively correct, and uses the theory of Heisenberg and Euler.[312]

I have told Oliphant and Cockroft[313] about what you found regarding the cyclotron, and the possibility of using deviations from cylindrical symmetry to focus it. Both were very interested by this, and wanted

---

[310]Meeting of the British Association for the Advancement of Science.

[311]Pièrre Auger (1899–1993), considered as the discoverer, in 1938, of giant air showers generated by the interaction of very high energy cosmic rays with the earth's atmosphere. See P. Auger and Thérèse Grivet, 'Analysis of deep rays', *Phys. Mod. Rev.* **11**, 323–54 (1939).

[312]W. Heisenberg and H. Euler, 'Consequences of the Dirac theory of the positron', *Z. Phys.* **98**, 714–32 (1936).

[313]John Douglas Cockroft (1897–1967), studied engineering at Manchester and mathematics at Cambridge, where he worked under Rutherford. During the Second World War, he worked on radar before taking charge of the Canadian Atomic Energy Project at Chalk River in 1944. In 1946 he returned to England to set up the Atomic Energy Research Establishment at Harwell.

to know whether you have done the general calculations and where the boundary is supposed to be (in principle). Could you tell me on some occasion?

Another piece of news is that Jones, who was my successor at the Mond-Laboratory, got a readership at the Imperial College. Cockroft is therefore looking for someone new. I mentioned Placzek, and he seemed to be very excited by the idea. We therefore decided to telegraph him to ask whether he is still free, as I had no idea how his negotiations with Cornell were going. I wanted to tell you, in any case. If he hasn't definitely decided what to do with Cornell, and if this new proposition has any influence on his decision regarding Cornell, I presume that he will immediately contact you. I will keep you informed about new developments in Cambridge in any case. Also, please tell me when you are coming and which ship you will take.
With best greetings from all of us, and also to your mother.
Yours,

[Rudi]

[49] Rudolf Peierls to Hans Bethe

Birmingham, 26.8.[19]38
(carbon copy)

Lieber Hans,
Vielen Dank für Deinen ausführlichen Bericht, nach dem ich mir die Situation sehr gut vorstellen kann. Ich bin in allem ganz Deiner Meinung, aber ich finde eigentlich gut, dass meine Eltern, wenn sie es überhaupt fertigbringen, sich zu entschliessen, zunächst mal nach England kommen wollen, insbesondere wenn, wie Du sagst, die USA Quote für 1939 ausverkauft ist. Weiss der Teufel, was in über einem Jahr noch alles passieren kann, und ob man dann überhaupt noch herauskommt, und ob man überhaupt noch seine Sachen mitnehmen kann u.s.w.

Ich habe daher jetzt direkt geschrieben, um ihnen sehr energisch zuzureden, wenn sie ein Affidavit von mir geschickt haben wollen, so kann ja ihre Absicht auszuwandern, kein sehr grosses Geheimnis mehr sein.

Es freut uns sehr, dass die Pläne für die Uebersiedlung Deiner Mutter anscheinend so gut funktionieren. Hoffentlich klappt es weiter.

Von Placzek habe ich auf ein am Dienstag abgesandtes Telegramm noch keine Antwort. Weisst Du irgendetwas darüber wo er sich befindet, und ob man erwarten sollte, dass ihm ein c/o Bloch, Physics Department, Stanford University adressiertes Telegramm nachgeschickt wird?

Physikalisch bin ich eigentlich ganz davon überzeugt, dass die Unendlichkeiten in der Elektrodynamik sich nicht durch geeignete Integration wegbringen lassen (sicher nicht in höheren Näherungen) und auch nicht durch die Subtraktionsterme, wenn man nicht den Differentialcharakter (bezüglich der Zeit) der Bewegungsgleichungen aufgeben will. Aber ein formaler Beweis dafür wäre doch sehr beruhigend.

Ich wollte Dich übrigens noch etwas bezüglich der Rechnungen über denjenigen Anteil der Kernkräfte fragen, der sich für kleine Abstände wie die Wechselwirkung zweier Dipole benimmt. Wenn ich mich recht erinnere, so sagtest Du, man könnte die Gleichung mit einem solchen Term nicht streng lösen, da die Winkelabhängigkeit der Eigenfunktion kompliziert würde, aber Deine Methode bestand darin, das Variationsintegral mit einer Vergleichsfunktion einfacher Winkelanhängigkeit aber beliebiger Radialabhängigkeit zum Minimum zu machen. Das erscheint mir nun, bei näherer Betrachtung komisch, denn die wirkliche Winkelabhängigkeit ist gar nicht so kompliziert. Es sind nämlich j, m und s noch gute Quantenzahlen. Für $s = 0$ gibt es nur eine Komponente (und da fällt die Anomalie natürlich heraus), für $s = 1$ kann $\cdots$[314] die Werte j $\ldots$[315] l und j haben. Davon kombinieren die beiden ersten miteinander, aber nicht mit der letzten. Es gibt daher im schlimmsten Falle zwei simultane Gleichungen zweiter Ordnung, die sich natürlich auf eine Gleichung vierter Ordnung zurückführen lassen. Diese ist zwar ein bisschen scheusslich, aber sie ist doch mehr oder weniger verdaulich. In jedem Fall kann sie numerisch behandelt werden (und für diesen Zweck ist sie

---

[314] Missing in carbon copy.

[315] Missing in carbon copy.

vermutlich nicht komplizierter als Deine vereinfachte Gleichung) und ausserdem lässt sich eine Menge allgemeines sagen. Sie wird besonders einfach, wenn man die Energie gleich Null setzen darf (was vermutlich für hinreichend kleine r erlaubt ist, falls die Energie endlich ist) und dann kann man die allgemeine Lösung in der Form $f(r) \ldots g(r) \log r$ anschreiben, wo f und g Potenzreihen nach fallender Potenz von r sind, deren Koeffizienten sich aus einer zweigliedrigen Rekursionsformel ergeben.

Möller hatte übrigens ein sehr hübsches Argument, warum man die Singularität dieses Terms bei kleinen Abständen nicht so ernst nehmen soll: Die Störungsrechnung, mit deren Hilfe man diese Terme ableitet, ist ja statisch, d.h. die schweren Teilchen werden als fest angenommen, und Retardierung (d.h. die endliche Geschwindigkeit der Yukonen im virtuellen Zustand) wird vernachlässigt. Das scheint vernünftig, wenn die Geschwindigkeit des schweren Teilchens klein ist. Aber man muss nicht vergessen, dass diese Kräfte eine Präzession der Spins der schweren Teilchen zur Folge haben, und dass die statische Rechnung nicht mehr erlaubt ist, sobald die Periode dieser Präzession kürzer wird als die Zeit, die das Yukon braucht, um von einem Teilchen zum anderen zu gelangen. Das ist aber gerade bei sehr kleinen Abständen der Fall. Er hat aber noch nicht ausgerechnet, was passiert, wenn man diese Retardierung mitnimmt.

Mit herzlichen Grüssen von uns allen
Dein

Rudi

Birmingham, 26.8.[19]38

Dear Hans,
Thank you for your detailed report; I can now imagine the situation very well. I agree on all points, but I think it would be a good idea for my parents to come to England first if they are able to decide at all. This is especially true if, as you say, the USA quota for 1939 is sold out. Who knows what might happen a year or more from now? Will you even be able to get out at all, take your belongings with you, etc.?

I therefore sent them a letter right away, in which I really tried to convince them to leave. If they want to get an affidavit from me, their plan to emigrate can't be a real secret anymore.

We are very happy that plans for the removal of your mother are going well. I hope this all works out fine.

I haven't received an answer from Placzek to a telegram sent on Tuesday. Do you know anything about where he is, and whether we can expect that a telegram addressed to "c/o Bloch, Physics Department, Stanford University" would be forwarded to him?

As for physics, I am quite convinced that the infinities in electrodynamics can't be solved just by an appropriate integration (surely not in higher-order approximations) or by subtracting the right values, assuming you don't want to give up the differential character (with respect to time) of the equations of motion. But a formal proof would be very comforting.

I also wanted to ask you something regarding your calculations of terms in the nuclear power series, which at small distances seem to behave like the interaction between two dipoles. If I remember correctly, you said that you couldn't solve the equation strictly with only one such value, as the angular dependence of the eigenfunction would be too complicated. But your method brought in a variational integral, using a simple function of the angle with an arbitrary radial dependence. On closer examination this seems a bit strange to me; the real angular dependence is not so complicated, because j, m, and s are all good quantum numbers. For $s = 0$ there is just one component (and there surely won't be an anomaly then), while for $s = 1$ you can have the values $j \cdots$[316] l and j. The first two components can combine with each other, but not with the third. In the worst case there are two corresponding quadratic equations, which can surely be derived from a fourth-order equation. This is a bit awful, of course, but it is still more or less tolerable. In any case, you can treat it numerically (and for this purpose it is presumably no more complicated than your simplified equation). And you can also learn a lot of general principles. It is

[316] Missing in carbon copy.

especially easy if you are allowed to set the energy to zero (which is presumably allowed for sufficiently small r, if the energy is finite); then you can write down a general solution in the form $f(r) \cdots$[317] $g(r) \log r$, where f and g are power series describing a potential that falls off with increasing r. Their coefficients can be determined using a two-part recursion formula.

Möller, by the way, has a very nice argument explaining why we shouldn't take the singularity at small distances in this term too seriously. The perturbation calculation which you can use to differentiate these values is static; i.e., the heavy particles are taken as solid, so their retardation (the finite velocity of virtual yukons, in other words) can be neglected. This seems to be reasonable if the velocity of the heavy particle is small. But you shouldn't forget that these forces also lead to a precession in the spins of heavy particles, and that the static calculation is not allowed if the period of this precession is smaller than the time the yukon needs to get from one particle to another. But this does happen, especially in case of small distances. He hasn't yet calculated what happens if you keep the retardation. With best wishes from all of us,
Yours,

Rudi

[50] Hans Bethe to Rudolf Peierls

Seelisberg, 27.8.[19]38

Lieber Peierls,
das ist ja alles sehr sehr interessant.
Vor allem das mit Placzek. Ich bin im Grunde sehr froh und beruhigt darüber. Natürlich hätte ich ihn, von meinem rein persönlichen Standpunkt aus, sehr gern in Cornell gehabt, aber —

[317] Missing in carbon copy.

Nr.1 er wäre in Cornell eigentlich am ganz falschen Platz, denn ein Theoretiker ist ja durchaus genug für ein Institut. Cambridge dagegen wäre ideal in dieser Beziehung und braucht einen wirklichen Theoretiker ja dringend. Nr.2 brauche ich eigentlich einen Sklaven, und Placzek ebenfalls, und es wären dann in Cornell keine weiteren Stellen für Sklaven vorhanden. Nr.3 wäre es wahrscheinlich schwierig mit den anderen Leuten in unserem Institut (verbesserte Auflage des Krachs Gibbs-Richtmyer und man sollte seine politische Situation nicht verschlechtern. Nr.4 würde Placzek sich wahrscheinlich in Europa wohler fühlen als in einem kleinen Städtchen in Amerika. Nr.5 ist zwar das Gehalt meines Wissens dasselbe ($ 2000 und £400), aber in Cambridge kann man eher was extra verdienen, und ausserdem sind dort die Gehälter der anderen auch nicht höher. Nr.6 war Cornell eigentlich nur als Übergang zu einer besseren Stelle in Amerika gedacht, und ob die Amerikaner je genug Verstand haben würden, über Placzeks "unamerikanische" Eigenschaften hinwegzusehen, ist fraglich. Übrigens Mond ist doch für 5 Jahre?

Also, alles in allem fände ich es eigentlich sehr viel besser, wenn P[laczek] nach Cambridge ginge. Das Einzige ist, ob man durch Einfürung eines weiteren Ausländers Ewalds Chancen in Cambridge noch verschlechtert? Ich glaube aber eigentlich Mond und Cavendish, und vor allem Kerne und Kristalle, sind linear unabhängig. Hat eigentlich Bragg beim Mond etwas zu sagen?

Ich schreibe Dir all diese vielen Gründe, die Du alle selber weisst, nicht nur zwecks Verminderung des Verdienstes der Post. Sondern es wäre vielleicht gut, sie Placzek selbst zu schreiben. Nun wäre es nicht sehr nett, wenn ich das täte, sondern sehr viel netter, wenn z.B. Du ihm die Vorzüge Cambridges gegenüber Cornell schilderst. Wäre das möglich? Ich glaube, es würde seine Entscheidung beeinflussen.

Vor etwa einer Woche bekam ich ein Telegramm von Placzek, er käme gerne nach Cornell, aber erst im 2. Semester. Im 1. Semester wollte er auf jeden Fall nach Europa. Ich telegraphierte zurück, das wäre recht, und telegraphierte in gleichem Sinne an Gibbs. Antwort erfolgte darauf noch nicht. Natürlich kann Cornell jederzeit rückgängig gemacht werden, besonders, da ein Anwärter da ist (Rose), der jederzeit auf Abruf zu kommen bereit ist. Rose würde ich übrigens nur für ein weiteres Jahr nehmen, und dann entweder Critchfield (von Teller) oder Schwinger (von Rabi), falls Placzek nicht kommt. Infolgedessen ist es möglich, Cornell sogar für Placzek zu reservieren, mindestens bis Februar, falls aus Cambridge nichts wird. Wie weit ist übrigens Cambridge gediehen? Hängt es hauptsächlich von Placzeks Entscheidung ab, oder muss noch jemand gefragt werden?

Bezüglich der anderen Dinge: Die Rechnungen über das nicht-cylindersymmetrische Cyclotron sind bisher nur bis zur der Näherung ausgeführt, in der der Focussierungseffect gerade auftritt (2. Näh[erung]), man kann also die Gültigkeitsgrenze nicht abschätzen. Es geht aber bestimmt solange der Relativitätseffekt noch als "Störung" betrachtet werden kann — ich würde sagen mindestens bis $\frac{1}{10}Mc^2$.

Die Höhenstrahlung ist sehr interessant. Wenn der Zerfall eine reine Frage der Zeit ist, anders als die ursprüngliche Blackett'sche Annahme der 200MV, glaube ich ihm gern. Nur schien es mir, dass die nötigen Zeiten ziemlich gross sind und z.B. nicht mit der extrapolierten β-Theorie stimmen (Rechnungen von Heitler und Nordheim). Ausserdem muss man dann erklären, wie die Barytronen in der Atmosphäre entstehen — dazu gibt's nicht nur keinen Mechanismus, sondern es führt (nach Nordheim) zu einigen direkten Widersprüchen, die ich allerdings vergessen habe. Aber wenn es so ist, dann muss es ja wohl so sein.
Dein

Hans

Seelisberg, 27.8.[19]38

Dear Peierls,
This is all very, very interesting, especially your news of Placzek.[318] I am very happy and relieved about this. Personally, I would of course have liked to have him at Cornell, but — 1. Cornell would have been totally wrong for him as one theorist is really enough for any institute. In this regard, Cambridge would be perfect as it urgently needs a real theorist. 2. I actually need a slave and Placzek does too; in Cornell,

[318]Reply to letter [48].

however, there wouldn't be any more jobs for slaves. 3. There would be difficulties with the other people at our institute (an improved version of the Gibbs-Richtmeyer conflict[319]), and one shouldn't make his political situation even worse. 4. Placzek would probably feel more comfortable in Europe than in a small town in America. 5. As far as I know the salary is the same ($ 2000 vs. £400), but in Cambridge it's easier to earn some extra money and the salaries of other jobs are no higher. 6. Cornell was just meant as a stepping-stone to a better job in America, but it is doubtful that the Americans would be reasonable enough to accept Placzek's "un-American" qualities. By the way, is the Mond position for five years?

So all in all, I would prefer it if P[laczek] went to Cambridge. The only thing is, would Ewald's chances at Cambridge be lowered because of another foreigner? But actually I think that the Mond and Cavendish, and especially nuclei and crystals, are linearly independent. Does Bragg have a say in things, there at Mond?

I'm not just telling you all these reasons you know already to reduce the post office's profit. It would probably be a good idea to tell them to Placzek himself. It wouldn't be very nice if I did this, but it would be much nicer if, for example, you explained to him the advantages of Cambridge as compared with Cornell. Would that be possible? I think this could influence his decision.

About a week ago, I received a telegram from Placzek in which he told me that he would like to come to Cornell but just for the second semester. In the first semester, he would definitely like to stay in Europe. I telegraphed him that this would be acceptable, and telegraphed the same to Gibbs. I haven't received an answer yet. Cornell can be cancelled at any time, of course, especially as we have an applicant (Rose) who is ready to come whenever we want. By the way, I would take Rose for one more year, then either Critchfield[320] (from Teller) or

[319] R. Clifton Gibbs and Floyd K. Richtmeyer, both professors at Cornell, were engaged in a continual feud, and as Bethe had been recruited by the Head of Department, Gibbs, his standing with Richtmeyer was on shaky grounds.

[320] Charles Louis Critchfield (1910–1994) received his Ph.D. from George Washington University, where he studied with Teller and Gamow. He joined the Manhattan

Schwinger[321] (from Rabi[322]), if Placzek doesn't come. It is therefore possible to reserve Cornell for Placzek at least until February, if Cambridge doesn't come through. How are things going in Cambridge, by the way? Does it only depend on Placzek's decision, or does anyone else need to be asked?

Now on to other things: my calculations for a non-cylindrical, symmetric cyclotron have only been done up (to the second order of approximation), which is just where the focussing effect appears, so you can't estimate the validity of the boundary. But it is surely acceptable so long as the effect of relativity can be regarded as a "perturbation" — I would say at least up to $\frac{1}{10}Mc^2$.

The result on cosmic radiation is very interesting. If the decay is just a function of time, as opposed to Blackett's first supposition of 200 MV, I will readily agree to it. Only it appeared to me that the necessary times are rather long, and do not (for example) correspond to the extrapolated β-theory calculations of Heitler and Nordheim.[323] Then you also have to explain how barytrons are made in the atmosphere — as for that there is not only no mechanism, but this also leads (according to Nordheim) to some direct contradictions which I have forgotten. But if that's how it is, that's how it is.

Yours,

Hans

Project in 1942/3. After a brief post-war spell at George Washington University he became professor at Minnesota. Julian Seymor Schwinger (1918–1994), received his doctorate from Columbia University in 1939, worked at Berkeley (1939–41) before moving to Purdue (1941–1945) and Harvard (1945–72).

[321] Julian Seymour Schwinger (1918–1994), obtained his Ph.D. from Columbia University in 1939. After research at Berkeley and Purdue, and war work at M.I.T. he accepted a position at Harvard in 1945, in 1972 he left Harvard for a postion at the University of California, Los Angeles.

[322] Isidor Isaac Rabi (1898–1988), studied at Cornell and Columbia University where he obtained his Ph.D. in 1927. After research at Copenhagen, Leipzig, Hamburg and Zurich, he joined the faculty at Columbia where he remained for the rest of his life.

[323] L.W. Nordheim and G. Nordheim, 'On the Production of Heavy Electrons', *Phys. Rev.* **54**, 254–65 (1938); W. Heitler, 'Analysis of Cosmic Rays', *Proc. Roy. Soc.* **A161**, 261 (1937); L.W. Nordheim, 'On the Absorption of Cosmic-Ray Electrons in the Atmosphere' *Phys. Rev.* **51**, 1110 (1937).

[51] Rudolf Peierls to Hans Bethe

Birmingham, 29.8.1938

Lieber Hans,
Vielen Dank für Deinen Brief. Mittlerweile habe ich auch von Placzek Antwort bekommen. Er ist im Prinzip sehr interessiert, sagt aber dass er den spring term in Cornell verbringen muss, und dass er sich ausserdem nicht sofort entscheiden kann, weil die Verhandlungen mit Paris schweben und bittet daher um eine Bedenkzeit von einigen Wochen. Die Mitteilungen habe ich natürlich sofort an Cockroft weitergegeben, und ich hoffe eigentlich, dass Cockroft auf ihn warten wird. Ich habe den Eindruck, dass in Cambridge alles in Ordnung ist. Bragg hat zwar im Prinzip etwas zu sagen, aber ich glaube, er wird sich in den reinen Cavendish-Dingen schon unbeliebt genug machen, und wird daher in the Mond-Geschichten nicht unnötig hereinreden. Ausserdem war es besonders günstig, dass Bohr in Cambridge war, denn Cockroft hetzte gleich Bohr auf Bragg, und das hatte wohl einen guten Effekt. Natürlich dauert es längere Zeit, bis die Sache offiziell entschieden ist, aber darüber mache ich mir keine sehr grossen Sorgen. Wenn Du ihm Cornell noch einige Zeit warm halten könntest, so wäre das natürlich äusserst nobel. Die Stelle ist offiziell jeweils für ein Jahr, aber das heisst nichts. Das Geld für die Stelle war von Anfang an für 5 Jahre gegeben worden, wovon drei schon um sind, und wie es mit einer Erneuerung des Grant steht, weiss ich nicht. Aber ich vermute, es wird verlängert werden, denn ohne Theoretiker (und ohne Allen, der eine parallele Stelle hat) kann das Mond Lab. kaum existieren. Jedenfalls scheint es mir sicher, dass, wenn Placzek erst mal hier ist, und wenn er sich nicht zu unbeliebt macht (was in England wohl viel schwieriger ist als in Amerika und so schlechte Manieren wie der Moritz hat er ja schliesslich nicht) so ist er bei dem hier herrschenden Mangel an Theoretikern auf die Dauer untergebracht.

Insbesondere freue ich mich, dass Du es “uns Engländern” nicht übel nimmst, dass wir versuchen, Dir den Placzek abzujagen. Ich glaube, es ist im Augenblick ‘nicht’ nötig, Placzek Deine 5 Punkte zu übermitteln, denn nach dem Telegramm scheint es mir dass er an sich selbst schon eingesehen hat, dass Cambridge für ihn besser wäre, und nur sein bereits gegebenes Versprechen halten möchte, im Spring Term in Cornell zu sein. Dementsprechend werde ich Placzek andeuten, ich hätte den Eindruck, Du würdest es ihm nicht übelnehmen, wenn er das rückgängig macht, falls das seine Chancen in Cambridge verbessert. Übrigens ist ja gerade das etwas, was Du ihm auch sehr gut selbst sagen kannst.

Ich werde auch Cockroft — in sehr viel vorsichtiger Formulierung — andeuten, dass es nicht ausgeschlossen sei, dass Placzek doch den ganzen Winter nach Cambridge kommt, falls er darauf Wert legt, und dass jedenfalls Cockroft dies nicht als Stein des Anstosses betrachten soll.

Ich glaube nicht, das das Ganze etwas mit Ewalds Chancen zu tun hat. Der einzige Mensch, der in Cambridge etwas für Ewald tun kann, ist Bragg. Was Bragg will, weiss kein Mensch. Aber entweder will er Ewald gerne haben, und dann wird er ihn bekommen, oder aber es ist ihm mehr oder weniger schnuppe, und dann kann man, jedenfalls in Cambridge, sowieso nichts machen. Diese Dinge sind ja nicht additiv.

Ich sollte eigentlich nach der British Ass[ociation] mit Bohr nach Kopenhagen fahren, dann sollte ich am 1.9. fahren, jetzt soll es der 7.9. werden. Du würdest sagen, ich fahre 1939. Jedenfalls schreibe vorläufig lieber hierher.

Mit freundlichen Grüssen
Dein

Rudi

Birmingham, 29.8.1938

Dear Hans,

Thank you for your letter. I have now received Placzek's answer. In principle he is very interested, but he told me that he had to spend the spring term in Cornell and that he couldn't decide right now as negotiations with Paris were still going on. He therefore asks for a few weeks to think it over. I immediately forwarded this message to Cockroft, of course, and I actually hope that Cockroft will wait for him. It is my impression that everything is all right in Cambridge. In principle Bragg has something to say about it, but I think he is sufficiently unpopular in the Cavendish itself and therefore won't try to interfere with matters at the Mond unnecessarily. It was also especially fortunate that Bohr was in Cambridge, as Cockroft immediately sent Bohr to speak with Bragg, which seems to have had a good effect. It will take a long time before the whole thing is officially decided, of course, but I'm not too worried about that. If you could keep him on the back burner at Cornell for a while, it would certainly be very generous. Officially the job is for one year, but that means nothing. In the beginning there was grant money for five years; three are already passed, and I don't know whether the grant will be renewed. But I'm guessing that it will be renewed, as without a theorist (and without Allen,[324] who has the second job) the Mond laboratory could hardly exist. Anyway, it seems certain that once Placzek comes here, provided that he doesn't become too unpopular (which seems to be much more difficult in England than in America, and his manners aren't as bad as Moritz's), he will be able to stay permanently, because Cambridge lacks a theorist.

I am especially pleased that you don't blame "us Englishmen" for trying to snatch Placzek. I think at the moment it is <u>not</u> necessary to forward your five points to Placzek, as his telegram seems to show that he has understood all by himself that Cambridge would be better for him, and that he just wants to keep his promise to Cornell for the spring term. I will therefore carefully tell Placzek that I have the impression you wouldn't blame him for cancelling Cornell, if it improved his situation at Cambridge. This is actually something you could tell him too, of course.

I will also tell Cockroft — with a much more cautious formulation — that it isn't impossible that Placzek would come to Cambridge for the whole winter if he really wanted him to do so, and anyway that he shouldn't take offence. I don't think that this whole situation has

[324] John F. Allen (1908–), a graduate from Toronto had come to Cambridge in 1935 to work with Kapitza. Instead, he continued Kapitza's work in Cambridge after the latter's detention in Russia in 1934. In January 1938, *Nature* contained two separate articles by Kapitza and by Allen and his student Miserer, which mark the discovery of superfluidity.

anything to do with Ewald's chances. The only one who can do anything for him in Cambridge is Bragg. What Bragg wants, nobody knows. But either he wants Ewald to come to Cambridge and then he will get him, or he more or less doesn't care and then there is nothing one can do about it, at least in Cambridge. These things are not additive. After the British Association I was supposed to go to Copenhagen with Bohr; then I was supposed to travel on September 1st, but now it has been postponed to the 7th. As you would say, I will go in 1939. It's better anyway if you send your letters to this address for the time being.
Sincerely yours,

Rudi

[52] Hans Bethe to Rudolf Peierls

Seelisberg, 6.9.[19]38

Lieber Rudi,
Programmmässig solltest Du zwar morgen nach Kopenhagen fahren, aber Du weisst ja selbst, dass das erst 1939 geschieht. Also:
Zunächst vielen Dank für Deine Informationen über Placzek. Ich habe inzwischen einen Brief von ihm, vor Erhalt Deines Angebots, in dem er zwischen Cornell und Paris im Zweifel war, falls Paris was wird und beides je 1/2 Jahr anschauen wollte. Inzwischen habe ich ihm geschrieben, was nach seiner positiven Einstellung zu Cambridge sehr einfach war. Ich schrieb, ich hätte von Dir gehört usw., und fände das sehr schön, ob er doch ein Semester nach Cornell kommen würde, aber er sollte das bleiben lassen, falls Cambridge irgendwie davon abhinge. Ausserdem Andeutung meiner 5 Punkte ··· übrigens auf jeden Fall Ende September auf ein paar Tage nach Cornell kommen, um alles zu besprechen. Meinem Direktor Gibbs habe ich geschrieben, was für eine herrliche Aufgabe auf den Placzek in Cambridge wartete; dass wir auf ihn verzichten müssten, falls Cambridge ihn verlangt, und dass wir uns andererseits die Finger lecken müssten, falls er nach einem Jahr aus irgendeinem Grunde doch zu uns käme. Ich glaube, das geht alles in Ordnung, und Cornell wird warmgehalten, bis Cambridge offiziell entschieden ist (was ja wohl vor 1939 zu erwarten ist?)

Ewald schrieb mir übrigens, dass er wieder 2 Hoffnungen in Amerika hat, von denen man aber nicht sprechen soll. Bezüglich der Physik: Es war natürlich sehr dumm von mir, nicht zu sehen, dass für $s = 1$ nur höchstens 2 Werte von l bei gegebenem j in Frage kommen. Dann kann man natürlich das Problem der wechselwirkenden Dipole viel hübscher exakt behandeln. Willst Du darüber noch was machen?

Möllers Agument mit der Retardierung ist natürlich hübsch. Ich glaube, es führt auch zu einem vernünftigem Abschneiden, nämlich etwa zu demselben wie wenn man dort abschneidet wo die 2. Näherung der Störungsrechnung gleich der ersten wird, d[as] i[st] r beinahe = Reichweite der Kernkräfte.

Nebenbei, kannst Du für den Spiegelberg'schen Namen Barytron wirken? Yukon geht ja wohl doch nicht, und Heavy Electron geht mir jedesmal durch Mark und Bein, bei einem Teilchen mit Bosestatistik, das ausserdem höchstwahrscheinlich auch noch eine neutrale Modifikation hat.

Mein Beweis, dass die Divergenzen der Quantenelektrodynamik nicht von der Integration herkommen (daran, dass es so ist, zweifele ich ebensowenig wie Du, aber es gibt immer noch Gelehrte, die es so rum versuchen), ist nicht weiter gediehen. Ich habe nur gelernt, dass die starken Divergenzen der gewöhnlichen (Nicht-Löcher-)Theorie eigentlich sehr wenig heissen, dass die logarithmische Divergenz der Löchertheorie sehr harmlos ist und das von mir geplante Variationsverfahren in beiden Fällen nicht anwendbar ist. Hauptsächlich habe ich mich aber mit der Arbeit für die Sommerfeld-Festschrift beschäftigt und die hat mir grossen Spass gemacht, denn sie ist ganz unbeabsichtigterweise sehr Sommerfeldisch geworden. Es kommt eine komplexe Integration und sogar eine Hankelfunktion mit imaginärem Argument drin vor. Das ganze ist eine Methode zur Behandlung sehr grosser Störungen — oder m[it] a[nderen] W[orten] zur Auflösung beliebiger Matrizen, deren Elemente nur langsam mit der Zeilennummer veränderlich sein müssen.

Wir werden bis 14. hier sein. Dass meine Britannic am 16. geht, habe ich wohl schon geschrieben. Hast Du von Deinen Eltern was gehört?

Schöne Grüsse an die ganze Familie

Hans

Seelisberg, 6.9.[19]38

Dear Rudi,
You are planning to go to Copenhagen tomorrow, but as you know this will happen in 1939.[325] So:

First, thank you for your information about Placzek. I have now received a letter from him, written before he received your offer, when he didn't know about either Cornell or Paris. If he got Paris as well as Cornell, he was hoping to try each one for half a year. I have now sent him a letter, which was easy because of his positive opinion of Cambridge. I wrote that I had heard from you etc., and that I would like him to come to Cornell for half a year but that he should cancel it if Cambridge appeared to rely on his coming earlier in any way. I also hinted at my 5 reasons ...[326] [and asked him] to come to Cornell at the end of September for a couple of days in any case, to talk about all this. I also sent a letter to Gibbs, the head of my department, to tell him about the great challenges waiting for Placzek in Cambridge, and that we would have to do without him if Cambridge asked for him, and that we should be grateful to get him if for any reason he decided to come back here a year later. I think that everything is all right here, and that Cornell is still a possibility for Placzek until Cambridge decides officially (which presumably can be expected before 1939?)

By the way, Ewald wrote to tell me that he had two other opportunities in America again, which we shouldn't talk about. Concerning physics: it was very silly of me not to see that for $s = 1$ there are at most 2 values of l if j is given. You can solve the problem of interacting dipoles in a much nicer way. Are you still planning to do something with that?

Möller's argument about the retardation is certainly very nice. I think this is a reasonable way of introducing the approximation. It's just as if you stopped with the second-order terms in the perturbation

[325] Niels Bohr was notorious for making last-minute requests for postponements of visits, and Peierls regularly had to re-arrange his Copenhagen visits in the mid- and late 1930s.

[326] See letter [50].

calculation, which is equal to the first. That means that r almost = range of the nuclear powers.

By the way, could you do something to promote Spielberg's name barytron?[327] Yukon doesn't work after all. 'Heavy electron' now sets my teeth on edge, especially for a particle obeying Bose statistics which is presumably very likely to have a neutral modification.

I haven't worked on my proof that the divergences of quantum electrodynamics do not result from the manner of integration (that this is the case I don't doubt at all, just like you, but there are still scientists who are trying to show this). I have just learned that the strong divergences in normal (no-hole) theories don't mean much, and that the logarithmic divergence in hole theory isn't very important, and that the variational method I wanted to apply can't be used in either case. Mainly, I have been working on Sommerfeld's festschrift, which was good fun, as it turned out, unintentionally Sommerfeld-like than was planned at the beginning. We have a complex integration, and even a Hankel function with an imaginary argument. This is a method of dealing with big perturbations — or in other words, of resolving any matrix whose elements change slowly with the line number.

We'll be here until the 14th. I think I told you that my ship the Britannic leaves on the 16th. Have you heard anything from your parents?
Best wishes to the whole family,

Hans

[327] Physicists disagreed about the naming of the particle. Initially it was called x particle; Niels Bohr called it the Yukon in honour of Yukon who first postulated its existence, and many American scientists, such as Bethe, preferred the name 'barytron' from the Greek meaning heavy electron.

[53] Rudolf Peierls to Hans Bethe

Birmingham, 5.11.1938
(carbon copy)

Lieber Hans,
Dies ist im Wesentlichen über Arbeitsprogramme.

Erstens. Du erzähltest mir in Cornwall, dass Deine Ansicht über den β-Zerfall nunmehr die wäre, dass wahrscheinlich der Fermische Ansatz der richtige ist, und dass, im Falle von erlaubten Übergängen das Auftreten von anscheinenden K.U.-Spektren immer durch mehrere Endzustände (und Strahlen) zu erklären ist. Diese Idee hatte ich schon vor längerer Zeit einmal mit Hoyle besprochen, aber wir hatten uns dann von den Experimentatoren einschüchtern lassen, die behaupteten, es gäbe Fälle, in denen es bestimmt keine β-Strahlen gibt. Hoyle hat sich weiter für diese Frage interessiert, und es gibt jetzt neue Experimente in Liverpool, die sehr stark für diese Erklärung sprechen. Ich möchte nun gerne wissen, ob Deine Bemerkung damals nur ein Stimmungsausdruck war, oder ob Du diese Frage ernsthaft untersuchst. In letzterem Falle würde ich Hoyle davon abraten, sich damit weiter zu beschäftigen, da Du das natürlich viel besser und schneller machst. Aber falls Du nicht die Absicht hast, es zu tun, so wäre es eigentlich sehr gut, wenn er das mal anständig diskutieren würde. (Eine Schwierigkeit liegt möglicherweise in folgendem: Wenn es auf der ersten verbotenen Sargent-Kurve viele Elemente gibt, die nachweislich keine β-Strahlen haben, und die alle ein U.K. Spektrum zeigen, so müsste dies auf die Verbotenheit zurückgehen. Nun folgt aber aus der Hoyleschen Arbeit, dass, falls das Elementargesetz das Fermische ist, zwar gewisse verbotene Übergänge ein U.K. Spektrum bekommen sollen, andere aber nicht. Es wird dann (da man ja über die Kernspins im allgemeinen zu wenig weiss) eine Frage der Statistik, ob die Erklärung plausibel ist.

Zweitens: (Ich weiss nicht, ob ich das nicht schon mal fragte) Wie steht es mit der Frage über die Elektronenpolarisation, die wir in Sandymouth besprachen. Ich würde an sich dieses Problem gerne Blackman vorschlagen, der darüber sicher sehr glücklich ist, da er bei G.P. Thomson sitzt und sich dort von Amtes wegen mit Elektronenbeugung befassen soll. Ich glaube, mit ein wenig Aufsicht, wird er das schon fertig bringen. Aber hast Du noch niemand für diese Problem in petto?

Auf diese beiden Fragen würde ich gerne bald eine Antwort haben. Hier ist inzwischen der Frieden ausgebrochen, wie Liebermann es einmal so schön in einem anderen Zusammenhang ausdrückte. "Soviel kann man gar nicht essen, wie man kotzen möchte!"

Placzek geht nun doch nach Paris (und dann nach Cornell) und Cockroft muss sich jemand anderen suchen. Von meiner Arbeit mit Bohr hoffe ich sehr, dass sie im Jahre 1939 erscheint, aber ich fürchte, es wird mindestens 1940.

Wie steht es mit der Übersiedelung Deiner Mutter? Meine Eltern haben natürlich so lange gewartet, bis es auch nach England fast unmöglich ist ein Visum zu bekommen, d.h. man bekommt es, aber wegen der Verstopfung des Home Office mit Anträgen von "modernen" Auswanderern aus der Tschechoslovakei und Italien u.s.w. dauert es mindestens drei Monate, bis irgendetwas passiert. Inzwischen scheinen aber die Auswanderungsbedingungen aus Deutschland immer schärfer zu werden. Wir leben in einer amüsanten Welt.
In diesem Sinne herzliche Grüsse
Dein

[Rudi]

Birmingham, 5.11.1938
(carbon copy)

Dear Hans,
This letter is basically about my plans concerning what to work on. First of all, in Cornwall you told me concerning the β-decay that you now believe Fermi's method was the right one,[328] and that in the case of permitted transitions the apparent occurrence of K.U. spectra can always be explained by supposing several final states (and rays).[329] I

[328] E. Fermi, 'Versuch einer Theorie der β-Strahlen. I', *Z. Phys.* **88**, 161–177 (1934).
[329] E. Konopinski and G.E. Uhlenbeck, 'On the Fermi Theory of β-Radioactivity', *Phys. Rev.* **48**, 7–12 (1935).

discussed this with Hoyle some time ago, but at the time the experimenters intimidated us a little, and claimed that in some cases there were definitely no β-rays. Hoyle never stopped being interested in this question, and we have new experiments at Liverpool which strongly support this explanation. I would really like to know now whether this comment was just an expression of your mood at the time, or whether you are doing serious research on this question. If this is the case, then I will recommend to Hoyle that he not deal with it anymore as you'll do it much better and faster. But if you're not planning to deal with it, it would be a good idea if he discussed it properly. One difficulty might be this: if on the first forbidden Sargent curve there are a lot of elements which demonstrably have no β-rays, and which all show a U.K. spectrum, then this has to be related to the forbiddance. If you apply Fermi's first law, then according to Hoyle's paper[330] some forbidden transitions are supposed to have U.K. spectra while others will not. Then whether or not it is plausible becomes a question of statistics (as we don't know a lot about nuclear spins in general).

Second: (I don't know whether I have already asked you this) What about the question of electron polarisation that we discussed in Sandymouth? I would like to suggest this problem to Blackman,[331] who will surely be happy about it as he is working with G.P. Thomson[332] and (officially) is supposed to deal with electron diffraction. I think that with some supervision, he will succeed. But do you have no one in mind for this problem?[333]

[330] F. Hoyle, 'β-transitions in a Coulomb Field', *Proc. Roy. Soc.* **A166**, 249–269 (1938).

[331] M. Blackman was working with G.P. Thomson on electron diffraction. See G.P. Thomson and M. Blackman, 'Theory of the width of rings formed by electron diffraction', *Proc. Phys. Soc.* **51**, 425–31 (1939).

[332] George Paget Thomson (1892–1975) studied physics and mathematics at Cambridge. He became Professor of Natural Philosophy at the University of Aberdeen where he stayed for 8 years. In 1930 he took up a professorship at Imperial College London. In 1952 he became Master of Corpus Christi Cambridge.

[333] Rudolf Peierls, Hans Bethe and Fred Hoyle published the results in 1939. See letter [54], note 335.

Could you answer these two questions quickly? We have had the outbreak of peace here. As Liebermann once said in another context: "You can't eat as much as you want to puke!"

Placzek has finally decided to go to Paris (and then to Cornell), and Cockroft now has to search for someone else. I am hoping that the paper I wrote with Bohr will be published in 1939, but now I'm afraid it will be 1940 at the earliest.[334]

How is your mother's removal going? Of course, my parents have been waiting so long that it is nearly impossible to get a visa for England; this means that you can still get one, but the Home Office is so busy with the applications of "modern" emigrants from Czechoslovakia, Italy, etc. that it takes at least three months before anything happens. And Germany's restrictions on emigration seem to be getting stricter and stricter. We're living in an amusing world.
In this spirit sincerely
Yours,

[Rudi]

[54] Rudolf Peierls to Hans Bethe

Birmingham, 10.12.1938
(carbon copy)

Lieber Hans,
Ich kann heute ein Kompliment erwidern, das Du mir vor einiger Zeit gemacht hast, nämlich Dir mitteilen, dass Du soeben eine Note an NATURE gemeinsam mit Hoyle und mir verfasst hast. Allerdings ist die Situation insofern durchaus anders, als damals die Frage war, ob ich überhaupt etwas dazu beigetragen hatte, während nun umgekehrt die Frage ist, ob Du das nicht alles schon selber weisst, und besser allein publiziertst. Hoyle und ich hatten die Frage schon einmal früher

[334] Peierls, Placzek and Bohr published a joint paper in 1939, R. Peierls, G. Placzek and Niels Bohr, 'Nuclear reactions in the continuous energy region', *Nature* **144**, 200–201 (1939).

diskutiert, aber wir hatten den Experimentatoren geglaubt, die die Existenz von β-Strahlen leugneten, und ohne Deinen Optimismus in Cornell hätten wir uns die Sache nie richtig überlegt.

Über die Richtigkeit des Standpunktes kann wohl kein Zweifel mehr sein, und da die Experimentatoren in Cambridge in heller Begeisterung sind, und es gern bald schwarz auf weiss haben wollen, so soll man ihnen den Gefallen wohl tun, und es publizieren. Ich weiss aber nicht, ob nicht die richtigere Form die wäre, dass Du selbst eine Note darüber schreibst oder einen Brief an Dich selbst (als Editor der Phys. Rev.) Dass ich bereits ein Manuskript beilege, soll Deiner Entscheidung in dieser Beziehung in keiner Weise vorweggreifen, sondern nur zur Beschleunigung beitragen. Bitte, teile mir Doch in amerikanischem Tempo telegraphisch mit, ob ja oder nein, und wenn nein, ob Du es selbst schreibst. Wenn Du kleinere Änderungen in der Form wünschst, so kannst Du sie brieflich mitteilen, da wir sie dann noch rechtzeitig für die Korrektur bekommen. Ich schlage daher folgenden Code bei Deiner Antwort vor:

- Accepted = Ihr könnt das Manuskript in der vorliegenden Form absenden, eventuelle kleine Änderungen brieflich.
- Amendments = Im Prinzip einverstanden, aber mir gefällt die Formulierung nicht, ich schicke postwendend ein abgeändertes Manuskript.
- Publishing = Ich schreibe selbst eine Note über diese Frage.

Neulich war Bohr auf ein paar Tage hier, und es war sehr interessant, obwohl unsere gemeinsame Arbeit in dieser Zeit nicht wesentlich fortgeschritten ist. In der letzten Zeit habe ich mich ein bisschen mit dem Beweis für die Vollständigkeit des von Kapur und mir verwendeten Funktionensystems beschäftigt, bei dem es mathematisch recht lustige Dinge gibt. Es stellt sich nämlich heraus, dass in den Ausnahmefällen, in denen das Quadratintegral einer Eigenfunktion verschwindet, das System in der Tat nicht vollständig ist, aber trotzdem gelten noch alle physikalisch interessanten Formeln, die mit Hilfe der Vollständigkeit abgeleitet sind, wie übrigens aus Stetigkeitsgründen sowieso plausibel ist. Es stellt sich heraus, dass wenn so ein Quadratintegral verschwindet, gerade zwei Eigenfunktionen identisch werden, was bei reeller Randbedingung nicht passieren kann. In dem Augenblick, wo sie gleich sind, hat man natürlich eine Funktion zu wenig zur Vollständigkeit. All das ist natürlich reine Mathematik, aber sehr amüsant.

Meine Eltern hoffe ich, in einigen Wochen hier zu haben. Wie weit ist es mit Deiner Mutter?
Mit herzlichen Grüssen von uns allen
Dein

[Rudi]

Birmingham, 10.12.1938
(carbon copy)

Dear Hans,
Today I can return a compliment you paid me some time ago, by telling you that you have just written a note for NATURE along with Hoyle and myself.[335] But whereas last time one could question whether I had contributed anything at all, this time the situation is just the opposite and one could question whether you should have published it alone, as you know everything about it anyway. Hoyle and myself had already discussed this question, but we believed the experimenters who denied the existence of β-rays so without your optimism at Cornell we would never really have pondered this question further.[336]

No one can doubt this point of view anymore. Since the experimenters at Cambridge are totally excited by this result and want to see it in cold print, we should do them the favour of publishing it. But it might be more appropriate for you to send your own note about this, or a letter to yourself (as the editor of *Phys. Rev.*). The attached manuscript shouldn't influence your decision in any way, just speed it up. Please reply with American speed by telegram to let me know whether or not you agree, and if not whether you will write it yourself. You can also

---

[335] R. Peierls, H. Bethe and F. Hoyle, 'Interpretation of β-disintegration', *Nature* **143**, 200–202 (1939).
[336] See letter [53].

send me a letter if you want to make any little changes in the form; we will receive it in time for corrections. I suggest the following code for your answer:

- Accepted = You can send this version of the manuscript, possible small changes by letter.
- Amendments = I agree in principle, but I don't like the formulation and I will send a modified manuscript by return mail.
- Publishing = I will write a note concerning this question by myself.

Some time ago Bohr was here for a couple of days, and it was very interesting even though we didn't make much progress on our collaboration. Since my last letter I have worked a little on proving the completeness of the function spaces that Kapur and I are using,[337] where I have been finding some mathematical oddities.[338] It turns out that in exceptional cases the square integral of one eigenfunction disappears, and as a result the system is not truly complete. All the physically interesting formulas derived by assuming completeness are still right, which is plausible for reasons of continuity anyway. It also turns out that whenever such a square integral disappears two of the eigenfunctions are identical, which is impossible under real boundary conditions. At the moment they become equal, you surely have one function too few for completeness. This is all just pure mathematics, of course, but it is still very amusing. I'm hoping my parents will be here in a few weeks. What about your mother?
With best wishes from all of us,
Yours,

[Rudi]

---

[337] P.L. Kapur and R. Peierls, 'The dispersion formula for nuclear reactions', *Proc. Roy. Soc.* **A166**, 277–95 (1938).

[338] The Peierls-Kapur paper discussed the response of a nucleus to neutron bombardment in terms of resonance. The paper takes for granted that the resonance states as defined form a complete basis. This is proved by Rudolf Peierls in a paper ten years later. R. Peierls, 'Expansion in terms of sets of functions with complex eigenvalues', *Proc. Camb. Phil. Soc.* **44**, 242–50 (1948).

[55] Hans Bethe to Rudolf Peierls

Ithaca, 12.12.1938

Deine Adresse habe ich wieder verdrängt
38 Calthorpe Road?
Lieber Rudi,
Dein Brief ist jetzt auch schon wieder über einen Monat alt, und Arbeitsprogramme sollte man eigentlich sofort beantworten.

Also erstens: Über die Elektronenpolarisation hat mein Rose viel, und ich glaube Vernünftiges, gerechnet. Das Resultat ist absolut Null. Er hat die Depolarisation durch die Stösse unter kleinen Winkeln berechnet, die zwischen den beiden Ablenkungen um 90° stattfinden. Nun gibt es zwar bei Dymonds Folien etwa 50 solche Stösse, aber die Depolarisation ist doch nur 3%, so dass sodass die theoretische Asymmetrie von 12% auf 11.6 reduziert wird. Berücksichtigt wurden elastische und unelastische Stösse, im ersten Fall durch Einfluss von Austausch mit einem Leuchtelektron entgegengesetzen Spins keine Bornsche Näherung. Er wird jetzt noch versuchen, ob vielleicht die Wechselwirkung mit Leitungselektronen was ausmacht, aber es scheint sehr unwahrscheinlich. Und ich bin überzeugt, dass die Kristallbeugung nichts tut.

Hingegen habe ich mich nicht mit den β-Spektren befasst, und meine diesbezüglichen Bemerkungen in Cornwall waren nur ein Stimmungsausdruck. Darum möchte es sehr schön sein, ob der Hoyle das anständig diskutieren würde. Im übrigen ist es ja inzwischen einem Californier gelungen, die obere Grenze des β-Spektrums von N^{13} unter die theoretische Grenze (von Kernreaktionen) zu bekommen. Was sind die neuen Experimente in Liverpool?

Wie geht es dem Compound nucleus? Placzek schrieb, trauriger-weise "hättest Du doch recht", nämlich damit, dass die schönen Formeln alle nichts nützen. Ich habe jetzt einen sehr gescheiten Mann, Groenblom, einen Finnen, der bei Heisenberg doktoriert hat. Er soll versuchen, eine genügend gute Näherung für schwere Kerne zu kriegen, so dass man wirklich sehen kann, wie weit die "Verschmierung" im Compound nucleus geht. Ob es z.B. wahrscheinlich ist, dass man irgendwelche Quasi-Auswahlregeln kriegt, (ausser den trivialen für

Drehim[uls] und Parität), wie sehr Matrix-Elemente schwanken etc. Ich glaube, wir haben eine Methode, die von den Fehlern der Hartree Methode frei ist, indem sie die E[igen]funktion bei kleinem Abstand zweier Teilchen exakt behandelt, und trotzdem hoffentlich noch übersehbar ist.

Persönlich habe ich mich mit den Sternen beschäftigt, bin aber, wie Ehrenberg zu sagen pflegt, jetzt ziemlich damit abgefüttert. Immerhin habe ich herausgefunden, dass praktisch alle Sachen, die die Astrophysiker annehmen, falsch sind. Ausserdem dass kein existierender Stern einen Neutronenkern haben kann, da der Stern dann kleiner sein müsste als die Erde. Ferner, dass es andererseits keine untere Grenze für die Grösse eines Neutronenkerns gibt, wie sie alle Gelehrten anzunehmen belieben. Und weitere derartige destruktive Dinge.

Wobei mir ein abstract von Houston im neuesten Bulletin of the American Physical Society einfällt, in dem er sich beklagt, dass Dein Erhaltungssatz für $\sum k_{Elektron} = \sum k_{Gitterwellen}$ in der Metalltheorie eine so katastrophale Wirkung auf die Entwicklung der besagten Theorie hatte. Er bringt es fertig, dann Dinge zu schreiben, die sowohl trivial wie falsch sind — das ist immerhin ein Erfolg. (Ich rege mich darüber glaube ich bloss auf, weil einige Leute hier den Houston als Institutsdirektor haben wollen).

Glaubst Du eigentlich auch, wie Møller, dass neutrale Teilchen eine reelle Eigenfunktion haben, d.h. $\psi^* grad \psi$ etc. den elektrischen und nicht den Teilchenstrom bedeutet? Das erschüttert meine tiefsten Überzeugungen — aber es mag ja richtig sein.

Speaking of Møller: Wie sollte doch die Retardierung eine Verminderung der Divergenz in der Barytron-Theorie bewirken? Ich vesuchte es Weisskopf zu erklären und brachte es nicht fertig.

Hast Du noch etwas über die Energie des Deuterons mit dem $\sigma_1 \cdot r \sigma_2 \cdot r$ Glied in der Wechselwirkung gerechnet? Ich will nächstens mal publizieren, was ich über diese Probleme gemacht habe und den Optimismus der Yukawisten etwas dämpfen durch Divergenzen.

Weisskopf und ich haben uns ziemlich viel mit den Divergenzen in der Elekrodynamik beschäftigt. Es scheint, dass die Löchertheorie wirklich viel besser ist als die Dirac-Theorie ohne Löcher, und dass es eigentlich überhaupt keinen Sinn hat, in der letzteren die Selbstenergie zu rechnen. Neuerdings behauptet Weisskopf, er könnte beweisen, dass die Selbstenergie in der Löchertheorie nicht stärker divergiert als $\frac{1}{r} \log r$ (in keiner Näherung!) aber das glaube ich vorerst noch nicht.

Es gibt schrecklich viele Leute hier zurzeit, und Bacher spricht nur noch von meiner production line. 5 Doktoranden, 3 zukünftige ditto, besagter Groenbom, Rose, und Ludloff (letzterer kein grosses Vergnügen). Es wird mir etwas zuviel, denn ich komme nicht mehr zum eigenen Arbeiten.

Placzek's Entscheidungen sind mir rätselhaft. Dieser Brief ist nur über Physik, denn wenn man an anderes denkt, wird man wahnsinnig. Meine Mutter ist leider (mit meinem Einverständnis) nach Ausbruch des Friedens wieder nach Baden-Baden zurückgefahren, um die Auswanderung vorzubereiten. Zwischendurch entschied sie sich mal wieder, sie wolle überhaupt in Baden bleiben. Inzwischen ist das trivialerweise vorüber, und ich habe ihr gestern das Affidavit geschickt. Aber — Und Deine Eltern?

Solange man nur an die nächste Umgebung denkt, ist es besser. Darum recht schöne Weihnachten und sehr viele Grüsse an Dich und Genia und Gaby und Ronny. Es ist wirklich schade, dass man mit der Queen Elisabeth (meiner) nicht mal über Weekend nach Birmingham fahren kann. So muss man sich mit dem New Hampshire zu Weihnachten begnügen.

Tausend Grüsse

Hans

[Ithaca], 12.12.1938

I have repressed your address again. 38 Calthorpe Road?

Dear Rudi,

Your letter is now more than a month old, and plans for what to work on should really be answered immediately.

So, first: on the subject of electron polarisation my Rose has been doing a lot of calculations, and I think they are quite reasonable. The result is absolutely nothing. He has calculated the depolarisation due to small-angle impacts, which takes place between the two 90° deflections.

Now in the case of Dymond's foils there are about 50 such collisions, but the degree of depolarisation is only 3%, so the theoretical asymmetry of 12% is reduced to 11.6%. Elastic and inelastic collisions were both taken into consideration. In the first case, this was calculated as an exchange between light electrons with opposed spins, even though there is no Born approximation. Next he will try to find out whether their interaction with conduction electrons has any effect, but this seems to be highly unlikely. I am also convinced that crystal diffraction is not at all important.[339]

But I have not been dealing with β-spectra, and my comments at Cornwall were just an expression of mood. It would therefore be very nice if Hoyle discussed this properly. By the way, a Californian was able to place an upper bound on the β-spectrum of ^{13}N that is below the theoretical predictions based on nuclear reactions. What are the latest experiments at Liverpool?

How is the compound nucleus coming along? Placzek wrote to me, to say that unfortunately "you were right", in other words that all these nice formulas don't help at all.[340] I now have a very smart man named Groenblom, from Finland, who did his dissertation under Heisenberg.[341] He is supposed to find a good approximation for heavy nuclei, so that you can really see how far the "luting" in the compound nucleus goes. For example, you could see whether the appearance of quasi-selection rules is probable (not counting the trivial ones for angular momentum and parity), to what extent the matrix elements are oscillating, etc. I think that we have found a way of doing this which is free from the errors of the Hartree method, as [our method] treats the eigenfunction exactly for two particles separated by a small distance, a case which easy to assess anyway.

---

[339] Bethe and Rose published the results shortly afterwards. M.E. Rose and H.A. Bethe, 'On the Absence of Polarization in Electron Scattering', *Phys. Rev.* **55**, 277–89 (1939).

[340] See letter of Placzek to Peierls, 4.9.1938, Lee, *Selected Correspondence*, 1, chapter 4.

[341] Berndt Olof Grönblom (1913–1941), had studied in Leipzig where he received his doctorate, and after a short spell at Cornell he returned to Heisenberg's institute. When Stalin invaded Finland, Grönblom returned to his home country and died in action in August 1941.

Personally, I have been dealing with stars, but now I am 'quite stuffed', as Ehrenberg used to say. I have at least found that almost everything the astrophysicists claim is wrong. I have also found that no existing star can be made of neutron nuclei, since then the star would have to be smaller than the Earth. Furthermore, conversely, that there is no lower bound on the size of a neutron nucleus as all scientists think. And other such destructive things.[342]

At this point an abstract by Houston comes to mind, which appeared in the latest Bulletin of the American Physical Society. He complains that your conservation equation $\sum k_{electron} = \sum k_{latticewaves}$ has had a catastrophic effect on the development of metal theory. He is silly enough to write things which are trivial as well as wrong — this is some kind of success, after all. (I guess I'm just upset by this, because some people want Houston to be the head of our institute).

Do you, like Møller, think that neutral particles have a real eigenfunction? I.e., do you think that $\psi^* grad \psi$ etc. is the electric current and not the particle stream? This is shattering my deepest convictions — but it might be true.

Speaking of Møller, how could retardation decrease the divergence of the Barytron theory? I tried to explain this to Weisskopf, but without success.

Have you calculated the deuteron energy from its interaction with the $\sigma_1 \times r \sigma_2 \times r$ matrix element? I will soon publish my work on these problems, and reduce the optimism of the yukawists with these divergences.[343]

Weisskopf and I have been working hard on the divergences of electrodynamics. It seems as if the hole theory is really much better than the Dirac theory without holes, and that it doesn't make any sense to calculate self-energy in the latter. Weisskopf now claims he can prove that the self-energy in the hole theory doesn't diverge more quickly

---

[342] Bethe had just submitted his first substantial paper on energy production in stars. H.A. Bethe, 'Energy Production in Stars', *Phys. Rev.* **55**, 434–56 (1939).

[343] Bethe published a paper on nuclear forces the following year. H.A. Bethe, 'The Meson Theory of Nuclear Forces', *Phys. Rev.* **55**, 1261–63 (1939).

than $\frac{1}{2} \log r$ (without any approximations!), but I don't believe that yet.[344]

There are a great many people here, and Bacher[345] has just started calling it my production line. 5 doctoral students, 3 future doctoral students, Groembom whom I mentioned, Rose, and Ludloff (this last is not a great pleasure). It's a bit too much for me, as I can't find the time for my own work anymore. I don't understand Placzek's decisions. This letter is just about physics, since thinking about anything else would just make one angry. Unfortunately, my mother returned to Baden-Baden after the outbreak of peace (with my agreement) to get ready for emigration. Since then she has decided to stay in Baden after all. By now this is over, and yesterday I sent her the affidavit. But —

And your parents?

It's better if you just think about your close surroundings. Therefore merry Christmas, and many greetings to you and Genia and Gaby and Ronny. It is a pity that I can't just go to Birmingham for a weekend on the Queen Elizabeth. So I will have to make do with New Hampshire for Christmas.

A thousand greetings,

Hans

[56] Rudolf Peierls to Hans Bethe

Birmingham, 25.12.1938
(carbon copy)

Lieber Hans,

Vielen Dank für Deinen Brief, der sich mit meinem gekreuzt hat. Ich nehme an, dass Du diesen Brief bereits als vorweggenommene Antwort

[344] Weisskopf published his results in V. Weisskopf, 'On the Self-Energy and the Electromagnetic Field of the Electron', *Phys. Rev.* **56**, 72–85 (1939).

[345] Robert Fox Bacher (1905–2004), studied at Michigan where he received his doctorate in 1930, after posts at CalTech, MIT, Michigan, and Columbia University, he became professor of physics and director of the Laboratory of Nuclear Studies at Cornell in 1935. Later he joined the Manhattan Project and after the war played a leading role in high energy physics at CalTech.

auf meine Anfrage bezüglich der Publikation betrachtest, und Dir daher das Telegramm sparen kannst, und ich interpretiere das als Einverständnis. Ich schicke daher heute die Note an NATURE ab. Sollte Dir das nicht recht sein, bitte ich um umgehenden Bescheid. Übrigens war Dee etwas böse, dass ich Dir das M[anuskript] geschickt habe, das die Feststellung enthält, dass man jetzt im Cavendish nach den β-Strahlen sucht. Er stellt sich offenbar vor, dass auf die Nachricht hin, dass dies im Cavendish gemacht wird, gleich ganz Amerika auch β-Strahlen suchen, und sie dann durch unzuverlässige Versuche finden werden. Also um ihn zu beruhigen, kannst Du vielleicht von der Information vorsichtigen Gebrauch machen (d.h. nicht von der theoretischen Feststellung, die Dir ja mehr als uns gehört, sondern von den unveröffentlichten Experimenten!)

Was die Polarisation betrifft, so würde ich es gerne sehen, wenn jemand mal anständig über die Frage diskutiert, wie weit man den Effekt einzelner Atome einfach suspendieren kann (Denn wenn ich recht verstehe, hat das doch wohl Rose auch gemacht, wenn er auch die Stösse weiterer Atome mit der bereits gestreuten Welle mitnimmt.) Vielleicht ist es trivial, aber dann muss man es auch in ein paar Zeilen anständig begründen können.

Betreffend Möller: Ich habe mich inzwischen auch davon überzeugt, dass es sehr unwahrscheinlich ist, dass die Retardierung in der Spin-Spin-Wechselwirkung der von Neutron-Proton in der Mesotrontheorie etwas retten kann. Möller hat natürlich Recht damit, dass gerade für kleine Abstände die Retardierung wesentlich ist. (Denn damit die vernachlässigbar ist, muss die Präzessionsperiode eines Spins in Felde des anderen lang sein gegen die Zeit, die ein Mesotron braucht, um von einem zum anderen zu laufen, und das ist gerade bei kleinen Abständen nicht der Fall). Andererseits kann man sich leicht überlegen, was passieren würde, wenn man grosse Spins (statt $\frac{1}{2}$) hätte. Dann würde der Zustand tiefster Energie, der der unendlichen Anziehung entspricht, gerade dem entsprechen, dass die Spins parallel zu ihrer Verbindungslinie eingestellt werden. In diesem Falle ist aber das "Feld" in der Richtung des Spins und daher gibt es keine Präzession und also macht die Retardierung nichts aus. Nun könnte es natürlich gerade noch sein, dass die Retardierung für Spin $\frac{1}{2}$ gerade noch die Unendlichkeit schwächer macht, aber das erscheint schon sehr unplausibel. Immerhin ist die Wechselwirkung zweier Spins mit Retardierung, aber ohne ihre Translation in Betracht zu ziehen, ein winziges Problem, und ich möchte gern jemanden das rechnen lassen.

Was die neutralen Mesotronen betrifft, so gefällt mir die Formulierung von Kemmer (Proc. Camb. Phil. Soc.) sehr gut, weil man dort in so zwangloser Weise die Gleichheit der Kräfte zwischen like und unlike particles bekommt.

Sonst habe ich über die Spindivergenzen nichts weiter gerechnet. Ich habe noch etwas vor; es erscheint mir nicht ausgeschlossen, dass man ohne den Spin-Spin-Term auskommen kann, d.h. ohne den zweiten Wechselwirkungsterm in der Hamiltonfunktion, der für die ganze Schweinerei verantwortlich ist. Man sagt immer, dass ohne diesen Term man keine Heisenbergkraft bekommen würde, und dass dann der Singlett- und Triplettzustand von ^{2}H dieselbe Energie haben sollte. Das ist aber ein fauler Zauber, denn um dieses Resultat zu bekommen, wird in der Ableitung die Geschwindigkeit der schweren Teilchen vernachlässigt. Nun benutzt man aber für die schweren Teilchen die Diracgleichung und die Geschwindigkeit ist daher immer gleich c (nur ihr Mittelwert ist sehr klein).

Dieses Argument ist also genauso schön als ob man sagen würde, dass die Diracgleichung kann die Feinstruktur der Atome nicht erklären, denn wenn das Elektron sich langsam bewegt, darf man die [...][346] gleich Null setzen, und dann gibt es keine magnetische Wechselwirkung, also muss man in die Diracgleichung noch die Neutrinoterme hinschreiben.

Nun ist es im Atom allerdings so, dass die Terme, die von der "Zitterbewegung" herrühren, immer um v^2/c^2 kleiner sind als die statischen. Ein entsprechendes Resultat wäre im Kern wohl zu wenig. (Ich bin nicht mal davon überzeugt.) Aber das Eintreten der Geschwindigkeit ist doch dort ziemlich zufällig, in Wirklichkeit steht natürlich da $\ldots/r^3$, und das r^3 hat eben im Atom so eine Grössenordnung, dass das ganze gleich $v^2/c^2 \cdot e^2/r$ wird. Das braucht aber alles im Kern nicht mehr zu

[346] Missing in manuscript.

stimmen, insbesondere weil ja dort die h/mc für die schweren Teilchen mit dem mittleren Abstand durchaus vergleichbar sind.

Aber all das hat wenig mit dem zu tun, was Du gerechnet hast. Es möchte nur schön sein, ob Du mir eine Kopie schicken könntest, wenn Du darüber etwas dichtest, dann könnte ich Deine Zerschmetterung der Spin-Spin-Wechselwirkung einfach zitieren (falls ich etwas herausbringe) ohne mich darauf näher einzulassen.

Vorgestern kam ein Telegramm von Teller mit (eventueller) Einladung zur Konferenz im Januar. Mit sehr blutendem Herzen musste ich absagen, denn ich habe hier für den nächsten Term viele Vorlesungen übernommen (11 pro Woche) dass es praktisch unmöglich ist, sie alle zu versäumen, und auch nicht, mich vertreten zu lassen, da ich vorläufig der einzige Theoretiker in Birmingham bin. Ich möchte gern einen Assistenten haben, und bemühe mich daher, zu beweisen, dass es soviel Arbeit für Theoretiker hier gibt, dass ich sie nicht allein schaffen kann (was übrigens stimmt). Dann wäre es natürlich sehr unpädagogisch, zu versuchen, ob es auch mal vier Wochen ohne mich geht. Die Schiffe gehen nämlich gerade so blöd, dass ich von hier vier Wochen weg sein müsste. (Im Summer Term, der im wesentlichen aus Examen besteht, wäre es natürlich leichter). Aber ärgerlich ist es doch!

Für meine Eltern habe ich das englische Visum bekommen, was ja jetzt gar nicht mehr so einfach ist, und sie bemühen sich jetzt schon seit sechs Wochen, ihre Unbedenklichkeitserklärung u.s.w. zu bekommen. Es ist sehr schwer, denn sie haben natürlich kein Bargeld, um das "Sünegeld" zu bezahlen, und Papiere darf man nicht ohne Genehmigung verkaufen, und die Genehmigung kriegt man nicht. Natürlich hat man das Auto weggenommen, was die Lauferei nicht gerade beschleunigt. Immerhin sind sie noch in ihrer Wohnung, und es sieht doch so aus, als ob sie in absehbarer Zeit alle Papiere zusammenbekommen werden. Es geht ihnen also noch wesentlich besser als den meisten anderen Leuten.

Schrödinger läuft hier in England herum (kriegt wohl eine Stelle in Dublin) und wundert sich sehr, wenn die Leute von ihm verlangen, er sollte doch von seiner Erklärung im Grazer Tageblatt abrücken. Das müsste doch jeder wissen, dass wenn er so etwas schreibt, es nicht ernst gemeint ist.

Im übrigen kriegt man natürlich massenhaft Briefe von Leuten, die nach England wollen, und den meisten muss man schreiben, dass es so gut wie ausgeschlossen ist. Natürlich gibt es unerwartete Glücksfälle, wie der von Kober, einem Mathematiker, der mit Watson befreundet ist. Er war früher Schullehrer und hat sich 33 zurückgezogen, um privat Mathematik zu betreiben. Jetzt hat sich ein früherer Schüler von ihm gefunden, der bereit ist, ihn und seine Frau vier Jahre lang hier zu erhalten, und bereits das Geld dafür auf den Tisch des Hauses gelegt hat. Kober selbst sitzt noch "im Konzert" wir hoffen aber, dass recht bald herausgelassen wird.

Aber Du hast recht, denken wir lieber an die nächste Umgebung (und auch das mit Auswahl, denn bei uns um die Ecke ist die Villa von Chamberlain). Wir haben eben die obligate Pute gegessen und dabei mit Rührung der Ente gedacht, mit der Genia und Du einen tug-of-war angestellt habt.

Beste Grüsse von und an
Dein

[Rudi]

Birmingham, 25.12.1938
(carbon copy)

Dear Hans,

Thank you for your letter,[347] which crossed mine[348] in the post. I suppose that you will regard this letter as an antedated reply to my questions concerning the publication. We can now do without the telegram, and I will interpret your letter as an agreement. I will therefore submit the note to NATURE today. If you don't want me to do this, please tell me immediately. Dee was a bit angry that I sent you the manuscript, by the way, which contains a note to the effect that they were searching for β rays in the Cavendish. He seems to think that (because of the note that this is being done in the Cavendish) all of America will now be searching for β rays, and that they will therefore

[347] Letter [55].
[348] Letter [54].

find them with unreliable experiments. To calm him down, maybe you can handle this information with caution (not regarding the theoretical conclusion, which is more yours than ours, but regarding the unpublished experiments!).

Concerning the polarisation, it would be great if someone properly discussed the question whether you can just suspend the effect of single atoms easily (As Rose did, if I remember correctly, although he considers collisions of the once-scattered wave with further atoms.) This may be trivial, but to do it properly you still ought to prove it with a few lines of maths.

As to Möller: In the meantime, I have become convinced that it is very unlikely that retardation of the spin-spin interaction between neutrons and protons might help the mesotron theory in any way. Möller is absolutely right that the retardation is especially important over small distances. (For it to be negligible, the precession period of one spin in the field of another has to be long compared to the time required for a mesotron to go from one to the other. This is clearly not the case over small distances). On the other hand, you can easily imagine what would happen if you had big spins (rather than $\frac{1}{2}$). Then the ground state, which corresponds to infinite gravity, corresponds to exactly the case where spins are parallel to the line connecting the particles. In this case the "field" is in the direction of the spin, so there is no precession and the retardation is not important. It would be possible that the retardation for spin $\frac{1}{2}$ would just weaken the infinity, of course, but this doesn't seem plausible. The interaction of two spins, with retardation but without taking into account their translation, is a tiny problem and I would like someone else to do the calculations. Concerning the neutral mesotrons, I like Kemmer's formulation (Proc. Camb. Phil. Soc.)[349] very much because you can use it to easily derive the equality of forces between like and unlike particles.

Aside from that, I haven't done any more calculations on the spin divergences. I have something else in mind. It doesn't seem impossible to do it without the spin-spin term, i.e., without the Hamiltonian's

[349] N. Kemmer, 'The charge-dependence of nuclear forces', *Proc. Camb. Phil. Soc.* **34**, 354–364 (1938).

second interaction term, which is responsible for the whole disaster. Everyone says that without this term you wouldn't get any Heisenberg force, and that the singlet and triplet states of ^{2}H would have the same energy. But this is a shady business, since to derive this result you need to neglect the velocity of heavy particles. For heavy particles you use the Dirac equation, so their velocity is always equal to c (only the average value is very small).

This argument is just as nice if you said that the structure of the atoms can't be explained with the Dirac equation; this follows because if the atom is moving slowly you can put [...][350] to zero. Then there is no magnetic interaction, so you also have to incorporate the neutrino values into the Dirac equation. In the atom it seems as if the v^2/c^2 terms resulting from the "oscillatory motion" are always smaller than the static terms. A corresponding result in the nucleus would be insufficient. (I am not even convinced of this myself.) But the velocity dependence seems to be quite arbitrary; actually, it must be $\cdots/r^3$, and because of the order of magnitude of r^3 in the atom the whole thing is equal to $v^2/c^2 \times e^2/r$. In the nucleus this might not be right anymore, especially because h/mc is comparable for heavy particles with an average distance.

But all this has little to do with what you calculated. It would be very nice if you could send me a copy anyway when you have figured something out; then I could just quote your shattering of the spin-spin interaction (assuming I discover something myself) without going into the details.

Two days ago I received a telegram from Teller with a possible invitation to the conference in January.[351] It was very hard for me to turn it down. Next term I will be holding a lot of lectures (11 per week), so it is almost impossible for me to miss them or to find a replacement, especially since I am still the only theorist in Birmingham. I would like to have an assistant, so I am trying to prove that this is too much work for one theorist and that it's impossible for me to do it alone (which is true, by the way). Of course, it would not be very pedagogical to see

[350] Missing in manuscript.
[351] Meeting of the American Physical Society, January 1939.

whether they could do without me for four weeks. The departure times of the ships are so stupid, that I would have to be away for four weeks. (In the summer term, when we just have exams for the most part, it would of course be easier.) But it is still annoying!

I received the English visa for my parents, which isn't very easy anymore, so they have been trying for six weeks now to get their certificate of innocuousness, etc. This is very difficult, since of course they have no cash to pay the "expiation fee". You are not allowed to sell shares without permission, and you can't get this permission. Their car has been taken away, of course, which doesn't speed them along at all. At least they are still in their apartment, and it seems possible for them to get all the necessary papers eventually. So they are still much better off than almost everyone else.

Schrödinger[352] is wandering around England (he will presumably get a job in Dublin) and he is surprised when people ask him to recant his declaration in the Graz newspaper.[353] Surely they must have known that, if he wrote something like this, it wasn't meant to be taken seriously.

Besides that, we are getting a number of letters from people who want to come to England and have had to tell most of them that it is nearly impossible. There are of course unexpected lucky cases like that of Kober,[354] who is a mathematician and friend of Watson's.[355]

[352]Erwin Rudolf Josef Alexander Schrödinger (1887–1961), studied in Vienna; after positions in Jena and Stuttgart, he became Professor at Breslau. In 1927, he moved to Berlin. Following short spells at Oxford and Princeton, he moved to Graz and later to Oxford and Ghent. In 1940 he became Director of the School of Theoretical Physics at the Dubin Institute for Advanced Studies where he stayed for 17 years. In 1956 he returned to Vienna.

[353]Between 1936 and 1938, Schrödinger taught at the University of Graz, where — on the advice of the new Nazi rector of the university — he wrote an open letter to the University Senate that he had 'misjudged up to the last the true will and the true destiny of his country'.

[354]Hermann Kober (1888–1973) studied at Breslau and Göttingen from where he received his doctorate in 1911. He taught at a Jewish school in Breslau but spent time in Cambridge doing research which helped him later to secure a grant to work in Birmingham in 1939.

[355]George Neville Watson (1886–1965), studied at Trinity College Cambridge, where

He was a schoolteacher, and retired in '33 to pursue mathematics in private. Now one of his former students here has agreed to support him and his wife for four years, and has actually put up the money for that. Kober himself is still sitting "in concert", but we are hoping that he'll be released soon.

But you are right in that it's better to think of one's close surroundings first (or rather, just a selected part since Chamberlain's villa is just around the corner). We have just had the obligatory turkey, affectionately thinking of the duck that you and Genia had a tug-of-war over.[356]
Sincerely yours

[Rudi]

[57] Rudolf Peierls to Hans Bethe

[Birmingham], 22.8.1939
(carbon copy)

Dear Hans,
Zunächst einmal von uns allen alle guten Wünsche für Deine, wie Rose uns anvertraut hat, so baldbevorstehende Ueberführung in den Dublettzustand. Wir hätten uns auf jeden Fall darüber gefreut, aber seit wir Rose in Cambridge kennengelernt haben, haben wir doppelten Grund, der Unternehmung Glück zu wünschen.

Die politische Situation in Europa ist ja im Augenblick nicht ohne interessante Seiten. Dass Russland voll Scham, seinen Fehler einsieht, sich selber einzukreisen und nun dem Antikomintern-Pakt beitritt, ist jedenfalls originell. Es gibt jetzt wohl nur zwei Möglichkeiten: Krieg oder Appeasement. Die erste Möglichkeit ist natürlich bitter, aber es ist dann jedenfalls die letzte Chance für Europa, und die letzte Hoffnung, dass die Diktaturen kaputtgehen. Es wird jedenfalls interessant,

---

he became a fellow in 1910. Between 1918 and 1951 he was Mason Professor of Pure Mathematics at Birmingham University.

[356]Refers to Genia's first attempt at roasting the Christmas turkey while Bethe and the Peierls were living in Manchester.

das mitanzusehen und vielleicht gibt es für uns auch Gelegenheit, in bescheidener Weise mitzuhelfen.

Es ist aber noch immer nicht ausgeschlossen, dass die Sache wie vor einem Jahre in Wohlgefallen und Appeasement endet, und dann hat Freund Adolf Europa in der Tasche. In diesem Falle würde ich auf das energischste versuchen, so schnell wie möglich Europa zu verlassen. Dieser Brief hat teilweise den Zweck, Dich hiervon zu informieren und davon, dass, immer unter der Voraussetzung, dass es ein weiteres Appeasement gibt, ich bereit wäre jede Stelle in U.S. anzunehmen. Mir ist aber klar, dass solche Stellen [nicht] herumlaufen und auf mich warten, und ich würde in diesem Falle Weihnachten (früher wird es wohl nicht gehen) hinüberfahren, um mich selbst zu informieren, und mich, falls möglich, auf einer Konferenz auszustellen. Falls ich dies beschliesse, wovon ich Dich natürlich noch informieren würde, evtl. telegrafisch, wäre ich Dir sehr dankbar, wenn Du mir eine Einladung formaler Art besorgen könntest, zwecks Erleichterung des Visums (Ich würde im übrigen natürlich auf eigene Kosten kommen.)

Das ist eine sehr ungewöhnliche Kombination von Gratulation und Geschäft, aber wir leben nun einmal in recht ungewöhnlichen Zeiten.
In Eile sehr herzliche Grüsse,
Dein

[Rudi]

[Birmingham], 22.8.1939
(carbon copy)

Dear Hans,
First, best wishes from all of us on your imminent transition into a doublet state, about which Rose has told us. We would have been very happy about this in any case, but since we got to know Rose in Cambridge we have two reasons to wish you the best.

The political situation in Europe is not without interesting sides at the moment. It is definitely a peculiar feature that Russia, full of shame, now accepts that it was a mistake to isolate itself and has joined the Anticomintern-alliance. Only two possibilities seem to remain: war or appeasement. The first is painful, of course, but would be the last chance for Europe and the last hope that the dictatorships might be destroyed. It will be interesting to observe in any case, and it might be possible for us to contribute in a modest way.

But it is still not impossible for the problem to vanish into thin air and end with appeasement, just as it did last year. Then friend Adolf will have Europe in the bag. If this happens I will definitely try to leave Europe as quickly as possible. One of the reasons for this letter is to inform you that if there were another appeasement, I would accept <u>any</u> job in the US. But of course I know that these kinds of jobs are [not] just sitting around waiting for me. In this case I would travel there at Christmas time (I fear it won't be possible earlier) to inform myself and if possible to present myself at a conference. If I decide to come, of which I would of course inform you right away (possibly by telegraph), I would be very grateful if you could provide a formal invitation to facilitate the visa procedures. (I would of course pay for the trip myself).

That was a very uncommon combination of saying congratulations and making a deal, but we live in uncommon times.

In a hurry, best greetings,
Yours

[Rudi]

[58] Genia Peierls to Hans Bethe

Birmingham, 17.6.1940
(carbon copy)

Dear Hans,
Things are moving so quickly now that I don't know where we shall be and what we shall be when you get this letter. At the moment, our children are with their school in the country, and we are both rather in the front line with the A[ir] R[aid] P[recaution] and would be in exposed position during a raid. It is quite possible that we may both get killed and that the children would then be left in a rather awkward situation.

If things should get very bad here I suppose the United States will organize some large scale scheme to get the children removed and if they are really left without us perhaps you could use your influence to get them taken along. If this country wins the war they will, of course, be all right. We have good friends who will be looking after them whatever happens to us. But if we should lose the war, their existence here would become impossible. In that case I imagine the Germans would shoot us, even if nothing happens to us during the air raids. But perhaps the children will be spared.

They are very nice children, very jolly, easy going and bright. They will get all the possible scholarships and so on and be a lot of fun to the people who take them. If they do manage to get taken into some kind of refugee camp in the USA perhaps you could occasionally keep an eye on them and bring them up in the good tradition of theoretical physics.

You know Rudi has a lot of relations in the USA, some of them even are not so bad. But I would much rather that the children would go to people who take them because they want to and not because they are the children of their brother-in-law or so. On no account would I let them feel as "poor relations" and that always happens in these conditions.

I am writing a similar letter to Fankuchen. We are very well, terribly busy and in good spirits. Qui vivra, verra, and if even "ne vivra pas", we had lots of fun and on the whole a long and interesting life. Nothing can be worse than death because you can always have that choice.

I hope to see you once more somewhere and some time.

Love to you all

[Genia]

[59] Rudolf Peierls to Hans and Rose Bethe

Birmingham, 12.7.[19]40
(carbon copy)

My dear Bethes,
I always knew you were friends but your telegram left us quite speechless.[357] This is really very nice of you. Meanwhile you will have got our cable explaining that the children have been invited by Toronto University and that we hope to send them there in a week or two. There is at present a great rush on steamer accommodation, some universities just managed to get their children away before the rush but I hope our party will be able to go quite soon. We thought for a moment whether we should now cancel this and accept your invitation instead, but we decided not to do this at the moment. The children have their passports and permits &c. ready to go to Canada, no visa is required there, they are travelling there as a party with other children (many of whom they know) and also as regards streamer accommodation the official party sent by the university is likely to get better treatment than people who try to book privately. Moreover, if we sent them now, this would probably mean that you would have to rush back from your holidays. We have made sure that the number of children sent is actually a little below that of the number of places offered to us by Toronto, so that there is no question of our children going to the exclusion of anybody else.

The main thing is to get them across to America first, and then there will be enough time to sort out everything. We shall certainly make use of your generous offer to the following extent: For one thing I shall arrange (or try to arrange) that in the event of our being killed Hans will be made guardian of the children. I shall have to see a solicitor about the best form of arranging this without getting into difficulties with any of the laws of the three countries involved. I shall also write to the people in Toronto with whom the children will stay, asking them to communicate with you in the event of our being killed, or if communication with us should become impossible.

[357] The Bethes had sent a telegram with an offer to provide a home for the Peierls children for the duration of the war.

Also, if sometime in the future you could go to Toronto and see for yourself how the children are getting on we would be extremely grateful. I believe Toronto is quite near from Rochester, where you will be now and then anyway. I cannot judge whether in the long run it will be better for the children to stay in Toronto or to come and stay with you, as this depends on so many things I cannot judge. I do not know how much they would be in your way, and whether they would fill your house beyond capacity. On the other hand, of course, I do not know, of course, what kind of people they will find in Toronto. Needless to say that we would be very happy to know the children with you. But in all probability they will be in good hands in Toronto.

Incidently among the mothers going to Toronto is Mrs Shapiro[358] whom, I believe, you have met in Manchester. She went ahead of the rest of the party, because she is again in the same position in which she was when you met her then, and is therefore in a hurry. She knows our children well and will keep an eye on them. We shall ask her to get in touch with you.

Well, in any case, whatever happens, and whatever you ultimately decide is best to do we shall not forget this telegram.

Yours sincerely

[Rudi]

[60] Genia Peierls to Hans and Rose Bethe

Birmingham, 16.7.[19]40
(carbon copy)

My dear Bethes,

I am not quite sure which you will get first, this letter or the one which Gaby will write to you from Toronto. After about a fortnight of red tape

[358]Isaac Avi Shapiro (1904–2004) and his wife Pauline had first met Rudi and Genia Peierls in Manchester, where Shapiro had taught English between 1933 and 1935, before returning to Birmingham where he spent the rest of his academic career and long retirement.

and bother I am now writing to you in a whirl wind of socks, jerseys, suits and dresses, coats and shoes. Everything is very small, everything is marked, some bags are "wanted", others "not wanted", but I think one really wants everything for a journey on the high seas starting with a cool English July and finishing on a Canadian August.

We are going round the children looking for some more suitable spots on which they can be labeled. They are wearing identity disks, University badges, school blazers, and I think I shall just stick some labels over them and varnish them over to make them waterproof.

In the meantime we are stuffing Gaby with wisdom and knowledge. They are just home from school, and the time is very short, and Gaby gets slightly confused. For instance after I warned her against the dangers of constipation and ordered her to go to one of the mothers travelling with the party and ask for a remedy, she was also warned about the dangers of falling into the sea and the necessity of listening to what the sailors told you. Next morning she told me that: "If I or Ronnie have not been to the lavatory I must go and tell a sailor." Such a lot of pitfalls occur in every problem.

Rudi is writing you all the business side, I will tell you about the children.

Gaby is very jolly, very quick and very unsentimental. I think she has what is called a masculine mind. She is absolutely logical, very clear and unemotional in all her thinking. She loves arithmetic and one can explain to her everything. She will understand a very abstract problem, and she thinks a lot in this direction. For instance she told Rudi already a long time ago that every book must have an even number of pages because every leaf has two sides. Nobody has asked her about it. She is already, and will be even more so, a book-worm. She reads everything and is quite capable of understanding easy geography and history books. She is mentally much older than her age. She actually enjoys learning and thinking and solving problems. As other people play tennis, she plays arithmetic, for the sake of mental gymnastics. But I do not think she will be able to "invent" anything. She has not much imagination and is terrible terre-à-terre. She is never afraid or abashed in the mental or sentimental sphere. She will go and talk to everybody and to any amount of people; she is very self-confident, (even cheeky) and free. But she is a coward in a lot of physical things, she does not believe in herself in that respect at all. She cannot ride a bicycle or climb, but loves water and probably will soon learn to swim. She is bad at sewing, knitting and all handwork, and her handwriting is positively inherited! She is very definitely my daughter in very many respects (bicycle) but much more a woman than I ever was. She loves frocks and ribbons, can spend hours in front of a mirror, and adores babies and small children. She is actually very good with them and very reliable.

She tremendously enjoys life and is very easily contented. She is a chatterbox, but has a lot of tact and savoir-vivre.

Now about Ronnie or, as he is called at school, Podgkin (he is still very fat). He is very much Hans's spiritual son. First of all, food is very important for him, and he can always eat as long as there is something to eat. But when it is good, he enjoys it tremendously. Ronnie is a real gourmand.

Then he has the same sort of reckless and senseless bravery as you have. He will climb the most impossible tree, he will try and get on a bicycle until he is dark and green and ultra-violet. He screams in between, but wipes his tears and goes on. He was swimming in Yorkshire last year in a real surf so that he was turning somersaults in the waves for minutes on end, but he never minded it. But if he is afraid of something (dogs!) nothing can be done.

And then he needs a lot of petting, and he adores it. He is very affectionate and every woman loves him. But he must be dealt with very firmly and he must be encouraged to consider himself a grown-up.

He was extremely good when his tonsils were taken out. He stayed for three days in the nursing home and never even cried or rang the bell. Just because he wanted to show off. He can very easily be sorry for himself. He is very jolly and has a big sense of humour. He has a lot of imagination, but he is not as quick or as logical as Gaby.

If he will ever stay with you, your mother will adore him and spoil him. And he will love her and tell her that "she is the best woman in the world, that he loves her more than anybody else, and that he will marry her when he grows up." He says that to a lot of people. He can read and write a little, and can do a bit of adding, &c. But he does not like arithmetic. He is a little lazy and easygoing.

The two are very fond of each other. When I told Gaby that in Toronto they might be separated, she went pale (you need a lot to make her feel anything) and said "but he is still so young and I must look after him!" I hope they will put them together. They both are quite independent, can dress, wash and Gaby can make her bed and wash up a bit or dust a room. They will eat everything, sleep everywhere, and play and work in the middle of the most terrible din (Gaby particularly) they learned that at boarding school.

They had whooping cough and were inoculated against diphtheria. Ronnie had his tonsils out. He must be stopped when he eats too much; he does not stand too much of eggs and fat things.

Now you know about them nearly as much as I do.

Gaby is a very simple problem. She is clever enough to understand another point of view and if she will not be frightened by something or somebody, she will always be happy and adapt herself to every situation. But she must be kept busy and set hard problems to think about. She did very badly at school when she was put in too low a form and work was too easy for her.

Ronnie is more complicated. He must be loved and petted or he will be unhappy, but not too much or he will not grow up. He must be treated and talked to as a "big boy" and he will learn to play up and live up to it.

Thank you very much once more.

Yours sincerely,

[Genia]

[61] Rudi Peierls to Hans and Rose Bethe

Birmingham 17.7.[19]40
(carbon copy)

Dear Bethes,

Genia has promised that I would write about the "business side", but the joke is that the business side is practically non-existent.

I am sending with the children a letter addressed to their hostess asking her to let you know her address and to approach you in the event of any difficulty. I have made a will by which, in the event of Genia's or my death you would be appointed guardian of the children if they are then overseas. The executor under this will is Prof. Roy Pascal, of Birmingham University. He has absolute discretion in the way he uses the money on behalf of the children (it was very hard to persuade my solicitor to draw up a will giving so wide powers, that was quite against his feelings). He would obviously try and get permission to transfer the money (consisting only of my superannuation insurance) to America and if he succeeds, would delegate his powers of absolute discretion to you so that you would not be tied in any way by regulations of trusteeship, or by restrictions as to what share of the money to use for which child, &c. But I do not see at the moment that permission for a transfer would be given and thus the question will not arise.

Meanwhile I also enclose an official authority for you which may be handy if anything has to be done for the children where it would cause too much delay to get our signature for, or if there is any delay in getting the will approved, or you appointed as guardian. The document has no actual legal value, but it may help in the case of an application on behalf of the children or any other papers which would normally have to be signed by a parent or guardian.

You will understand that the intention of all this is not to give you more responsibility than you wish to take, but merely to avoid technical complications and bother.

I still hope that there will be a chance of showing you our gratitude. Yours very sincerely,

[Rudi]

[62] Hans Bethe to Rudi and Genia Peierls

[location unspecified], 15.8.1940

My dear Peierls',
Our telegram to you was so much the obvious thing to do that we hardly needed to think before we sent it.[359] It might just as easily have been the other way round, and you would have done the same thing.

So far, we have not yet heard from the children. We wrote a letter to Toronto University to tell them that we would take Gaby and Ronnie if there were any difficulty in finding a suitable family for them to live in Toronto. We have not yet had an answer from Toronto, but our letter was written only about 2 weeks ago, and it takes quite a long time from here to Toronto. But if there is no answer by the beginning of next week, and if we do not hear from Gaby in the meantime, we shall write again to Toronto.

Of course we shall go and see how the children are taken care of as soon as we are back in the East — i.e. probably early in October. I believe it may be best to convince ourselves personally whether all is well, and then decide whether it would be better to take them to Ithaca. For the purpose of bringing them over to America, it was certainly the best to have them go with the big transport of the other university children — everything goes so much more smoothly if there is an official organization behind it.

I remember enough of the children to know that they are about the nicest children one could acquire. All the real education, it is said, is done between 0 and 6 years; what follows is mostly the consistent development of the character built up in the first six years, and any outside influence contrary to the child's character provokes only antagonism and has an effect opposite to the intended one. Well, if this is true, nothing can spoil Gaby and Ronald — the first six years were too good. I have hardly ever seen children who were at the same time so natural and happy and so well-balanced and understanding as they were in 1938. It

---

[359] See letter [59].

will not be a hard task for us — or for anybody else — to bring them up, and I know it will be a lot of fun.

Your description of them, Genia, was as good as if they had been with us for months. We are repeating all the time: "If I have not been to the lavatory I tell a sailor"; and it is if they were playing here in this room. I am already giving Gaby a problem in arithmetic and climbing up the big tree in our lawn with Ronnie.

Thanks very much for the business arrangements. I hope they will never become effective but if the situation arises the free hand arrangement will simplify everything tremendously.

We are practically living with you all this time. Whenever there are news in the radio we listen anxiously. There is seldom much reason for jubilation. Still, we are happy enough to know that so far the German planes have been repulsed every time with great losses, and that the R.A.F. obviously has learnt a great deal from the mistakes in the Belgian and French War. We are happy to know that British ships still are everywhere on the sea and that England's sea power has not been seriously challenged, that the French fleet did not fall in Hitler's hands and many other things. And we hope for the best.

Yes, this life was beautiful and full, and we are enjoying it at the moment in spite of everything. The last year alone was worth living for (this concerns my personal affairs only), and if there should be an end soon, we still have had our life. If England should be defeated, our turn will be next — everybody knows that here. And I am very much surprised at how indifferent I am to the possibility of death. I dislike the idea of dying mostly because there would then be one person less to fight Hitler. If somebody had told me that 3 years ago, I should have considered him crazy.

Good luck — and I hope we can soon write to you something more positive about the children.
Yours,

Hans

3. The Birmingham-Cornell Pipeline

After the end of the war, Rudolf Peierls and Hans Bethe returned to their homes at Cornell and the UK. Although Peierls had attractive offers from a number of universities, including Oxford, Manchester, London and Cambridge, he chose to remain at Birmingham.[360] He had come to Birmingham in 1937 as the first professor of Mathematical Physics and had set himself the task of establishing a school devoted to both first-class research and first-class teaching. The war had put the effort on hold, but as soon as Peierls returned to Birmingham he re-engaged in the process. Virtually from scratch he built a school of mathematical physics, or theoretical physics as it would be called later, which was arguably the best in the country and which could compete with any in Europe and with most others globally. He had clear ideas of what he regarded as important for a prosperous theoretical physics community in the UK: a balanced flexible system that provided good training and high standards without prejudicing against students outside Oxbridge.

Hans Bethe, after settling at Cornell in the mid 1930s, had a similar vision. Starting in 1933, the Physics Department at Cornell had made plans to enlarge its activities which had been focussed on teaching with research primarily serving to provide theses topics for Masters and Ph.D. students. The new Chairman of the Department, R.C. Gibbs and one of Bethe's acquaintances from Munich days, Lloyd Smith, conceived a very different model for the department, and Hans Bethe had been one of four new appointments who were to turn this vision into a reality.[361] On his

[360] See letters [63–64].

[361] The other new staff members were Lyman Parratt, in X-ray spectroscopy, and two other physicists in the very new field of nuclear physics: Stanley Livingstone, who had assisted Ernest Lawrence in building the world's first cyclotron at Berkeley, and the young yet experienced experimentalist Robert F. Bacher.

return from Los Alamos, Bethe and his colleagues restarted where they had left in 1939. Hans Bethe, as head of the theoretical division at Los Alamos, had had to direct many of the brightest scientific minds. This experience of dealing with idiosyncratic experts in an effort to achieve a common goal proved a useful experience when he returned to Cornell. Together with his former colleagues and some reinforcement recruited at Los Alamos, such as Richard Feynman, the pre-war programme of building a strong nuclear physics laboratory and a strong theoretical division at Cornell was continued.

The early post-war letters exchanged between Peierls and Bethe touch on many of the key concerns of academics in the UK and the US who were engaged in rebuilding their academic communities after the end of hostilities: lack of resources and facilities, lack of staff and lack of qualified research students.[362] The one thing nobody on either side of the Atlantic was lacking was a variety of interesting new research topics.

Collaboration with the US throughout the war and close contact with many friends and colleagues across the Atlantic had sharpened Rudolf Peierls' awareness of the role reversal which had occurred with regard to academic physics. As early as September 1945, he expressed, in a letter to Raymond Priestley, the Vice Chancellor of Birmingham University, that 'American universities [had] matured a great deal and contact with this country [was] now less important to them, and more important to us'.[363] The consequence of this, in Peierls' view, had to be regular academic exchanges which would allow the UK to benefit from scientific achievements of colleagues in the US. Therefore, his attempts at putting Birmingham firmly on the academic map in theoretical physics meant that he was keen to secure a sizeable fraction of the exchange for this institution. Since his Munich days he had been establishing contacts with colleagues all over the world, and his work at Los Alamos added more depth and breadth to his international links.

[362] Letters [63], [66], [68].

[363] See letters exchanged between Rudolf Peierls to Raymond Priestley, September to December 1945, *Peierls Private Papers*, Ms.Eng.misc.b198, B.9.

First and foremost within the post-war collaborative network in and out of Birmingham was the link to Cornell University.

Perhaps the most influential of the exchanges orchestrated by Bethe and Peierls was based on the recommendation of Hans Bethe to Freeman Dyson,[364] in early 1949, to spend some time at Peierls' institute.[365] Rudolf Peierls and Robert Oppenheimer, at the time director of the Institute of Advanced Studies at Princeton, where Dyson, the rising star of theoretical physics at the time, held a research position, arranged a flexible fellowship. Dyson would be based at Birmingham, but it was agreed that he was at liberty to spend time at Princeton regularly, as long as it fitted in with departmental requirements at Birmingham. This resulted in Birmingham being in direct contact with the development of quantum field theory, which at the time was worked on by Schwinger, Tomonaga, Feynman and Dyson. The arrangement demonstrated two essential ingredients that promoted the success of the Peierls School at Birmingham. Firstly, Peierls was excellent at spotting talent, and secondly he was flexible enough to make Birmingham an attractive option for young scholars and therefore ensure that they chose his institute despite stiff competition from Cambridge, Oxford, Liverpool, Manchester, Bristol and other universities.

For both Bethe and Peierls the exchange of ideas with colleagues and the encouragement of academic discourse among junior members of the department were among the keys in their endeavour to advance research in theoretical physics. This explains why they were equally interested in establishing what they would later refer to as the Birmingham-Cornell pipeline, an 'exchange' of post-doctoral researchers and young colleagues.[366] Many scholars crossed the Atlantic and moved between Birmingham and Cornell at some stage in their career, for instance

[364]Freeman Dyson (1923–), studied mathematics at Cambridge between 1941 and 1943. After war service in the research division of the R.A.F., Bomber Command, he undertook research at Cambridge (1946–47) and Cornell (1947–1948) before becoming research fellow at Birmingham. In 1951 he took up a professorship at Cornell; in 1953 he became Professor of Physics at the Institute for Advanced Studies at Princeton.

[365]See letters [95–99].

[366]Letter[146], [149–150].

McCauley,[367] Nina Byers,[368] Elliot Lieb,[369] Jim Langer,[370] Dick Dalitz,[371] Ed Salpeter,[372] Jeffrey Goldstone[373] and Donald Sprung[374] moving between Birmingham and the US.

Another example of Peierls' excellent intuition in spotting scholars of promise and great potential was his slightly unconventional recruitment of Gerry Brown.[375] Brown had studied at Wisconsin and Yale, where he obtained an M.S. and a Ph.D. A short-lived membership of the Communist Party from which he was eventually expelled, put his academic career in the US at risk, despite his outstanding doctoral work with Gregory Breit. Various enquiries to universities in England led to the by now famous three-penny folded airmail return from Rudi Peierls saying: 'Come ahead.'[376] In February 1950 Gerry Brown arrived as a political refugee from pre-McCarthy anti-communist America; he stayed at Birmingham for a decade and, as is evident from the correspondence, he had a significant impact on the physics department.[377] In 1960 he left to take up an appointment as full professor of Theoretical Physics at Niels Bohr's Nordic Institute for Theoretical Physics (NORDITA).

While not many of Peierls' students arrived as refugees in the same way as Gerry Brown did, many moved on to take up distinguished positions. The Birmingham department itself was seen as an exceptional

---

[367] See letters [149], [151–53], [160], [163].

[368] See letter [144].

[369] See letter [140].

[370] See letter [143].

[371] See letters [122], [125].

[372] See letter [95].

[373] See letters [132], [135].

[374] See letter [161].

[375] Gerald E. Brown (1926–), studied at the University of Wisconsin to be trained as an electrical engineering officer for the Navy. At the same time he took physics courses with Gregory Breit and later moved to Yale with Breit, where he obtained his Ph.D. in theoretical physics in 1949. After research and teaching at Birmingham, he became professor at NORDITA in 1960, at Princeton in 1963 and moved to Stony Brook where he still works as Distinguished Professor of Physics. See letter [99].

[376] G.E. Brown, 'Flying with Eagles', *Annu. Rev. Part. Sci.* **51**, 1–22 (2002), here, p. 6.

[377] See letters [124], [137–140], [149], [161].

training ground for young scientists well beyond the United Kingdom. Many of those who came to Birmingham as students, graduates or research fellows in the 1950s later filled lectureships and professorships around the globe: Freeman Dyson, Dick Dalitz, Sam Edwards, Nina Byers,[378] Brian Flowers, Stanley Mandelstam,[379] John Bell, Denys Wilkinson, Elliot Lieb, Jim Langer to name but a few. Hans Bethe, similarly, was a 'magnet that attracted a world-class faculty to the physics department at Cornell',[380] among them Richard Feynman, Freeman Dyson, Robert Wilson and Ed Salpeter.

Hans Bethe was awarded the Nobel Prize in 1967 for work done just before the war, but many argued that while his understanding of the nuclear reactions that make the stars shine was significant, his calculations of the Lamb Shift, done in the early post-war years, were the more profound contribution to physics. They provided the insight that opened up a new way of thinking about quantum electrodynamics, they opened, as one colleague eloquently described it, the way to the modern era of particle physics.[381] Bethe's correspondence with Peierls contains few references to this work, as much of it was done during and after the Shelter Island Conference of June 1947 on the Foundation of Quantum Mechanics, and Bethe and Peierls saw each other intermittently between the summers of 1947 and 1948.[382] The Lamb Shift calculations were a fine example of that Gerry Brown would later call the "H.A. Bethe way".[383] When confronted by any problem Bethe would sit down with his stack of paper and fountain pen and his slide rule

[378]Nina Byers had obtained her Ph.D. at Chicago in 1956 before coming to Birmingham; she continued her research at Stanford (1958–61) before joining the faculty at UCLA as assistant professor and later professor of physics.

[379]Stanley Mandelstam (1928–) studied at Witwatersrand and Cambridge and obtained his Ph.D. from Birmingham in 1956. He went on to do research at Birmingham, Columbia, and Berkeley where he eventually became Professor of Physics in 1963.

[380]See http://www.news.cornell.edu/stories/March05/Betheobit.deb.html.

[381]Freeman Dyson, 'Hans Bethe and Quantum Electrodynamics', in Brown and Lee, *Hans Bethe and His Physics*, pp. 157–163, here p. 161.

[382]See letters [77–78], [82–83], [87], [91], [93].

[383]G.E. Brown, 'Hans Bethe and His Physics', in Brown and Lee, *Hans Bethe and His Physics*, pp. 1–23, here p. 16.

and calculate the problem in the most obvious way. Most importantly, he would not to be deterred by anyone who would tell him — as most colleagues would — that the situation was much more complicated than you thought.

Having helped to determine the new direction of QED, Bethe continued to work on related or consequential problems. Together with Ed Salpeter he formulated a relativistic wave equation for bound states of two particles,[384] and later the two physicists published a greatly revised version of Bethe's 1933 Handbuchartikel on quantum mechanics of one-electron and two-electron systems.[385] This work, through the 'Cornell-Birmingham pipeline' permeated into Peierls' department and had significant impact on the research there.[386]

While scientists returning from their war-related work were primarily concerned with rebuilding their institutions and re-igniting peacetime research and teaching, many of them also engaged in activities aimed at dealing with the consequences of their war work. This was true in particular for many of the scientists who had helped develop atomic weapons. Hans Bethe and Rudolf Peierls were among those who played a prominent role in discussions on nuclear energy, disarmament and control of nuclear weapons.[387] They applauded a report on international control of Atomic Energy, issued by the Department of State Committee on Atomic Energy and drafted largely by Robert Oppenheimer.[388] The report echoed the views expressed by scientists on both sides of the Atlantic in the newly formed *Atomic Scientists' Organization* and the *Federation of Atomic Scientists* in the UK and the US respectively. Bethe and Peierls were active in these, and they both published prolifically in journals committed to Arms Control and Nuclear Disarmament for decades to come. In 1950 Peierls, together

[384]E.E. Salpeter and H.A. Bethe, 'A Relativistic Equation for Bound-State Problems', *Phys. Rev.* **84**, 1232–1242 (1951).

[385]H.A. Bethe and E.E. Salpeter, 'Quantum Mechanics of One- and Two Electron Systems', *Handbuch der Physik*, **35**, 88–436 (1957).

[386]See for instance letters [70], [77], [91], [95], [97].

[387]See letters [161], [163], [165], [208], [210–211].

[388]Chester I. Barnard, J.R. Oppenheimer, David E. Lilienthal at al., 'Report on the International Control of Atomic Energy', Washington, USGPO, 16.3.1946.

with J.L. Crammer edited a popular science volume about Atomic Energy, to which Hans Bethe contributed. This elucidated some of the controversial issues dealing with atomic energy such as public health, civil defence, industrial potential, and also questions of raw materials or reactor design, detailed discussions of which had not previously been happening in the public domain.[389]

While both Bethe and Peierls were active in non-governmental organisations aimed at nuclear disarmament, arms control and better East-West dialogue, their roles within the political establishments of their chosen countries of residence was rather different. Peierls ceased his involvement in weapons production as such; his expertise was enlisted primarily by way of consultancy work for the Atomic Energy Establishment at Harwell. Among the friends and colleagues from the Manhatten Project who were on the staff at Harwell was Klaus Fuchs. His arrest, in 1950, on charges of passing secret information to the Soviet Union was a severe blow to the British scientific community as a whole, and it was a particularly traumatic experience for Rudolf Peierls and his family. Fuchs had been a close friend of Rudolf and Genia Peierls', he had lodged with them when he first came to Birmingham, and he had collaborated closely with Rudolf Peierls who had not only hired him into his department at Birmingham but had also been instrumental in securing his appointment at Los Alamos.[390]

Rudolf Peierls never shied away from expressing his views in public. He did so regardless of the effect this would have on is own position. He defended civil liberties in the aftermath of the Fuchs affair in his memorandum 'Lesson of the Fuchs Case',[391] although his close association with Fuchs had made him a prime target of suspicion. He was never secretive about his friendships with people from communist countries and of communist persuasion, he argued for the re-establishment of

[389]R.E. Peierls and L.J. Cammer (eds.), *Atomic Energy*, Harmondsworth: Penguin, 1950. See also H.E. Wimperis, 'Science and Social Order', *International Affairs* **27**, 72 (1951).

[390]See letter [102].

[391]S. Lee (ed.) *The Private and Scientific Correspondence of Sir Rudolf Peierls*, vol. II, World Scientific, 2007, chapter 6.

scientific exchange with the Soviet Union and its satellites. He rejected the idea of oppressing the voices of dissenters by arguing that this totalitarian measure would bring security at the expense of values that any democracy had to fight to retain. In the aftermath of the Fuchs arrest, Peierls' overt expression of these views led some to question his reliability, especially in view of the fact that he had access to sensitive and secret information in connection with the UK nuclear programme. However, at that time, as on many other occasions during the subsequent decades, it was recognised by people in authority that the views may have been uncomfortable at times, but at no point did they undermine the security and values of democracy in the UK, and at all times did Peierls prove loyal to the UK national interest.[392]

Hans Bethe's impact on nuclear weapons policy in the United States was more direct. He continued to consult for Los Alamos and he attempted to influence policy decisions. Alongside other scientific leaders, he advised against a crash programme to develop a hydrogen bomb as a response to the first Soviet nuclear test in August 1949, and he had hoped to be able to prove that an H-Bomb would be technically infeasible. But when he realised that this was not the case, and when a decision had been made to develop such a device, he decided to return to Los Alamos, in an attempt to make his influence felt within the establishment.[393] Following his initial work on the atom and hydrogen bombs, he subsequently advocated and worked tirelessly to create effective tools for verifying and validating negotiated agreements to slow down the arms race between the two super powers.[394] His continuing disagreement with his old friend Edward Teller about the wisdom of manufacturing the H-Bomb put severe strain on their relationship. But even more severe was the effect of their diametrically opposed positions in the Oppenheimer Case. While Teller with whom Oppenheimer had

[392]See S. Lee, 'The Spy That Never Was', *Intelligence and National Security* **17**, 77–99 (2002).

[393]Lee Edison, 'Scientific Man for All Season', *New York Times Magazine*, 10.3.1968, p. 125.

[394]S. Drell, 'Shaping Public Policy', in Brown and Lee, *Hans Bethe and His Physics*, p. 255.

disagreed on the hydrogen bomb testified against Oppenheimer at his security hearing in 1954, Bethe, at the time President of the American Physical Society, was one of Oppenheimer's staunchest supporters. He never made a secret of his abhorrence of the treatment Oppenheimer received.[395] Like Bethe, Peierls — though further removed from the the scene — tirelessly spoke out in support of Robert Oppenheimer, and his many letters to 'Oppie'[396] are evidence of the deeply-felt indignation at the attacks launched against his friend and colleague.

Despite their significantly increased administrative duties within and without the university, the 1950s were a very productive period for Peierls and Bethe. More and more, Peierls' impact on the discipline was mostly through teaching and collaboration with younger colleagues rather than through independent contributions. This is evident as much in his detailed correspondence with his students abroad[397] as it is in the exchange of letters with Bethe about the Birmingham-Cornell pipeline. At the same time, it is clear from the correspondence, that throughout the 1950s and even more so during the 1960s, his focus was shifting further towards political work and publications in the area of arms control.[398] Having been offered the Wykeham Chair of Physics in 1961, Rudolf Peierls, after lengthy negotiations decided to accept in early 1962 and took up his appointment in the autumn of 1963. When asked about the reasons for his decision to move from Birmingham to Oxford, he would later refer to the need for change after quarter of a century at the same university. But it was more than simply the desire of change. Peierls liked the challenge. After successfully building a school of theoretical physics at Birmingham, he wanted to achieve something similar in Oxford.

In contrast, Hans Bethe not only remained at Cornell for the remainder of his career; furthermore, despite his at times extremely onerous consultancy and political advisory roles, he remained prolific in his own research. A sabbatical at Cambridge in 1955 triggered a significant new

---

[395] See letter [127].
[396] See *Peierls' Private Papers*, Ms.Eng.misc.b211, C.227–28.
[397] See for instance Lee, *Peierls' Correspondence*, vol. II, chapters 6–8.
[398] See letters [161], [163–164].

phase in this research work. It involved the use of the G-matrix theory developed by Keith Brueckner[399] in order to "tame" the extremely strong short-range repulsions entering into the nucleon-nucleon interaction.[400] Brueckner's pioneering approach to solving the two-body scattering problem in the nuclear medium by rearranging perturbation theory in such a way that the contribution to the total energy at each order was proportional to the number of particles allowed a better understanding of nuclear properties. Energies per particle were manifestly finite. The challenge of calculating these nuclear properties was yet another one to which Bethe's abilities, experience and knowledge were ideally suited. In one substantial paper[401] Bethe gave a self-contained and largely new description of Brueckner's method for studying the nucleus as a system of strongly interacting particles with the aim to develop a method that was applicable to a nucleus of finite size while at the same time eliminating any ambiguities of interpretation and approximations required for computation. Thus Bethe — using the work of Brueckner and collaborators — produced an orderly formalism which the evaluation of the two-body operators G would form the basis for calculating the shell model potential $V(r)$. As in other scientific contexts, in this field too, analytic solutions to specific problems were a source of additional insight for Hans Bethe, and he turned to one such specific

[399]Keith Brueckner (1924–), obtained his Ph.D. from Berkeley in 1950. He joined the physics faculty at Indiana University in 1951, and in 1956 he moved on to the University of Pennsylvania before he taking up a Professorship for Physics at the University of California, San Diego in 1959.

[400]K.A. Brueckner, C. A. Levinson and H.M. Mahmoud, 'Two-body Forces and Nuclear Saturation. I. Central Forces', *Phys. Rev.* **95**, 217–228 (1954); K.A. Brueckner, 'Nuclear Saturation and Two-Body Forces. II. Tensor Forces', *Phys. Rev.* **96**, 508–516 (1954); K.A. Brueckner and C.A. Levinson, 'Approximate Reduction of the Many-Body Problem for Strongly Interacting Particles to a Problem of Self-Consistent Fields', *Phys. Rev.* **97**, 1344–1352 (1955); K.A. Brueckner, 'Two-Body Forces and Nuclear Saturation. III. Details of the Structure of the Nucleus', *Phys. Rev.* **97**, 1352–1366 (1955); K.A. Brueckner, 'Many-Body Problem for Strongly Interacting Particles. II. Linked Cluster Expansion', *Phys. Rev.* **100**, 36–45 (1955).

[401]H.A. Bethe, 'Nucelar Many-Body Problem', *Phys. Rev.* **103**, 1353–1390 (1956).

problem.[402] With Jeffery Goldstone[403] he went on to investigate the evaluation of G for extreme infinite height hard core potential,[404] and he encouraged his graduate student David Thouless to investigate the problem that, given the empirically known shell potential $V_{SM}(r)$, what properties of G would produce it.[405]

[402]Letter [135].

[403]Jeffrey Goldstone (1933–), studied at Cambridge where he obtained his Ph.D. in 1958. After research at Cambridge, CERN, Harvard and Copenhagen he became a faculty member at Cambridge in 1962. In 1977 he was appointed Professor of Physics at MIT.

[404]Letter [132].

[405]Letter [147].

[63] Rudolf Peierls to Hans Bethe

[Birmingham], 14.3.1946
(carbon copy)

Dear Hans,
Thank you very much for your article, which I think is admirable.[406] I believe it will fit into the series without altering a single word. This completes the series, except for the article from Phil Morrison from whom I have not heard yet.

As what you sent me was a carbon copy, I assume that you have sent the original to Groves' office. They have in the past been quite good about clearing things quickly, in particular when reminded from time to time, and Mr. MacMillan in Chadwick's office is looking after that. If, by any chance, you should not have sent your article to Groves' office, McMillan will get in touch with you directly.

I am in your debt in many ways, owing you (a) a letter, and (b) a nickel. On the latter, will you please begin to charge interest.[407] In other words, I have decided to stay here, largely because the spirit of the place is more attractive than it appears to be now in Cambridge, and it is hard to predict how Cambridge will change in the near future. It is, of course, possible that a good man will succeed Cockroft, but even then the whole administrative machinery is very cumbersome and conservative. Birmingham is rather fun and I have the ambition to put Birmingham on the map and get a good team of research workers started here, which would, I think, be a healthier thing from a wider point of view than to begin concentrating everything at Cambridge. People in Cambridge were naturally disappointed and they are now trying to find an alternative solution. There are, I think, one or two possibilities which would work quite well; if they are reasonable about it, but I know that amongst others, your name is on their list and I am trying to tell them

[406] Peierls helped put together a series of articles on atomic weapons as part of the Penguin News Series. *Science News* **6**, Penguin, London, 1948.

[407] Bethe and Peierls had had a bet over Peierls' future career. The latter had had several offers of chairs, among others from Cambridge University. He decided to remain at Birmingham among others for the reasons outlined in this letter.

that to attempt to offer this job to you would be a waste of time and of a postage stamp.

Even in Birmingham, with all the help from the University, it takes, of course, time to build up. So far I have 1 (one) research student,[408] with two more expected in the near future and one apparently good undergraduate finishing in the summer who might stay on. There is practically any amount of money for fellowships, so that I could get more people if I could find them. In particular I am trying to induce Skyrme to leave Los Alamos, but have not heard from him as yet.

The process of settling down and getting rid of excess papers etc. is a very painful one, as no doubt you have found yourself, and it is not made easier by the fact that a week or so after my return one of our lecturers fell ill and I had to take over most of his work, which, for a month or so, involves me in about thirteen lectures a week, mostly to large classes of engineers. I do not regret this, because I always meant to try these engineer courses in order to see whether they cannot be given in a more reasonable way and it is rather fun. Anyway, there are only two more weeks of this and, on the principle of the old story with the pig and the goat, etc., I shall then find that never before in my life did I have so much free time.

My one research student is now tidying up the question of the integral theory of the electro-magnetic field. As you know I was going to try this out. So far it appears that the classic side of this is perfectly straightforward and contains no snags. We are trying to apply this to some specific problem, such as the emission of radiation by an electron, to see how it works, but the real question is, of course, that of quantization and I have no confidence that this can be done.[409]

You will by now have received an invitation to the conference in Cambridge in July.[410] It is, I think, unfortunate that the organisers are not able to pay people's fares but only their stay in Cambridge. Even

[408]Hugh McManus. See letter [66], note 430.

[409]H. McManus, 'Classical electrodynamics without singularities', *Proc. Roy. Soc. A* **195**, 323–36 (1948).

[410]Between 22 and 27 July 1946, a Physical Society Conference on fundamental particles and low temperatures took place in Cambridge.

so, I very much hope that you will be able to come. It goes without saying that once you are here, and if you are not too hard pressed for time, we would count on seeing you in Birmingham.

We are still living in a boarding house, but we have bought the tail end of the lease of quite a nice house and hope to move in on April 15th. With very best wishes to all of you,
Yours sincerely,

[Rudi]

[64] Hans Bethe to Rudolf Peierls

Ithaca, 8.4.1946

Dear Rudi:
Thanks a lot for your letter of 3rd April.[411] It is unfortunately true that I had not sent my manuscript to Washington, but I am doing this today. It had weighed on my conscience for about a week past, but I was away all the week.

I am very happy to have won the nickel. I think in the long run you are going to be happier at Birmingham than at Cambridge. Still, it is a pity that Cambridge has gone so much to the dogs as you indicated by your refusal. Bretscher is apparently so disgusted that he will not come back to Cambridge.

I am just having a visit with Wentzel for two days. He mentioned that there were several young theorists in Zurich who took their degrees in the last two years. He wanted to send one over to Cornell with a Fellowship, which we like very much. But I thought also of your problems and thought you might be interested in getting one or two to England. Wentzel is going back to Zurich at the end of this week. You might like to write to him there as to what kind of people you would be interested in.

I got the official invitation from Cambridge. It is very tempting to go, but I had promised myself to take a really long vacation this

[411] Bethe almost certainly refers to letter [63] dated 14 March 1946.

summer. So it is probably "no", but this is certainly the most difficult invitation to refuse.

I got settled here quite nicely and find some time to study mesons and nuclear forces and to do a little bit of work myself. Physics is, after all, a fascinating subject.

You may have seen the report of the State Department on international control of atomic energy of which Oppie is one of the authors.[412] I think it is very good. Greetings from Rose who is enjoying the comforts of civilisation and from myself to you and the family.
Yours sincerely,

Hans

[65] Rudolf Peierls to Hans Bethe

[Birmingham], 27.4.1946
(carbon copy)

Dear Hans,
Thank you for your letter of April 8th. I have immediately written to Wentzel and I hope something will come out of this.[413]

I do hope you will eventually decide to come to Cambridge. You deserve a long holiday, but I believe the meeting is going to be very nice and except for time taken up by actual sessions not so far from a nice holiday.

The report of the State Department Committee arrived here only a few days ago,[414] but everybody including myself, is very impressed with

---

[412]United States Department of State Committee on Atomic Energy, 'Report on the International Control of Atomic Energy' (Washington, 28.3.1946).

[413]Peierls was hoping to get Wentzel to join him at Birmingham on a Fellowship for Swiss students.

[414]In 1946 scientists and politicians debated at length the various issues arising from the development and use of atomic weapons during the Second World War. One of the key issues was the control of atomic energy. See Barton J. Bernstein, 'The Quest for Security: American Foreign Policy and International Control of Atomic Energy, 1942–1946' *Journal of American History*, March 1974, pp. 1003–44. See also Donald A. Strickland, *Scientists in Politics: The Atomic Scientists Movement, 1945–46* (Lafayette, Indiana: Purdue University Studies, 1968), pp. 125–35.

it and we were particularly gratified to see that on general principles it is very close to a report we drew up on behalf of the new Atomic Scientists' Association[415] which we are forming here, for submission to the United Nations Atomic Energy Commission. This similarity in some places even goes so far as to make certain sentences similar almost word for word. Our document, on the other hand, was very much more modest and rather more like a skeleton on which the American report has put some very appropriate flesh. People here were so satisfied with this report that we did, in fact, consider withdrawing our own document entirely and in place of it backing the American report. I did decide eventually to go ahead with ours, mainly because it strengthens the case to show that a completely independent group has reached so similar conclusions and also because the wording of the American report makes it a national rather than international document which it would not be quite correct for us to back as a substitute for a report of our own.

I have not been as lucky as you in that I have, as yet, had very little opportunity to learn some physics again, but things are clearing up and life is getting easier particularly now that we have moved into our newly acquired house. This is by no means perfect, but in the circumstances quite satisfactory. And it was a great pleasure to unpack our twenty-nine pieces of luggage for the last time (some of them for the sixteenth time), and forget their existence.

Yours sincerely,

[Rudi]

[66] Rudolf Peierls to Hans Bethe

Birmingham, 24.1.1947
(carbon copy)

Dear Hans,

There are many things to write about after such a long time but for the moment I shall leave the domestic matters and write only about shop.

[415]The Atomic Scientists' Association was set up in 1946 primarily to educate British public opinion about nuclear matters and to make a case for international control of atomic energy.

My department has got going and we have started on quite a number of problems, although we have not as yet produced much in the way of results. I would like, in this letter, however, to tell you what we are doing, for two reasons: firstly I would be glad of your comments on any of these points and, secondly, many of the problems are of such a type that I rather expect someone in America is also thinking about them. In that case it is quite likely that you may know and, while it may be quite worth while to go ahead tackling such problems in several independent ways, it would in any case be a comfort to us to know.

I think I can best describe our work by listing the people who are here and what they are doing.

> Jahn[416] is a man whose name I am sure you know. He has been here since August and in connection with his knowledge of group theory he was particularly interested in Wigner's theory of symmetries in nuclei. His ultimate aim is to see which of Wigner's results, in particular about the most reasonable coupling to assume, have to be modified in the light of the existence of non-central forces. No doubt this problem is also being thought about by other people, but we want to know the answers and want to learn the techniques. For the moment he is trying to alter the presentation of the Wigner theory by bringing in the explicit representation of the permutation group to see why the results appear only to depend on whether the nucleus is of the type $4, 4r+1$ etc., although for the derivation one has to use explicitly the permutation groups of a higher order. What he is trying to do is analogous to Slater's method in the case of the atom. This, I think, would be a very useful step.[417]

[416]Hermann Arthur Jahn (1907–1979), studied under Heisenberg and van der Waerden in Leipzig, before working at the Royal Institution (1935–41) and the Royal Aircraft Establishment (1941–46); he joined Peierls' department in 1946 and later became professor of applied mathematics in Southampton (1949–1972).

[417]See among others H.A. Jahn and J. Hope, 'Symmetry Properties of the Wigner 9*j* Symbol', *Phys. Rev.* **93**, 318–21 (1954).

Preston,[418] a pupil of Infeld's from Toronto, came here with an almost completed paper on the theory of alpha-decay.[419] This carried the one-body-problem to a logical conclusion, fitting the phases of the wave function inside the nucleus and using an exact analytical solution, even for the case $l \neq 0$. It is doubtful whether it makes sense to carry the one-body model that far but in the course of his calculations two interesting points have come out: (1) the formula given by Gamov from the transparence barrier in the case $l \neq 0$ is completely wrong[420] and based on an inaccurate evaluation of the integral, which, of course, with some trouble can be done in closed form. Apparently this wrong formula has been used ever since, e.g. by you. (2) In the one-body model, the transparency of the barrier as a function of l first increases and then decreases. This surprising result is due to the fact that if you always assume a rectangular potential barrier, but fit its eigenvalue correctly, higher l forces you, for the same total energy, to assume a deeper well (because of the effect of the centrifugal term) and therefore both the density and velocity near the edge of the nucleus increases. It is, of course, very questionable whether this feature of the one-body problem has a real meaning, but it might have and there seems some evidence that the assumption of the decay constant behaving in that way allows a somewhat better fit for the experimental data.

Preston is also working on an explanation of Chang's peculiar results on alpha fine structure assuming that the nucleus is excited by long-range Coulomb forces after passing the potential

[418]Melvin A. Preston, studied at Toronto before joining Peierls' research team in Birmingham where he completed his Ph.D. In 1953 he returned to McMaster, where, apart from a few years at Saskatchewan, he has worked since.

[419]M.A. Preston, 'The Theory of Alpha-Radioactivity', *Phys. Rev.* **71**, 865–77 (1947).

[420]G.A. Gamow, *Constitution of Atomic Nuclei and Radioactivity*, Cambridge, 1937, pp. 91 and 103; G.A. Gamow, 'Zur Quantentheorie des Atomkernes', *Z. Phys.* **51**, 204–212 (1928).

barrier.[421] A simple calculation regarding the alpha-particle as a classical point charge gives a perfectly reasonable estimate for the probability. We felt, however, that this should be done more decently and we struck some snags in that calculation which have dissolved themselves now, leaving him only with somewhat nasty integrals to evaluate. We have seen now a reference to work by Dancoff[422] on the same idea but, as from this it is quite impossible to understand how Dancoff has done the calculation or what the answer is, we propose to continue. In any case, it is a fascinating problem.[423]

When this is finished he will tackle the question of the effect of non-central forces on proton-proton scattering. This, of course, appears only at higher l and therefore at higher energies than the p-n scattering, but it is likely that such energies will be, or in fact have been, reached. In this connection we computed for the experimentalists some curves for the angular distribution of the p-p scattering at about 13 MeV and the curves are very entertaining. They mean, in particular, that in order to sort out the contributions of s, p, etc. waves, one has to measure the scattering with quite a fine angular definition. Experiments along these lines are planned here now, but as the cyclotron is not working yet they will have to wait.

Skyrme is studying the problem arising from the dispersion formula (Bethe-Placzek,[424] Kapur-Peierls etc.[425]) with a view to

---

[421] W.Y. Chang, 'A Study of Alpha-Particles from Po with a Cyclotron-Magnet Alpha-Ray Spectrograph', *Phys. Rev.* **69**, 60–77 (1946); W.Y. Chang, 'Low-Energy Alpha-Particles from Radium', *Phys. Rev.* **70**, 632–39 (1946).

[422] S.M. Dancoff, Metallurgical Project Report, *Short Range Alphas in Natural Radioactivity.*

[423] Preston published his results two years later in M.A. Preston,'The Electrostatic Interaction and Low Energy Particles in Alpha-Radioactivity', *Phys. Rev.* **75**, 90–99 (1949).

[424] H.A. Bethe and G. Placzek, 'Resonance Effects in Nuclear Processes', *Phys. Rev.* **51**, 450–89 (1937).

[425] P.L. Kapur and R. Peierls, 'The dispersion formula for nuclear reactions', *Proc. Roy. Soc.* **A166**, (1938), 277–95.

seeing what one can get out about potential scattering and the like. We have just received Wigner's latest papers on this subject[426] but do not appreciate yet to what extent they dispose of the problem. Skyrme has also given a tidier proof of the completeness of the complex eigenfunctions used in my paper with Kapur.

Salpeter,[427] an Austro-American, is tackling the problem of self energy and has succeeded in proving that (contrary to the idea of Peng's paper[428]) the infinities are not due to using perturbation theory, at least in the following sense: if the integrals are cut off at a certain maximum we have factor k_0 and the wave equations are then solved rigorously for any k_0: then in the limit $k_0 = \infty$ the energy tends to infinity. This proof is now complete for all one particle problems. In the relativistic one-electron problem it leads to a divergence as $\int dk$ rather than $\int k\, dk$, as in Waller's case. This, however, we have convinced ourselves, justifies no hope of a similar improvement in the Weisskopf case.

The next step would be, of course, to try and generalise such a proof to Weisskopf's case of pair theory.[429] In attempting this, however, we discovered that Weisskopf's method is wrong, since the subtraction of the infinite charge density due to the electrons in negative states is done by reference to the states in zero field comparing the two infinite sums, term by term, for the same momentum. Since, however, momentum is not gauge invariant, the whole procedure is not gauge invariant, as I have shown years

---

[426]Eugene P. Wigner, 'Resonance Reactions', *Phys. Rev.* **70**, 606–18 (1946).

[427]Edwin E. Salpeter, a graduate from the University of Sydney, completed his Ph.D. under R. Peierls in 1948. He then moved on to Cornell, where he became professor of physics in 1957.

[428]H.W. Peng, 'On the representation of the wave function of a quantized field by means of a generating function', *Proc. Roy. Ir. Acad.* **51**, 113–22 (1947).

[429]Weisskopf had published several papers on self-energy. See, e.g., V. Weisskopf, 'On Self Energy and the Electromagnetic Field of the Electron', *Phys. Rev.* **56**, 72–85 (1939). See also R.E. Marshak and V.F. Weisskopf, 'On the Scattering of Mesons of Spin 1/2 $\hbar$ by Atomic Nuclei', *Phys. Rev.* **59**, 130–35 (1941). For a survey of developments in electron theory see V. Weisskopf, 'Recent Developments in the Theory of the Electron', *Rev. Mod. Phys.* **21**, 305–15 (1949).

ago in the Royal Society Proceedings. This makes the whole theory inconsistent, since in the ordinary quantum electro-dynamics used by Weisskopf the relation $divE = 4\pi\rho$ can hold at all times only if all interactions are gauge invariant. We are trying to put this right, but cannot say yet whether this will affect Weisskopf's result.

McManus[430] is still working on the field theory for finite size electron, of which you know the general idea. There were many snags in the mathematics but as far as the classical theory is concerned everything straightened itself out, with one exception: we have obtained the equation of motion of an electron without external field which, in this case, is an integral equation. We can easily prove that this has a trivial solution corresponding to motion at constant velocity and also that if you add external fields you get the usual radiation terms up to the normal radiation damping, with corrections only of higher order in the frequency than the Lorentz damping. It seems very plausible that these should be the only solutions of the type of Dirac's runaway solutions. We are still struggling, however, with a reasonable proof for this. As to the quantisation of such theory we need a brainwave but we have not yet given it up. An attempt to do this is also being made by Bleuler,[431] one of Wentzel's pupils whom I managed to get largely through your good offices and who also is tackling one or two more formal problems in the theory of quantised fields.

A young research student is just starting work on an attempt to calculate internal pair creation in the case of a gamma-ray of arbitrary multipole character.[432] I do not believe this has ever been done but in the case of gamma-rays of energies well above 1MeV

[430]Hugh McManus who had been a research fellow at Birmingham before the war, returned to Birmingham from his war-time assignment at Chalk River, where he had been working at the Canadian atomic energy laboratory. He continued his work as a research fellow at Birmingham until 1951 and eventually became professor of physics at Michigan University.

[431]Konrad Bleuler (1912–92), completed his Ph.D. under Wentzel's supervision in Zurich; he later became Professor of Theoretical Physics at Bonn University.

[432]The student was R.A. Fatehally.

internal pair creation should be quite well observable and experiments on the sharing of energy between electron and positron or on their relative angular distribution ought to give a convenient way of determining the order of the multipole. This, of course, can also be done with internal conservation, but as for high gamma-ray energy and low atomic number the distinction between K, L, etc. electrons may not be easy, this alternative method seems of interest. This, in particular is a problem which one would hardly like to duplicate, so if you know of anyone else doing this I would be glad to know.

Krook,[433] a South African who formerly worked on astrophysics, is trying to calculate the continuous gamma-ray spectrum emitted in a proton-neutron collision when the neutron is not captured. For low energies, this will, of course, be quite a small effect although it should be observable. For high energies, it will be more likely than capture and may there help to get information about forces. What is planned at present is to calculate it for some simple type of force so as to have this available for comparison when we get experimental data.

Another research student is looking into Landau's theory of liquid helium.[434] We are rather dissatisfied with that theory because, while it makes out an excellent case for the existence of what Landau calls "rotons", he assumes that there is a lower limit to the roton spectrum from dimensional arguments, leaving out of account that there is a dimensionless number, namely the number of helium atoms, which may enter into this result. We think it should, in fact, be possible to prove that this is the case and that for a reasonable amount of helium the lowest roton level is practically zero. This, however, is not quite an easy problem. It

[433] Max Krook (1913–1985), had studied astrophysics at Cambridge before becoming research fellow at Birmingham. Eventually, he returned to astrophysics, becoming professor at Harvard.

[434] After the discovery of superfluidity and liquid helium in 1938, Lev Landau constructed a theory of 'quantum liquids' at very low temperatures. His papers between 1941 and 1947 are concerned with the quantum liquids of the 'Bose type' such as the superfluid liquid helium (^{4}He).

is mathematically connected with the old problem of the rotation of nuclei, on which I am dissatisfied with the paper by Teller and Wheeler.[435] Probably if one can decently solve one of the two problems one can also get the solution to the other.

Another research student is learning all about luminescence in solids in order to assist Garlick[436] (whom I believe you met at a recent meeting at Cornell) in the analysis of the rather interesting experiments he is doing here.

Kynch,[437] who is rather busy with teaching, is working on the non-central forces in the nucleus. There seems, in particular, a peculiar discrepancy between your result and a similar result of Schwinger and Rarita on the one hand[438] and the result of Rosenfeld and Møller on the other.[439] You find that in order to account for the observed quadrupole moment of deuteron one has to assume a very strong tensor force, comparable in order of magnitude to the central force. Møller and Rosenfeld start with a tensor force which, coming only from the non-static terms, is about ten times smaller, treat it with perturbation theory and get the right quadrupole moment. One would say, at first sight

---

[435]E. Teller and J.A. Wheeler, 'On the Rotation of Atomic Nucleus', *Phys. Rev.* **53**, 778–89 (1938).

[436]G.F.J. Garlick, was experimenting with phosphors and phosphorescence. See G.F.J. Garlick, 'Phosphors and Phosphorescence', *Rep. Prog. Phys* **12**, 34–55 (1948). He later published a joint paper with Fatyehally: G.F.J. Garlick and R.A. Fatehally, 'Measurement of Particle Energies with Scintillation Counters', *Phys. Rev.* **75**, 1446 (1949).

[437]J.G. Krynch (1915–2003) had studied theoretical physics at Imperial College, London, joined the Birmingham department initially to take over Rudolf Peierls' teaching, while the latter was engaged in work for the M.A.U.D. committee. His temporary appointment was extended several times and he stayed until 1952 when he accepted the Chair of Applied Mathematics at University College of Aberystwyth. In 1957 he became professor of mathematics at UMIST.

[438]W. Rarita and J. Schwinger, 'On Neutron-Proton Interaction', *Phys. Rev.* **59** 436–52 (1941); W. Rarita and J. Schwinger, 'On the Exchange Properties of Neutron-Proton Interaction', *Phys. Rev.* **59** 556–64 (1941).

[439]C. Møller and L. Rosenfeld, *Kgl. Danske Vid. Math.-Fys. Med.* **17**, No. 8 (1940); C. Møller and L. Rosenfeld, *Kgl. Danske Vid. Math.-Fys. Med.* **23**, No. 13 (1945).

that this simply means the perturbation theory is not justified, but it is the same kind of perturbation theory that one uses for calculating the fine structure in the atom and there we are reasonably confident that it works. It might, in fact, be that because of the peculiar nature of the problem the perturbation theory is more trustworthy than the rigorous solution. In this connection Kynch is also exploring the use of the Svartholm method, which treats the Schrödinger equation as an integral equation in momentum space and uses iteration. It appears that this works also for tensor forces and is, in fact, quite convenient.

This, I think, completed the list, except that I have not said what I am doing, but that is easy — I sit on committees and write letters.

I shall be writing about other things soon.

Yours sincerely,

[Rudi]

[67] Hans Bethe to Rudolf Peierls

Ithaca, 8.7.1947

Dear Rudy,

The rumors about my visit were in a still different category, (d) change of mind. I had made some plans to come to your conference as a representative of the Brookhaven Laboratory, and to bring Rose with me. Unfortunately I then changed my mind for two reasons: first because Rose and I had a rather hectic year, with much running around. Second, the preliminary agreement of the Brookhaven Laboratory and the Atomic Energy Commission came so late that it seemed most doubtful that we could get a passport for Rose in time, and I did not want to go without her.

I am very sorry about this, because I really should have liked to come, and to see you and discuss physics with you. I am enclosing a little note instead of myself. We discussed a lot about these problems of self-energy at a small conference in Long Island a month ago.[440]

Instead of coming this summer for a short time, I am making some very tentative plans to come next year for a longer time. There is a chance that I can get a sabbatic leave from here for the Fall semester, 1948 (September 1948 to January 1949). If this materializes, I should very much like to spend this semester in England. Is there any chance of getting a guest professorship at some university?

The family is very well, but the part under four years is very noisy. How are you? We heard Genia was ill. Is she better?
Sincerely yours,

Hans

[68] Rudolf Peierls to Hans Bethe

Birmingham, 28.7.1947

Dear Hans,
Thank you for your letter. Our conference has just finished[441] and it was a lot of fun, though no doubt it would have been more fun if you had managed to get here. We discussed a number of problems, including the one related to your note.[442] My personal impression is that physically you are almost certain to be right, i.e. that the observed shift is due to self energy, but that this will not come out of the present theory

[440]Between 2 and 4 June 1947, the first of three conferences on the foundation of quantum mechanics was held at Long Island's Shelter Island. Among the 24 attendees were Edward Teller, J. Robert Oppenheimer, John van Neumann, John A. Wheeler, I.I. Rabi, Richard Feynman and Julian Schwinger, Victor Weisskopf, H.A. Kramers and K.K. Darrow.

[441]Peierls had organised a small theoretical conference at Birmingham which dealt with the fundamental difficulties and with elementary particles. This took place between July 23rd and 26th. See also letter [67].

[442]Bethe had just submitted a note on the electromagnetic shift of energy levels which was to be published later that year. H.A. Bethe, 'The Electromagnetic Shift of Energy Levels', *Phys. Rev.* **72**, 339–41 (1947).

and that, in particular if one applies present theory to this problem the reduction in the order of magnitude due to pair theory and that due to taking the difference between two states will not be cumulative, so that one will get at least still a logarithmic infinity in this result. One may not even get a definite result at all because, as far as I know, no formulation of pair theory exists which is consistent and Lorentz invariant beyond the first approximation. I do believe, however, that in a future theory in which one has eliminated the infinities the result for the level shift will look very much like yours.

We have discussed further the paper by you and Oppenheimer[443] on the Heitler theory and Ferretti has traced the trouble to the result that in the damping theory light signals do not propagate with light velocity.[444] This, of course, is the effect of the reinterpretation of the theory in which one gives up the ordinary space-time description. We have all come to the conclusion that this is a fundamental feature of all those theories in which one tries to throw out all self-energy terms and therefore one should now consider only theories involving a fundamental constant of the dimensions of a length in which only the contributions from very short waves are neglected but the others, which, as you point out, are needed, are retained. Your letter, of course, tends to strengthen this view.

How about plans for next year? I think it should not be difficult, if you can get leave of absence, to find a suitable position for you for a few months. Needless to say that it would appeal to me most if that position could be found in Birmingham, but it depends somewhat on what you want. There are a number of places which, at the moment, have no decent theoretician — for instance, Dee in Glasgow has only some rather junior men and a new Chair has just been created at Liverpool which I imagine it will be hard to fill. In one or the other of these places, you would therefore be carrying out a valuable job of putting the experimentalists on the right lines.

[443] Hans A. Bethe and J.R. Oppenheimer, 'Reaction of Radiation on Electron Scattering and Heitler's Theory of Radiation Damping', *Phys. Rev.* **70**, 451–58 (1946).

[444] Ferretti's results were the basis of a paper published in *Nature*. See letter [69], note 447.

On the other hand, I take it your point in taking sabbatical leave is to get away for a time from administration and teaching duties and to sit in a place with the right atmosphere in which you could do your own research and where there would be enough younger people to pick up any spare problems you happen to scatter. From my point of view Birmingham might be the best place, though this would, of course, not exclude your meeting the people from other universities.

Another alternative would be a kind of general job, sponsored, perhaps, by the Royal Society, in which you would spend periods at all the places that want to see you, but if you want to sit down and get some work done this is hardly to be recommended. Please let me know what your ideas are on this subject so that we can set the wheels turning.

Another question is about the financial side of the arrangements. The new scheme of sabbatical leave that we are going to institute here provides that the full university salary continues and may be supplemented by a grant to cover travel expenses and higher cost of living. In trying to organise something here, one would have to know whether your regulations are similar, because if one of us went on study leave and got a grant by the university he was visiting, this would merely serve to reduce the cost to this home university, and naturally nobody here would be anxious to make a grant merely to save expenses to Cornell. On the other hand, we would, of course, be anxious to make sure that you are not out of pocket as a result of the transaction. Would you want to bring your family and, if so, would they be with you or perhaps in Belfast?[445]

There is no doubt whatever that we can arrange for you to get the necessary status and facilities and very little doubt that we can arrange an adequate grant.

Yours sincerely,

Rudi

[handwritten addition]
I hear you are getting Dyson for next year, he seems quite an outstanding man. I hope you will look after him well!

[445]Paul Ewald, Hans Bethe's father-in-law had settled in Belfast.

[69] Rudolf Peierls to Hans Bethe

Birmingham, 6.8.1947
(carbon copy)

Dear Hans,
We have made a little further progress on the line of thought that you and Oppie have started,[446] and you may be interested in the note which we are sending to Nature.[447]
Yours sincerely,

[Rudi]

[70] Rudolf Peierls to Hans Bethe

Birmingham, 14.8.1947
(carbon copy)

Dear Hans,
Thank you for sending me a draft of your paper on the two-meson hypothesis.[448] This is all very interesting as Uncle Nick would say, but I do not believe that the second kind of meson which you postulate can have anything to do at all with the Bristol photographs.[449] The chief

---

[446]See letter [68], note 444.

[447]R. Peierls and B. Ferretti, 'Radiation damping theory and the propagation of light', *Nature* **160**, 531–34 (1947).

[448]Hans A. Bethe and R.E. Marshak, 'On the two-meson hypothesis', *Phys. Rev.* **72**, 506–509 (1947).

[449]At Bristol, photographs had shown the development of light mesons from heavy ones. C.M.G. Lattes, H. Muirhead, G.P.S. Occhialini and C.F. Powell, 'Processes involving charged mesons', *Nature* **159**, 694–97 (1947). The results were reported, among others, at a conference at Harwell, where Powell explained the experiments in detail. See also C.M.G. Lattes, G.P.S. Occhialini and C.F. Powell, 'Observations on the tracks of slow mesons in photographic emulsions', *Nature* **160**, 453–56 and 486–92 (1948). Further developments are described in S. Schweber, 'Shelter Island, Pocono, Oldstone. The emergence of American quantum electrodynamics after the war', *Osiris* **2**, 265–302 (1986).

reason for this is that in the Bristol technique for any heavy meson stopping in the plate, the probability of the track of the light meson, if any, also lying within the plate is very small. In other words, for any track which they have found in which a light meson is visible there must be many others which also exist, but do not appear. This means that the heavy mesons do not represent, as you say, something like 6% of all mesons stopping in the plate, but from the latest Bristol figures something like 50%. In the light of this, the discussion on your page two is somewhat misleading. Incidentally, the photographs were not taken at 30,000 feet as you say on page two, but on top of a mountain and your guesses on page five about the dimension of the plane do not make much sense, instead there was a lot of snow around.

Similarly, your footnote seventeen is hardly justified since it is very likely that there was in that case a secondary meson present but that it went out of the plane of the emulsion. Equally, of course, this may have been a light meson.

I cannot help feeling that all the facts we have at present, if they are right, cannot even be understood with the help of two kinds of mesons.

I enclose a re-draft of our note to Nature which I hope is more intelligible than the first.[450]

Yours sincerely,

[Rudi]

[71] Hans Bethe to Rudolf Peierls

Ithaca, 3.9.1947

Dear Rudi:

Thanks very much for your many letters, scientific and otherwise. I liked your paper about Heitler's Theory very much.[451] It is certainly the most striking argument against the theory that it does not give the

[450] See letter [69], note 447.

[451] See letters [68–70].

correct velocity of light. It ought to be possible to show this not only by indirect argument but also by direct calculation. You are probably doing this right now.

I did a little more on the electro-magnetic line shift. I have done it relativistically and it is indeed finite. I do not have the numerical result because there are too many terms, each of which has to be evaluated. But the result is essentially the same as in the unrelativistic ca[l]culation. There are some interesting points in it, concerning the subtraction of the electron-magnetic mass, and also concerning the cut-off procedure for the self energy. But the main point is that it is finite and depends essentially on $\psi^2(0)$.

Thanks for your criticism of our note on the heavy meson. I am afraid the note was written in a great hurry, and what is worse, it will be printed by "Physics Review." without our getting any proof.[452] So the large density of mistakes will stay in. I am particularly unhappy about the mess I made concerning the altitude in which the measurements were made; as they were made on a mountain, our statement is not correct that most of the mesons are produced in the neighborhood of the measuring apparatus. You are also at least qualitatively right that there must be many heavy mesons which do not appear as such because they escape from the plate. However, I believe that quantitatively this is not quite as important as you say, because the emulsion is rather thick and amounts to about 40 * percent of the range. I think some change in ionization density should be noticed in that distance. The famous Footnote 17 is of course unintelligible, but the meson in question originates in a nuclear disintegration in the plate itself, and should therefore be a heavy meson according to our theory. Since this statement was omitted in the footnote, your objection that it might be a light meson will occur to most readers, but really it cannot be.

Now about the plans for the future. It was awfully nice of you to write immediately about the possibility for my possible visit. I have not yet got my leave approved but I am working on it and I hope it will be definite in about a month or so. The financial arrangements are

[452] See letter [70], note 448.

that the University pays my regular salary but no expenses. I am losing a considerable amount in consultation fees from the General Electric Company which of course I cannot visit while I am in England. None of the money which I might get from an English University will go to Cornell. So it would be nice if I could get some sort of position; in this case I will probably come out just about even, that is I will be able to pay the travelling expenses and replace the loss of GE fees.

I should like to go to a place where I am really useful and should like to give some regular lectures, as long as there is not too much of this (let us say no more than six a week). I think at the present time, with physicists so scarce, I should not completely retire to the luxury of pure research. Concerning places I want to try — first of all to make a deal with Mott.[453] Our department here has invited Mott previously to come for a semester and would still be very anxious to have him. There is a lot of work in the solid state going on here, and I am too much out of this to give any advise. So the present plan is to ask Mott to change places with me for one semester. Please keep this to yourself for the time being, because I have not yet approached Mott, since my leave has not definitely been approved.

If this does not work out, I think Liverpool would be a very attractive place. Also, now with Frisch there,[454] Cambridge would be interesting. But I think I should let fate take its course and not put too many boundary conditions on this problem. Any University where there is an attractive program in experimental physics and where a theoretical physicist is needed, will be fine. I should like to stay at one place and not travel around too much. Of course I would want to see you frequently, but distances generally are not very great.

The plan is to take the whole family and this might make a difficulty from the standpoint of housing. We would visit Rose's parents during the summer and would expect to leave the children there while Rose and I go to the Continent for a few weeks. But when term begins, it would be nice to have the family happily reunited. The time would be approximately from July to the end of January. I do not know how this

[453]Nevill Mott held the Chair of Theoretical Physics at Bristol at the time.
[454]Otto Frisch had accepted the Jacksonian Chair in Cambridge in 1947.

fits in with your terms; as I remember the fall term ends at Christmas. If the job I am going to get is just for the fall term, it might be very nice if I could spend a month with you in Birmingham.

This is just to give you a vague idea, I hope it won't make too much trouble for you to look for a suitable place. But both Rose and I are dreaming very much of this visit.

We had a very nice time together in southwestern Colorado. At present, Rose's mother is visiting us and it is really very nice. She has hardly changed at all in the nineteen years since I know her.
Best regards to the family.
Yours sincerely,

Hans

*I did not have time to look up the numbers, so this may be wrong.

[72] Hans Bethe to Rudolf Peierls

Ithaca, 23.9.1947

Dear Rudy,
One of my students, Edwin Lennox,[455] is very much interested in spending a year or two in Europe. He will probably get his Ph.D. next June and would like to come over after that. He would like best to come to you if you would like to have him, and if you could give him some research position.

You may remember Lennox from Los Alamos while he was in Vicky's group. In the meantime, he has developed very well and had learned a lot of physics. When he was at Los Alamos, he had had only a very small amount of physics courses. I think he is very good; at least he shows great interest in all problems and tries to get a real understanding of everything in physics.

[455] Edwin Lennox completed his Ph.D. with Hans Bethe and later took up a post at Ann Arbor before moving to the Department of Chemistry at the University of Illinois, Urbana.

I remember that you had some research positions available when you went back to Birmingham. Would any of these be available to Lennox for next year? What would be the salary that he could get? He just encumbered himself by marrying a widow with three children; however, they get some pension from the U.S. Government and possibly some financial support from the grandfather of the children. So the situation is not as desperate as it sounds and you may consider him simply as married without any children, as far as financial needs are concerned.

If it is possible for you to take Lennox, he would of course also be interested in the problems of life, especially whether there is any chance to find a place for him to live. If this looks very black, he may be better advised not to come. He would also be interested in knowing for how long your research appointments would run.

Physics continues to be very exciting especially the electromagnetic shift of the energy levels and the mesons. At a conference the other day, we admired the latest pictures from Bristol, especially the one in which a meson is produced in a star and then produces another star.[456]

I hope to hear from Vicky and Placzek very soon on what happened in Copenhagen. I assume you were there.[457]

Best regards to the family.

Yours sincerely,

Hans

[456]Prior to the advent of large-scale accelerators in the mid 1950s, Cecil Powell and his collaborators at Bristol (Occhialini, Lattes, Camerini, Muirhead, and many others) made a number of important discoveries in the study of cosmic radiation, so much so that Louis LePrince-Ringuet referred to Bristol as 'the big sun surrounded by little satellites'.

[457]Rudolf Peierls attended the latter part of the conference organised by Niels Bohr in late September 1947.

[73] Rudolf Peierls to Hans Bethe

[Birmingham], 27.9.1947
(carbon copy)

Dear Hans,
Thank you for your letter. I still do not agree with your interpretation of Powell's experiment.[458] The point is not that if the track goes at an angle to the plate it cannot be recognized as that of a meson, but that it is not detected at all. The plate is always horizontal under the microscope and if the track has a strong "dip" the focal plane will only contain one or a few grains at a time and, therefore, no recognizable track will appear. No doubt a more elaborate inspection technique would disclose such tracks, but there would be no point in trying this since all the information can be obtained by observing only the tracks which are nearly in the plane of the emulsion.

In any case, the others have not been looked for, and all the dozen or so cases found at Bristol are cases in which both primary and secondary mesons lie in the plane of observation. Powell is going to publish his conclusions about the statistics.[459]

Now about your plans, I am not taking any action at present since the question of Bristol is best explored by your approaching Mott directly.

Here are, however, a few points for your information; the winter term lasts in all universities from early October until Christmas. The term is, however, not an important unit for teaching; all courses are planned on the basis of a whole session and as far as I can see it would make no difference if your visit did not cover an integer number of terms. (A whole session, October to June would be different but I take it that is too long for you).

[458] Cecil Frank Powell (1903–1969), studied natural sciences at Cambridge where he gained his Ph.D. working under Wilson and Rutherford. He moved to Bristol where he eventually (1948) became Melville Wills Professor of Physics.

[459] C.M.G. Lattes, G.P.S. Occhialini and C.F. Powell, 'Observations on the tracks of slow mesons in photographic emulsions', *Nature* **160**, 453–56 and 486–92 (1948).

In Bristol, the teaching of Applied Mathematics is in the hands of the Mathematics Department. The theoretical physicists do not do the teaching. I do not think that Mott himself gives as many as six lectures a week.

At Birmingham (and in some other places) people with our background do lecture on mechanics, hydrodynamics, electricity etc. to undergraduates including engineers, and we have now a staff of three besides myself, for this purpose. It would be hard to justify a temporary appointment for this kind of work, particularly if it is only for part of the session, so that someone else would have to finish the courses which you start. Here, at any rate, it would be much easier to build a case on the stimulating effect your presence would have on the research in both experimental and theoretical physics. I cannot speak for Bristol, but I imagine things would be rather similar. If you just swap places with Mott, your main function would be to replace him in the administration of the department and in the supervision of research.

Housing is, as you know, not easy. One does, however, see from time to time furnished houses or flats at not unreasonable rents, and there are also boarding houses that might take you with the family. (Best of all close to somebody's house where Rose could go for washing or similar activities not tolerated in a boarding house).

Liverpool are trying to fill a new chair in theoretical physics.[460] This is likely to be filled before your visit. If they are unsuccessful and the chair is still vacant in 48–49 they would, of course, jump at the chance of having you for a short period and it would there be particularly easy to organize funds since the job exists.

Failing this, I think Cambridge should also not be difficult.

Next come some plans of mine. I find I shall have to attend a meeting in Washington on 14th and 15th November, though there is some talk of it being a week earlier. It comes at an awkward time from the point of view of my duties here and I do not want to stay away longer than necessary. But I might get in an extra day or two and in that case would,

[460] The Chair was filled in 1948 by Herbert Fröhlich who remained at Liverpool until 1973.

of course, very much like to come and see you. As we are very limited on dollars now, this would only work, if my fare from Washington to New York could be covered somehow. What are the chances of this? I could, of course, give a lecture on any subject.

I shall let you know more about dates as soon as possible.

Greetings to everybody.

Yours sincerely,

[Rudi]

[74] Hans Bethe to Rudolf Peierls

Ithaca, 3.10.1947

Dear Rudy,

This is very fine that you will come to this country. Of course we want to see you and of course we can pay your travel expenses from Washington and back to New York or any other linear combination. In addition, we can pay you $40.00 for each day you are here. I hope they will be many.

It would be nice if you gave us a talk in the Colloquium on Monday the 17th. Either your work on the Heitler theory or something about the meson experiments would be fine. We should also like to hear about your work on the relativistically invariant cut-off. Please talk about whatever subject seems most interesting to you. Let me know as soon as possible on what date you expect to arrive and what the title of your talk will be.

If you need any travel advance sent to Washington, please let me know.

Hoping to see you in November.

Yours sincerely,

Hans

[75] Rudolf Peierls to Hans Bethe

[Birmingham], 8.10.1947
(carbon copy)

Dear Hans,
Thank you for your letter about Lennox. I am naturally very interested in the possibility of having him here. He must, however, realise that things are not going to be easy at all from his own point of view.

Our research fellow-ships vary in seniority and salaries range from £450 a year to £700. This is subject to the usual 5% deduction for superannuation with the University adding 10%. The superannuation is optional and would not apply to anyone who was not intending remaining in academic life in this country.

In addition there is a family allowance of £50 p.a. for each child of school age or below. I would, of course, have to know more about his progress and ability before expressing an opinion where his salary would lie within the range I have quoted, but unless he has developed in quite a spectacular way, it is hardly likely that he would find himself at the top of the grade. My present guess would, therefore, be that he might well expect £500 to £550 and more if we can make out a case. I cannot also at this moment give an assurance that there will be a vacancy for him, we are likely to have a vacancy next session for which there may also be competition, but more vacancies may arise.

The terms of these fellow-ships are yearly appointments up to a total tenure of five years. This last limit is not likely to be serious since, if he wants to settle down in this country, it is quite likely that he will find a teaching job that would attract him either in Birmingham or elsewhere.

He would, however, have to pay his own passage as well as that of his family, since we have no grant that could be used for this purpose. I do not know whether it would be possible to find some outside grant to help him, but I am sure that at most his own passage could be covered in this way.

As regards living conditions here, the salaries I have quoted are adequate though not generous for young married men. To support three children on them would be very tough. Housing is still very short here and it is virtually impossible to rent a house or flat unless he is exceedingly lucky or unless he is allocated one of the very few houses under the control of the University for which, however, there is very strong competition. One can occasionally rent furnished accommodation, but that would be rather expensive and it may not be easy to find anything satisfactory for the size of his family.

It is possible to buy houses, a small modern house is likely to cost about £1,500, of which two thirds, or with luck a little more, could be covered by a mortgage. This would have to be furnished unless he can bring furniture from America, this again is difficult and expensive.

It should also be born[e] in mind that the style of dress, especially for the children, is rather different here from what it is in America, and he would, therefore, probably have to get almost completely new outfits before he came over.

It looks, therefore, as if the whole project was extremely difficult unless he was able to supplement his income at least for the initial expenses and during the first year quite substantially from private means. If he feels, provided things turn out the right way, he might settle in Europe for good, it may be worth going through all this trouble; if what he has in mind is a year or two on this side, it would seem too big a sacrifice.

Perhaps you would discuss these facts with him and if you would let us know a little more both about his progress in research and about his other plans, I may be able to give more definite advice.

The meeting to which I have to go in Washington has now definitely been fixed for 14th–16th November. I intend to leave here about the 10th and to start back on the 19th. If during that time there could be a chance of seeing you, that would, of course, be very nice.

Yours sincerely,

[Rudi]

[76] Rudolf Peierls to Hans Bethe

[Birmingham], 1.11.1947
(carbon copy)

Dear Hans,
My movements are changed again, since the only flight I could get leaves London on the 5th and should arrive in New York on the 6th. I am planning to call at Chicago, M.I.T. and Princeton in that order if it will suit my learned colleagues in those places and if it suits you I would like to leave our date unchanged, spending the 17th and 18th at Cornell.

I shall let you know as soon as I make contact with the others what my movements will be, so you can get in touch with me if you want to rearrange the date.
Yours sincerely,

[Rudi]

[77] Rudolf Peierls to Hans Bethe

Birmingham, 24.11.1947

Dear Hans,
On my return here, I found a list of people who have been invited to write reports for the Solvay Conference,[461] and I am glad to see your name. I had already promised to write a report on self-energy problems, not knowing, of course, the rapid progress that is now taking place in the United States in this field. Definitely we should try and avoid too much overlap and we ought to agree what is the best division of labour, if any. Could you let me know as soon as possible whether you are proposing to come to the conference, whether you are agreeing to write a report and what your views are about the best division.

[461]The 8th Solvay Conference, the first post-war conference, took place in September 1948 in Brussels.

For example, I might try to summarise the account of the difficulties of the old theories including a criticism of the attempts by Dirac, Gustafson, Heitler etc.,[462] and deal with the position of theories not including perturbation. This would include your own proof that the divergence does not depend on perturbation theory. It would also include the classical theory of McManus and the general method of Feynman to the extent to which I have understood it, or shall understand it. It would leave you all those theories in which finite results can be obtained by an intelligent application of perturbation theory without modification, including your own work, the recent Princeton results,[463] Schwinger etc.[464] This, however, is only a tentative suggestion and any division would be all right with me provided it does not involve my writing about the Princeton results of which I have only heard in conversation, and Schwinger's calculations which I do not know at all.[465]
With best wishes,
Yours sincerely,

Rudi

---

[462]At the Shelter Island Conference of June 1947 (see letter [68], note 440), Willis Lamb and Robert Retherford had presented their results of measurements of the fine structure of the energy levels of the hydrogen atom, which were in disagreement with older theories of the hydrogen atom, e.g. Dirac's theory, which had predicted that the lowest two excited states of the hydrogen should have equal energy. The discrepancy between Lamb's experimental data and theory was explained by some as resulting from quantum fluctuations of the electromagnetic field acting on the electron in the atom, which would have given the electron additional energy, the so-called self-energy. However, in the existing theory of QED the value of self-energy was infinite.

[463]Hendrik Kramers who had spent a term at Princeton as a visiting scientist introduced the idea of renormalisation to the discussion. He argued that according to QED, the observed energy of an electron is the sum of two unobservable energies, the bare energy and the self-energy. Through renormalization bare energies are replaced by observed energies. Hans Bethe carried out the necessary calculations. H.A. Bethe, 'The Electro-Magnetic Shift of Energy Levels', *Phys. Rev.* **72**, 339–41.

[464]After Bethe had completed non-relativistic calculations about the Lamb Shift and magnetic anomalies of the electron momentum, Weisskopf and Schwinger had done some relativistic calculations. See S. Schwinger, 'On quantum electrodynamics and the magnetic moment of the electron', *Phys. Rev.* **73**, 416–17 (1948).

[465]Schwinger submitted his results to *Physical Review* in December and they were published in February 1948.

[78] Hans Bethe to Rudolf Peierls

Ithaca, 4.12.1947

Dear Rudi,
Thanks for your letter of 24th November. It tells me, among other things, that you arrived safely back in England which is fine. In the meantime, we also saw Fuchs whose visit was very nice.

For the Solvay Conference, I would have proposed exactly the same division of labor as you. I wrote Bragg that I would talk about level shift and related problems. By this I mean the Princeton results, Schwinger's results and any other results we might obtain in subtraction physics.[466] I would leave everything about the free electron and the relation to classical theory to you. So in short, I agree.

I[t] was very fine to have you here and we all got a lot of ideas from your visit.[467] Unfortunately, Feynman[468] has not yet quantized your theory but got sidetracked into some invariance problems. He and Lennox have shown tentatively that Pais' method for getting the a finite self-energy does not give an invariant result.

The last two days Vicki was here,[469] so we got still more ideas and still less work done.

I will let you know if anything new develops.

Greetings to everybody, especially to Genia.

Yours sincerely,

Hans

---

[466]See letter [77].

[467]See letters [76–77].

[468]Richard Feynman had taken up a professorship at Cornell. He stayed there until 1950 when he accepted a post at Caltech where he remained for the rest of his career.

[469]Refers to Viktor Weisskopf.

[79] Rudi Peierls to Hans Bethe

Birmingham, 5.2.1948

Dear Hans,
This is mainly about the Solvay Conference. The present position seems to be that the committee is unable to pay transatlantic fares. I have pressed them strongly to approach the London representative of the U.S. Navy, who I imagine would be delighted to take this on. Somehow this has not been done. Instead Blackett, who is now in the U.S. has been asked to contact the Navy people there. This seems a rather complicated way of doing things, and I hope no confusion will come from it. I thought you ought to know this in case you have the opportunity to catalyse things a little.

In confidence I may mention that Blackett also has a list of further names to be added, if the limits on local and transatlantic finance make this possible. I hope that this will lead to the most glaring gaps in the list being filled.

Will you be very rushed for time on your trip, or if not will you be able to visit Birmingham? We cannot treat you quite as generously as you seem to be able to treat visitors to Cornell, but we can take care of your extra expenses and we shall, of course, be delighted to have you here for whatever time you can spare.

Our official Easter vacation extends from 20th March to 20th April, but most of the physicists will, of course, be here anyway, particularly after the Solvay.
With best wishes,

Rudi

Skyrme is applying for a Commonwealth Fellowship and would like to come to Cornell, if he gets it.

[80] Hans Bethe to Rudolf Peierls

Ithaca, 12.2.1948

Dear Rudy,
Thanks for your letter of February 5. There seems to be quite a mess about paying the fares for the Solvay Conference.[470] I saw Blackett for a few minutes and he told me he would talk to the U.S. Navy. At that time he did not seem to want any help, but if he does I shall of course stand by.

Bragg wrote me recently that "he was sure they could arrange to pay for my fare" I am not sure how much that means, but I shall wait for developments.

I am sorry I will be very much in a rush around the Solvay Conference because it is in the midst of our term. Moreover, I am still planning to come over for a more leisurely visit this summer. So I would rather not visit you in connection with the Solvay Conference, unless it just happens that my return plane is delayed.

It seems that our summer trip will also be made my plane so that we shall arrive quite early in June. My plans are to spend the time from then until July 15 visiting various places in England, except for about two weeks that I want to spend with my parents-in-law. That can easily be arranged at my convenience. When will it be best to come to you and to go to other places in England? Will you have a conference during this time? This might be a great advantage to me because it would probably facilitate getting my trip paid.

Skyrme[471] is of course most welcome. The only boundary condition is that I will be in Columbia the first semester next year and at Cornell the second. I do not think that the Commonwealth people and Skyrme will mind this.
Yours sincerely,

Hans

---

[470]The 8th Solvay Conference was to take place in October 1948. Bethe did not take part in the conference after all. See letters [80–84].

[471]Tony Skyrme, at the time university research fellow at Birmingham, spent the academic years 1948–50 in the United States, as research associate at MIT and as a member of the Institute for Advanced Study at Princeton.

[81] Rudolf Peierls to Hans Bethe

[Birmingham], 3.3.1948
(carbon copy)

Dear Hans,
Thank you for your letter of 12th February. I have now at last got my Solvay report finished and, while you will receive a copy through official channels, I thought I had better let you have one as soon as possible for your information.

In most English universities term finishes about 1st July, in Birmingham in particular it finishes with the Degree Ceremony on the 3rd. Between the 24th and 29th of June are our examiners' meetings when there is one or sometimes two meetings almost every day. I am hoping to go on a holiday about the middle of July, so that the best time to see you here would be either before 22nd June or between 30th June and about 12th July. I shall be away for three weeks and then again here from early August, though naturally, of course, during August other members of my department may be away at various times. Lastly, we shall have a conference here from the 14th to 18th September. I hope this is not too late for you. It will this time be a bigger affair covering experimental as well as theoretical problems, but we hope nevertheless to preserve the informal character of our last conference, at any rate in the theoretical meetings. You will receive an official invitation to that conference in a day or two.
Yours sincerely,

[Rudi]

[82] Rudolf Peierls to Hans Bethe

Birmingham, 12.3.1948

Dear Hans,
Enclosed the formal invitation to our conference. I do hope this is not too late, but if you are staying over until the Solvay it will just fit in.

Invitations are being sent mostly to heads of departments and other VIP's, and we hope to get further names from them. We would like to know who could come from your department (Feynman?).

The total number (experimental and theoretical) is limited to 250, so we have to select carefully, but we would be glad to consider any recommendations from you.

Yours sincerely,

R.P.

Are you accepting your own invitation to Jamaica?

[83] Hans Bethe to Rudolf Peierls

Ithaca, 18.3.1948

Dear Rudy,

I suppose the enclosed letter to Bragg is self-explanatory.[472] I should really like very much to be back in time. This of course also applies to your conference.

In view of the new date for the Solvay Conference, I wonder whether your Theoretical Conference would not be more convenient in June or early July. I do not know whether that is possible. If not, could it be early in September? Otherwise the program outlined in your letter suits me very well. I intended to be in Switzerland and possibly Germany from the middle of July to the end of August which overlaps with the time you intend to be absent from Birmingham.

I shall probably come to see you in the beginning of July but maybe I shall stop for a day or two in early June.

Yours sincerely,

Hans

[handwritten addition]

Thanks a lot for your Solvay report!

---

[472]Letter Hans Bethe to W.L. Bragg, 18.3.1948 asking Bragg to take into account the American academic calendar when rescheduling the Solvay Conference. Carbon copy in *Peierls Papers*, Mc.Eng.misc.b202, C.16.

[84] Rudolf Peierls to Hans Bethe

[Birmingham], 23.3.1948
(carbon copy)

Dear Hans,
Thank you for your letter. I am very sorry that the date of our meeting is too late for you. Unfortunately, it is impossible to change it now since the invitations have gone out, the accommodation has been booked (this is very short as you may imagine) and particularly as there are a number of other events in September. It begins with a[n] International Conference on Applied Mathematics and Mechanics from 6th–11th September which will overlap to a slight degree in membership with ours. There are also likely to be some business meetings involving a number of our people as well as possible foreign visitors during the later half of that same week. In the week following our conference is the Cosmic Ray Conference at Bristol which was the first one to get fixed and, in fact, our date was partly chosen to enable the overseas visitors to attend both conferences. Following that, of course, is the Solvay. I am surprised at what you say about the starting date of the American universities since last year the majority of visitors seemed to remain in Europe until the end of the Copenhagen Conference in the very last days of September. I very much hope you will be able to delay your return to take in our conference at least, if not Bristol and the Solvay.

I understand the meeting sponsored by the American Emergency Committee of Atomic Scientists will now not be held in Jamaica, but in the United States. I have said that I accept in principle, but if the date remains in June I shall probably not get away in time. I would be grateful to have your views as to whether this meeting is likely to serve a useful purpose and whether it would be worth the effort to go there, particularly if it means missing important University meetings.
With kindest regards,
Yours sincerely

[Rudi]

[85] Marc Oliphant and Rudolf Peierls to Hans Bethe

Birmingham, 24.3.1948

Dear Dr. Bethe,
We very much hope that your will be able to join us at our conference in September and we hope, in that case, you will let us look after your hotel expenses while in Birmingham.
Yours sincerely,

M.L. Oliphant
R.E. Peierls

Please read this in conjunction with the notice sent to you on 12th March 1948 of our conference on "Problems of Nuclear Physics".

[86] Hans Bethe to Rudolf Peierls

Ithaca, 7.4.1948

Dear Rudy,
Thanks for your letter of March 23. I am rather sad that the conference cannot be moved. The starting of American Universities has always been around September 20. The trouble was only that Uncle Nick has always disregarded university schedules and chosen the dates for his Copenhagen Conference at random.

Some visitors, like Vicky, were too anxious to go to the grand re-opening of the Copenhagen meetings[473] last year and therefore did not return in spite of the semester. (I appreciate that the Bristol meeting

[473]The conference had attracted a large number of physicists who spent some time in Copenhagen during the last two weeks of September 1947. Among them were Kramers, Weisskopf, Pais, Rosenfeld, Peierls, Blackett, Placzek, Wheeler, Rossi, Ferretti, Klein. See letter Pauli to Stern, 19.8.1947, *Pauli Briefwechsel*, III, p. 471.

was the first to get fixed and I protested immediately to Powell about his date, but without success.)

I have pretty much decided to go to your conference but none of the later ones.

I hope that I will hear everything in Birmingham then and earlier in the summer.

The meeting sponsored by the American Emergency Committee[474] is not likely to serve a useful purpose. It was Szilard's idea to have such a meeting to bring together scientists from all countries east and west of the iron curtain. Those from the east of the curtain have already refused to attend, as was to be expected. In the present situation, it is my opinion, that nothing useful can be done from the side of atomic energy or of scientists in general and that any stress on atomic energy can only deepen the international conflict. I was very much against holding this conference and this opinion is shared by other people such as Oppenheimer and Weisskopf.

By the way, I resigned from the Emergency Committee,[475] not because of the conference but because of my conviction of the futility of these endeavors at the present time. I have also heard some rumors that the Federation of American Scientists[476] is giving up the idea of the International Conference. This seems sensible to me, but of course the Emergency Committee is an independent agency. Anyway, I would much rather talk physics with you than have you go to that conference.

Thanks very much for the parliamentary debates of the House of Lords.[477] I am of course very proud.

Yours sincerely

Hans

---

[474]See letter [84].

[475]Emergency Committee of the Atomic Scientists. See above letters [84], [86].

[476]The Federation of American Scientists was formed in 1945 by atomic scientists from the Manhattan Project to address a broad spectrum of national security issues of the nuclear age and to promote the humanitarian uses of science and technology.

[477]The Archbishop of York had launched a debate on 18 February 1948 about the 'Perils of Atomic Warfare'. Hansard, House of Lords, vol. 152 (1948), vols. 1178–213.

[87] Hans Bethe to Rudolf Peierls

Ithaca, 14.4.1948

Dear Rudy,
I am now making more detailed plans about my trip to England. As I remember it, you were expecting to be busy during the last week of June. Would it be all right if I spent the week from June 13 to 19 with you? I suppose this will be better than to come very shortly before your Conference. (I had in mind some time around September 1?)

I have now decided to come to your Conference and to go back to this country immediately thereafter. Thanks very much for your offer to pay my expenses during that time which I gladly accept.

We had a very fine Conference in the Pennsylvania Mountains[478] around Easter, during which Schwinger gave us a one and a half days lecture on his complete theory in which all observable effects are finite and relativistically invariant.[479]
We are beginning serious preparations for our trip in the summer.
Hoping to see you.
Yours sincerely,

Hans

[88] Rudolf Peierls to Hans Bethe

[Birmingham], 22.4.1948
(carbon copy)

Dear Hans,
The week from 13th to 19th June sounds admirable. I have so far only one (two-hour) senate meeting for that week which I may even be able to skip. I shall try to avoid further engagements, but I may have to mark

[478] This refers to the Pocono Manor Conference between 30 March and 2 April 1948.
[479] See report by Richard Feynman. R.P. Feynman, 'Pocono Conference', *Physics Today* **1**, 8–10 (1948).

some exam papers and the like. This is certainly better than September, since the number of assorted conferences in September already makes my flesh creep. It is fine also to know that you will be here again for our conference. The acceptances so far received sound very encouraging.
Yours sincerely,

R.E. Peierls

[89] Hans Bethe to Rudolf Peierls

Ithaca, 4.5.1948

Dear Rudy,
My plans are pretty well settled. I shall arrive in Belfast on June 4 and then go first to Cambridge and then to you. I shall probably arrive some time on June 12, depending on train connections from Cambridge. I would plan to stay until the 18th and go to Bristol either on that day or early the following.

I am looking forward very much to seeing you and the many brilliant young men who must be working with you.
So long.
Yours sincerely,

Hans

[90] Rudolf Peierls to Hans Bethe

[Birmingham], 6.5.1948
(carbon copy)

Dear Hans,
You may be interested in the enclosed note which we have sent to Nature. The results bear out qualitatively what you have found independently with a square well. We were afraid of a square well, since at very short wavelength the sharp edges of the well are liable to give serious

interference effects. We started, of course, like most people with the idea that 100 MeV is a high energy. The results show that it isn't, and so probably the square well is as good as anything else. It is, however, desirable that at these energies the results are insensitive to the nature of the forces, no doubt this would change at still higher energies, but even if these could be attained, they will be very much harder to interpret because of relativistic effects and the like. What is more pleasing is the rapidity with which it is possible to calculate phases by a method of Ferretti and Krook.[480]

The trick is as follows:

You choose a point $r = 2$ beyond the range of the forces and you imagine that you integrate the equations inwards from this point. In general, i.e. choosing the wrong slope, the answer will be singular at the origin and its Taylor series about a will have a radius of convergence at most equal to a. For the correct slope the radius of convergence will be larger and may even be infinite. Accordingly you can get the correct slope by imposing the condition that the Taylor series should converge as well as possible. As an approximation you make the nth term in these series as small as possible, i.e. equal to zero, and for quite reasonable n this gives a good approximation to the slope and hence to the phase. Since in this operation only successive differentiations are required, it is quite easy to go to, say $n = 10$, without much trouble if your potential is of a reasonably simple analytical form. The method works amazingly well and can be generalised to tensor forces. We are getting rather excited about the self-energy problem. You remember Salpeter here having a proof for the one-body problem that the self-energy if calculated exactly and without perturbation theory was really infinite.[481] In trying to generalise this to the case of hole theory, he struck some snags, however, and there seems now the possibility that

[480]B. Ferretti and M. Krook, 'On the Solution of Scattering and Related Problems', *Proc. Phys. Soc.* **60**, 481–90 (1948).

[481]Ed Salpeter (1924–) was completing his Ph.D. at Birmingham in 1948. After a brief period he went to Cornell to work with Hans Bethe where, apart from short research spells elsewhere he stayed for the remainder of his career.

the statement is not in fact true in hole theory, but that the logarithmic infinity disappears if one calculates exactly. We are very far from having proved this, but the mere possibility is, of course, very important.

Another point that agitates me is the self-energy of the photon. In the ordinary way this, of course, is again infinite and one would expect with the usual tricks or, for example, with the kind of modification that McManus or Feynman are playing with,[482] to make it finite. However, the rest mass of the photon is not merely finite, it must be very exactly zero, and the only way to get this in any of these theories is to assume a finite mechanical rest mass for the photon which would then exactly cancel the finite (negative) dynamic mass. This is rather interesting formally and it means, amongst other things, that perturbation theory applied to this must always give utter nonsense since you can never get a finite rest mass to rest mass zero by a small perturbation.
See you in June.
Yours sincerely,

[Rudi]

[91] Rudolf Peierls to Hans Bethe

Birmingham, 10.5.1948
(carbon copy)

Dear Hans,
Thank you for your letter. On 12th June so it happens there will be in Birmingham a meeting of the Council of our Atomic Scientists' Association which starts at 11.30 am and goes on until about 5 o'clock. I do not think you will want to take part in all the routine administrative business, but I am sure that the members of the Council will be delighted if you could arrive either in time to have lunch with us, or if you could come to the meeting for a time and chat about the general

[482]R.P. Feynman, 'Relativistic Cut-Off for Quantum Electrodynamics', *Phys. Rev.* **74**, 1430–38 (1948).

problems. If you come for lunch, you will have to excuse me for the first half of the afternoon, but if you want to start talking shop at once, I am sure the rest of my department can at a moment's notice produce a queue of people anxious to talk to you.

Skyrme has now had his application for a Commonwealth Fund Fellowship turned down for reasons only known to the Board of the Commonwealth Foundation, if to them. This is a pity since a change of scenery would do him good, and we are now considering Copenhagen or possibly Zurich as alternatives. From your knowledge of the general situation, would you think there is any chance of getting a grant from American sources in the time now remaining for next session?[483]

In view of your absence from Cornell,[484] it would probably be better to go for a grant (if such a thing can be done) at Cambridge (Harvard or M.I.T.) or Princeton. Cornell is not such an attractive place without you, and at Columbia you will not have a coherent team.

What I am afraid of, however, is that if we try now, the approval would at best come so late that he would have had to make other arrangements for the meantime. Any help or advice from you on this would be most welcome.

Yours sincerely,

R.E. Peierls

[92] Hans Bethe to Rudolf Peierls

Ithaca, 13.5.1948

Dear Rudy,

Because of G.I. Taylor, I am slightly postponing my visit.

I shall arrive in Birmingham probably with the first manageable train on Sunday, June 13.

See you then.

Yours sincerely,

Hans A. Bethe

---

[483]See letters [79–81].

[484]Bethe was planning to spend half of the following year, a sabbatical, at Columbia.

[93] Hans Bethe to Rudolf Peierls

Ithaca, 20.5.1948

Dear Rudy,
Thanks for your various communications. I have slightly changed my travel plans and now want to do everything a day later in order to catch G.I. Taylor at Cambridge. So I will come to you late on Saturday, June 12. This means that we do not need to make any arrangements about Saturday noon and afternoon. I will let you know more accurately when I have the English timetables.

Thanks for sending me your note to "Nature" about neutron-proton scattering.[485] It is very nice that these results are so invariant against change of shape of the potential. Your method of determining the phase shift from the convergence seems very nice.

We have thought some about Skyrme. As it happens we have a vacancy in a research associateship. If he wants it, we would be very glad to consider him for next year or possibly part of the year. He could either come for the entire year and work with Feynman and Morrison while I am at Columbia, or he could come only during the second semester when I shall be back here. The salary would be in the order of $3500 for the year. Please let me know right away whether he is interested. I do not want to give a definite promise because we might decide to take someone who is prepared to stay for a longer time. It is more likely, though that we shall prefer to have Skyrme.

Very soon now we expect to make our crossing of the Atlantic and be much closer to you.
Until then,
Yours sincerely,

Hans

---

[485] Peierls was working on this question together with F.C. Barker. See also: F.C. Barker and R.E. Peierls, 'On the definition of the "effective range" of nuclear forces', *Phys. Rev.* **75**, 312–13 (1949).

[94] Rudolf Peierls to Hans Bethe

Birmingham, 19.11.1948

Dear Hans,
We have noticed that the very convenient equation defining the effective range of the force for an arbitrary shape of well has never been stated in print. Blatt does not seem to be aware of it or at least not of the fact that it can be derived in a few lines.[486] He refers to the famous unpublished paper by Schwinger.[487] It seemed worthwhile to write it up unless the paper by Schwinger is, in fact, in press and contains this elementary argument. I am afraid of writing to Schwinger as the letter would probably join the pile of unanswered letters on his desk, but perhaps you could help us as you are likely to know what is going on and I am therefore enclosing the letter. If you think it is a sensible idea to publish it, would you pass it on to the Physical Review. I also enclose a carbon copy in case that is of help.
Yours sincerely,

Rudi

We are well. Enjoying a visit from Linde[488] just now!

[95] Hans Bethe to Rudolf Peierls

New York, 10.12.1948

Dear Rudy,
Thanks for your several letters. I shall be very happy to have Salpeter[489] and I have just written to Wilson[490] about giving him an appointment.

---

[486] John M. Blatt, 'On the Neutron-Proton Force', *Phys. Rev.* **74**, 92–96 (1948).
[487] J. Schwinger, 'On the interaction of several electrons' (unpublished, 1934).
[488] Linde Davidson, Rose Bethe's younger sister.
[489] Ed Salpeter moved to Cornell after completing his Ph.D. at Birmingham.
[490] Robert R. Wilson (1914–2000), who had been heading the Manhattan Project's Experimental Nuclear Physics Division at Los Alamos, had moved to Cornell from Harvard in 1947, where he eventually directed the construction of a particle accelerator, the Cornell Electron-Positron Storage Ring.

I am reasonably sure that we can give him a position as a Research Associate at a salary of $3500 to $3800. This, I think, should be adequate and will correspond approximately to his current salary plus traveling expenses to this country and back. We cannot make any separate arrangement for travel expenses.

I hope Dyson will actually decide to come to you. He gets better every day. However, he has so many offers that he has not yet made up his mind where to go in England. He has some excessive loyalty to Cambridge; I told him he would waste his time if he went there.

I have sent your Letter to the Editor to *The Physical Review.*[491] I knew this derivation or a rather similar one; in fact, I am planning to write a paper in which a still more general and exact relation is derived and used.[492] Blatt[493] has used still another derivation, but for practical purposes has used essentially your relation. Everything about the scattering of neutrons and protons by protons is getting to be exceedingly simple. Especially the complicated calculations by Breit[494] are now quite unnecessary. I have also taken care of your application for membership in The Physical Society.

I am enjoying my stay at Columbia.[495] I find that I have much more time than usual to do some work. There is quite a lot of progress in quantum electrodynamics and the people most concerned, namely

---

[491] F.C. Barker and R.E. Peierls, 'On the Definition of the "Effective Range" of Nuclear Forces', *Phys. Rev.* **75**, 312–12 (1949).

[492] H.A. Bethe, 'Theory of the Effective Range on Nuclear Scattering', *Phys. Rev.* **76**, 38–50 (1949).

[493] John M. Blatt and J. David Jackson, 'On the Interpretation of Neutron-Proton Scattering Data by the Schwinger Variational Method', *Phys. Rev.* **76**, 18–37 (1949).

[494] G. Breit, H.M. Thaxton and L. Eisenbud, 'Analysis of Experiments on the Scattering of Protons by Protons', *Phys. Rev.* **55**, 1018–64 (1939); L.E. Hoisington, S.S. Share and G. Breit, 'Effects of Shape of Potential Energy Wells Detectable by Experiments on Proton-Proton Scattering', *Phys. Rev.* **56**, 884–890 (1939), and G. Breit, L.E. Hoisington, S.S. Share and H.M. Thaxton, 'The Approximate Equality of the Proton-Proton and Proton-Neutron Interactions for the Meson Potential', *Phys. Rev.* **55**, 1103 (1939).

[495] Hans Bethe was visiting professor at Columbia University in the autumn of 1948.

Schwinger, Feynman and Dyson, have been busy writing up their knowledge. At the moment, Rose is visiting me but without the children. Regards to all of you.
Yours sincerely,

Hans

[96] Rudolf Peierls to Hans Bethe

[Birmingham], 25.1.1949
(carbon copy)

Dear Hans,
I am writing again, this time about a number of assorted points, instead of sending you three of four letters in succession.

The first point is that Dyson has now decided to come to Birmingham next session, which, of course, cheers me very much. It remains now to get the authorities here to fix his salary and I am trying to get for him a reasonably high salary which may be a little difficult in view of his age. Would you be good enough to write me a letter with a frank statement of your opinion of him; this should cover not merely his ability and promise but also his actual achievement to date, since the point at issue is the extent to which he can be regarded as of fairly senior standing. For example, it would help, if you could state what sort of job he could expect to have in an American university if he was an American, or if you were able to compare him to the general run of people holding senior jobs in this country (if this can be done without being rude).

The next point is a question of physics. I don't remember whether I have mentioned to you before some work which Swiatecki[496] here has done recently trying to get to the bottom of the magic numbers, 20, 50 and 82, in nuclear structure. The occurrence of these numbers smells of

[496] Wladek J. Swiatecki (1926–), completed his Ph.D. under Peierls in 1949 working on nuclear structure. He later went to Copenhagen, Uppsala, and Berkeley where he spent the rest of his career.

the existence of "shells", i.e. single particle states to some approximation, but there are two factors against this. (1.) that on a shell model you would have to explain why you get only the breaks at these numbers and none at the intermediate numbers which should correspond to closed shells. (2.) you cannot get a complete shell at 50 unless you omit the 2s level, whereas you cannot complete 20 unless you include it. Both these difficulties can be met in principle by observing that with increasing weight the shape of any potential well would have to change both on account of the predominance of the surface effects for light nuclei and because of the increase in the Coulomb forces which tends to decrease the density near the centre as compared to the edge. In particular, the latter effect will make the sequence 2s, 3p $\cdots$ go up with increasing weight relative to 1s, 2p $\cdots$ It is, therefore, not impossible that when Y or $N(=A-2)$ is about 20 the 2s level should still be fairly low, whereas near Z or $N=50$ it might have moved up above 3d. At the same time, this would explain why intermediate shells do not appear to be marked when two levels are just about passing each other one would not get any clear break. On this view one would be dealing with a situation analogous to the rare earths in atomic structure.

The question whether the reduction in the density near the centre is big enough to do this depends on what one assumes for the compressibility of nuclear matter. There exists on this only a paper by Feenberg[497] in which one of the two required parameters is fixed from the observed mass defect curve, whereas the other is fixed from fishy theoretical treatment. We have been unable to rule out the possibility that the compressibility might be somewhat larger than is usually assumed and it might even be possible that heavy nuclei are almost hollow. We have thought of means of deciding this question experimentally; in principle it might be done by observing nuclear radii, but this is very insensitive. One interesting possibility is to use the diffraction of electrons by nuclei using energies of 30MeV or more. Now the diffraction patterns to be observed have been calculated by Rose using the Born approximation and if one does this both for a solid and a hollow nucleus

[497] E. Feenberg, 'Nuclear Shell Structure and Isomerism', *Phys. Rev.* **75**, 320–22 (1949).

the difference is small but not altogether negligible. Now Born approximation is, of course, very poor, particularly for heavy nuclei and, while it can hardly be expected to give a long order of magnitude, it would be nice to have a better theory. I believe this is in principle contained in the work of Smith[498] which you mentioned when you were here and the main object in telling you this long story (apart from getting your general comments) is to see whether Smith's work, when complete, will cover the answer to this question, or if not whether his method can be used for the purpose. If you do not intend to have this done we would very much welcome a chance to see Smith's work in advance so that we could adapt it for this purpose.

My third problem arises from the planning of the 1950 volume of the Physical Society Progress Reports which you know, as you wrote an article for it in the past.[499] We were wondering whether we could have an article on nuclear forces; would you be interested in writing this? The date of publication is not fixed yet, but probably the MS. would have to be received during the late summer for publication early next year. You will remember that these articles are not generously paid but serve a very useful purpose. Compared to the Review of Modern Physics they are rather less specialised and less high-brow. If you cannot do this, can you suggest someone else? Equally, of course, other suggestions about articles in the series would be welcome.

Lastly, we understand that the Fulbright Scheme is coming into effect next session and that, as you probably know this will allow research students and more senior people to come and work here. We have been asked to say whether we would accept research students, which, of course, we would, if they are suitable, and what kind of more senior people we would like. This information was required at such short notice that it was impossible to find out who would be interested. I have mentioned a few names at random including Feynman and Placzek, though these are long shorts, but if you hear of anybody who might like to spend a year or so in Europe, will you let them know or get directly

[498]Presumably Jack H. Smith at Cornell.

[499]H.A. Bethe and R.E. Marshak, 'The physics of stellar interiors and stellar evolution', *Reports on Progress in Physics* **6**, 1–15 (1939).

into touch with the Fulbright people. I have also mentioned to them that I would quite like to have someone who works on solids and/or metals to arouse some interest in this here, though of course he could not learn very much on this subject from us. Under this heading I have not mentioned any names at all, but would be grateful for suggestions.
Yours sincerely,

R.E. Peierls

[97] Hans Bethe to Rudolf Peierls

Ithaca, 2.2.1949

Dear Rudy,
I think the enclosed letter on Dyson is strong enough but the strange thing is that it is all true and sincere. He is really incredibly good.

Concerning your calculations on nuclear structure, I just have on my desk 2 papers, one by Nordheim[500] and the other by Feenberg.[501] Feenberg especially uses a model similar to yours in which he takes into account that the nucleus may be something like a hollow shell.

Having just returned to Cornell I have not had time yet to find out about the recent calculations of Jack Smith. I think he has calculated the effect of the distribution of protons on the nucleus and came to the conclusion that the effect is hardly observable of the distribution is that assumed by Feenberg in previous papers. However, if the nucleus were actually a hollow shell, this should be observable in electron diffraction. Nobody as yet has experimented on this but this will probably soon be

[500] L.W. Nordheim, 'On Spins, Moments, and Shells in Nuclei', *Phys. Rev.* **75**, 1894–1901 (1949).

[501] Eugène Feenberg had published a paper on nuclear shell structure in early 1949 (see letter [96], note 497) and had submitted another paper on nuclear shell structure to be published later in 1949. E. Feenberg and K.C. Hammack, 'Nuclear Shell Structure', *Phys. Rev.* **75**, 1877–93 (1949). A few weeks after Bethe's letter to Peierls, Feenberg, Hammack and Nordheim submitted a joint note on the same scheme. E. Feenberg, K.C. Hammack and L.W. Nordheim, 'Note on Proposed Schemes for Nuclear Shell Models', *Phys. Rev.* **75**, 1968–69 (1949) about the same subject.

done. The Berkeley synchrotron is going and ours is beginning to go, and electron diffraction is one of the points on our program. I will write more about this when I have checked with Smith.

Concerning the Physical Society progress reports, I should like to think it over a little more. At the moment I would rather say no, because I still have a good deal of writing to do for the books of Segre[502] and Schein[503] on nuclear physics and cosmic rays respectively. I will ask Phil Morrison if he wants to do it; he would do it very well. Weisskopf may be another possibility.

Concerning exchange students, I had a very good boy at Columbia by the name of Slotnick[504] who would like to go to Europe next year. His first choice is Pauli, but he is also very much interested in coming to you. Possibly he will want to divide his time between Birmingham and Zurich. He will write you directly. I am writing to Eyges[505] about possible jobs for next year for him.

I had a rather fruitful semester at Columbia and will soon send you a paper on the effective range of nuclear forces.[506] I am also sending you a copy of the first paper by Dyson which will soon appear in print.[507] With best regards to you and the family.

Yours sincerely,

Hans

---

[502]Emilio Segrè (ed.), *Experimental Nuclear Physics*, Wiley, New York, 1953.

[503]D.J.X. Montgomery (ed.), *Cosmic Ray Physics: based on lectures given by Marcel Schein at Princeton University*, Princeton University Press, Princeton, 1949.

[504]Murray Slotnick, in fact went to Princeton in 1950, before moving to the University of Michigan, Ann Arbor.

[505]Leonard Eyges went to Berkeley, M.I.T. before joining the Air Force Cambridge Research Laboratory in Massachusetts.

[506]H.A. Bethe, 'Theory of the Effective Range in Nuclear Scattering', *Phys. Rev.* **76**, 38–50 (1949).

[507]F.J. Dyson, 'The Electromagnetic Shift of Energy Levels', *Phys. Rev.* **73**, 617–26 (1948). The paper was not Dyson's first paper; he had already published twelve papers between 1943 and 1948, but all of these were in mathematical journals.

[98] Hans Bethe to Rudolf Peierls

Ithaca, 2.2.1949

Dear Rudy,

I am very happy to write about Dyson. In my opinion Dyson is one of the very best theoretical physicists now living regardless of age. He is probably the best English theorist since Dirac. In this country I would rate him on the same level with Schwinger who is a full professor at Harvard University and is considered the leading theorist under 40 years.

My high opinion of Dyson is shared by Dr. Oppenheimer who invited him to come to the Institute of Advanced Study at Princeton whenever he likes and for whatever period he likes within the next 5 years. Whenever Dyson chooses to come to the Institute, he will receive a regular member's salary from the Institute. The only other foreign scientist who has a similar invitation at present is Niels Bohr.

During the last five months, Dyson has completely reworked the Schwinger theory of quantum electrodynamics. He has proved its identity with the theory of Feynman and has obtained a theory which combines the thorough foundation of Schwinger with the ease of calculation of Feynman's theory. With this new formulation it is possible to attack the arbitrarily complicated problems in arbitrarily high order and write down the result in terms of an integral in one-half hour. Furthermore, Dyson together with Feynman, has succeeded in proving that all effects in electro-dynamics are finite, except for those which can be interpreted as renormalizations of the mass or the charge of the particle. This, you will agree, is a tremendous progress in the theory.

At the Institute of Advanced Study in Princeton, Dyson is the undisputed expert on all questions of advanced quantum theory. At the recent meeting of the Physical Society, many much older men with permanent professorships crowded around him to learn about the latest advances. Many universities including Columbia, have tried to attract him by offering him good positions. Columbia, for instance, would have offered him an Associate Professorship which would carry a salary in the neigh-

bourhood of $6000 to $7000. He has refused these offers because he feels under obligation to return to England.
Yours sincerely,

Hans A. Bethe

[99] Rudolf Peierls to Hans Bethe

Birmingham, 27.7.1949

Dear Hans,
It looks almost as if I have not written to you since your letter of 7th February.[508] Meanwhile your letter about Dyson has done just what I wanted it to do, namely not merely persuaded the university to offer him a good job, but also got the Royal Society to appoint him to a Warren Fellowship which has advantages. He has been here already for two days and it will make all the difference to the department to have him here next year.

I assume that you have not done any more about the Physical Society Progress Reports and that is just as well, because with all the experiments going on now it is likely that a report written now would be obsolete before it appeared (the volume now in preparation will not come out before July 1950). Things may have settled down a little more in a year's time. The closing date for manuscripts for the following volume is going to be about September 1950 and it is much too early to talk about an article for that volume. However, I need hardly say that if you actually have started to write an article and if you are prepared to let us have it by September, everybody will be delighted to see it go in to the forthcoming volume.

On the question of nuclear structure it seems indeed that Feenberg and Swiatecki here have been thinking along very similar lines, but for some time, Swiatecki has been working on a decent derivation of the surface tensions of a nucleus which is essential for the question under

[508] Refers to letter [98], dated 2nd February 1949.

discussion and he had discovered that nothing said on this subject in the literature is any good. He has now developed a generalization of the Thomas-Fermi method which is suitable to treat potential gradients and, while the numerical work is not complete as yet, he seems to be able to derive a fairly good value for the surface term in the semi-empirical formula and one can then have some confidence in applying the methods to internal density gradients. I have not heard anything from Slotnick,[509] but I had some correspondence with Gerald Brown of Yale [510] who makes a very good impression and who is coming here probably with a Fulbright Grant. He was coming in October, but he has to finish some job for Breit and will not be free before January.

The last session has not been very productive, largely, though not entirely, because we were always trying to catch up with the work on field theory by Schwinger, Feynman and Dyson etc. We are now preparing for a determined attack on discussing meson theory by the new techniques without the use of perturbation theory; most of what has been said in the literature seems just to be perturbation theory run riot.

Dyson told us the sad story of your car; I hope it is making a good recovery and that in spite of it you are by now all reunited.

Yours sincerely,

Rudi

[100] Hans Bethe to Rudolf Peierls

[location unspecified], [August 1949]

Dear Rudy,

de Hoffman tells me you will probably come to a declassification meeting in Canada the end of September. I should very much like to see you then! Please come to Cornell, if you can at all arrange it! I shall be

[509] Murray Slotnick, at the time at Columbia University then moved on to Princeton postdoctoral fellow.

[510] Gerry Brown had worked with Gregory Breit at Yale. He joined the Birmingham group in 1950.

here presumably from the 25th Sept. on. As usual we can pay you $40 a day (up to 5 days) plus travel expenses from Chalk River and back to New York, or some similar places of your choice. Also, of course, we should like you to stay in our house. Please come as long as you can.

No, I have not written anything for the *Progress of Physics.* About the other things in your letter later.

I have not yet congratulated you and Genia to Nr. 4. We were very happy! Especially that it came, apparently, so easily (after the false alarms). Rose sends her regards.

Please let me know as soon as possible (a) whether you can come, (b) when and for how long, (c) whether you can give a colloquium and what the title would be. My address until Sept. 10 is P.O. Box 1663, Los Alamos, N.M., then Ithaca.

Regards!

Yours,

Hans

[101] Rudolf Peierls to Hans Bethe

Birmingham, 20.8.1949

Dear Hans,

Many thanks for your warm invitation. I would have got in touch with you earlier if I had seen any chance of getting to Cornell on this visit. Unfortunately it does not look possible because I do not get to New York until the morning of 23rd September (or rather that is when I am due, the rest depends on the plane) and I am due in Chalk River on the evening of the 25th.[511] Since I can hardly pass through New York without calling on my relatives there,[512] it would be a mad scramble to

[511] Frequently, meetings between representatives of the United Kingdom, Canada and the United States about questions of nuclear research and policies took place at the Chalk River Laboratories northwest of Ontario, Canada.

[512] Rudolf Peierls' stepmother, his sister's family and his father's brother with family lived in Montclair, New Jersey.

get to Cornell. Even if I arrive on time, and after having spent Thursday night all over the Atlantic, I do not relish the prospect of spending Friday and Saturday nights on the Leigh Valley. I may go to Princeton for half a day or so. The answer to the obvious question why I cannot leave here a couple of days earlier is that I only get back from holiday on the 19th and it would be difficult to alter the date of that, being connected with the reservation on the cross-channel steamer for our car. I am rather reluctant to cut my holiday altogether. At the other end I expect to leave Chalk River on the evening of the 29th and have a booking to fly back from Montreal on the 1st of October and I have promised to spend the 30th in Toronto. Here, the time limit is set by the beginning of term on 3rd October. Since I have lost two lecturers and for one of them the successor has not even yet been appointed, it would make things awfully complicated if I were not here at the beginning of term. I had been imploring the authorities to put the Chalk River meeting at a different time but without success.

It looks, therefore, as if I shall not see you on this occasion, unless you should happen to be in New York on the 23rd or 24th September or in Toronto on the 30th. I hope I may come again some time.
Yours sincerely,

Rudi

[102] Rudolf Peierls to Hans Bethe

Birmingham, 15.2.1950

Dear Hans,
I feel I should write to you about Fuchs,[513] though I cannot say much that you do not know already from the papers. The whole affair came

[513]Klaus Fuchs (1911–1988), had been working with Peierls initially at Birmingham and later at Los Alamos. After lodging with the Peierls family when he first came to Birmingham, he had become a close family friend. In 1950 Fuchs was arrested on suspicion of breaking the Official Secrets Act by passing military secrets (about nuclear weapons) to the Soviet Union, at the time a friendly nation. He was later sentenced to fourteen years in prison of which he served nine years and four months before being released. He then emigrated to the GDR.

as a complete shock to me; I had not known that there was anything wrong until the day of his arrest. We have spent many hours trying to understand and trying to remember and it is still as unbelievable as on the first day. The only evidence so far produced in court, apart from some vague reference to suspicions about contacts (which might mean anything) rests on his own statements. This, in fact, has led us to question whether possibly the whole story comes from his imagination and whether, as a result of some mental strain, he is accusing himself of things he has never done or, at any rate, is greatly exaggerating some minor indiscretion. The answer, of course, depends on what other evidence is there to corroborate his story and nobody, of course, will tell me at this stage whether such evidence exists or what is its nature. It is evident, in any case, that his mind is in quite an unbelievable state and must have been all along, if, indeed, the story is true. If you have seen in the papers what Counsel of Prosecution have said about him, it is clear that the authorities also have noticed his very strange state of mind. They must have asked themselves the question whether or not the story is true and must have chosen to believe it.

If it is true the picture comes out that all the time we knew him he was hiding his allegiance, with an almost religious conviction, to the communist cause and one can imagine this. But what makes i[t] so unusual is that at the same time he should allow himself to get so closely attached to people personally and let people become so attached to him and equally that he should have been so genuinely fond of his work. I think you will have the same impression as I that his devotion to his work and his passionate interest to see matters done the right way must have been genuine. This became even clearer during his later work at Harwell where he had more administrative responsibility and when his whole heart was not merely in the technical side of the work but in looking after his staff. There was no problem in the running of the whole place in which he would not intervene with energy if he felt that someone was not being fairly treated or that the work was not being done in the best possible manner.

He himself says that it was these personal attachments which broke him in the end and which made him see that he was backing the wrong cause, but I believe he has not yet understood what he has done to people and I wish he could now be allowed to go around Harwell and see the faces of his friends and members of his staff. It is amazing that with all his consideration for people he should not have realised all along what he was doing to them. This, of course, is quite apart from what this whole business has done to the scientists in both countries, everyone of whom will be again under suspicion and to the political atmosphere. I gather, in particular, that this has led to an anti-British outburst in the American press. This was bound to result and there is not much one can do about it. Still one should remember that the major charges are based on actions during the time when he was at Los Alamos and therefore happened right under the eyes of the American security people who must have (or should have had) considerable interest in the rare journeys of Los Alamos staff. However, logic is of no great use for public opinion and I imagine we are all in for a pretty tough time. His trial is likely to take place in a few weeks' time and I hope the facts will come out a little more clearly. Some of the features are still completely fantastic. For instance, if one remembers how much he could drink on occasion at Los Alamos, several times a little more than he could stand, this one would have thought is the last thing a man in his position would have dare[d] to do, unless he had an almost crazy self control. Another small fact is that on his return from America he brought as a present for Genia the book by Kravchenko.

It does not do much good to go on speculating, but I thought I should write to you about it and I would be very glad to know how things look from your end.

I shall not write about physics today, but hope to do so very soon.
With kindest regards,
Yours sincerely

Rudi

(handwritten addition)
Thanks for your warm letter about Joanna.[514]

---

[514]The youngest daughter of Genia and Rudi Peierls, Joanna, was born in 1949. She was soon diagnosed to be an achondroplastic dwarf, and as a result of this condition had some health troubles in early childhood. See Peierls, *Bird of Passage*, p. 221.

[103] Rudolf Peierls to Hans Bethe

Birmingham, 28.3.1950

Dear Hans,

It seems a firm tradition that about this time of year I am writing to you about one of my collaborators who wants to go to America. This time the question concerns McManus and in some way he represents a more difficult problem than the previous ones.

I don't know whether you remember meeting him here. You will certainly remember his paper.[515] The paper, of course, was based on an idea of mine, but there is a lot of his imagination in the execution. However, he has not published, or indeed completed, any work since then. For about a year after completing the work contained in that paper he explored various possibilities of quantizing this theory. This seemed worth the effort and he should certainly not blamed for failing in this. However, since then he has not really got properly started on any concrete problem. For this I feel to a large extent responsible myself. As you know I am stronger on criticism and critical understanding than on constructive work, and this tendency seems to affect my department. I think McManus has learnt from me to understand rapidly the essence of a new piece of work, and place it in relation to existing knowledge and he has probably also learnt from me to give more weight to such appreciation than to working things out in detail. For the last year he has been on our teaching staff, which meant giving seven undergraduate lectures a week in addition to helping with tutorials, setting and marking papers, etc. He has done this well and particularly so with the advanced work (he gave our main introductory course to quantum theory); his elementary courses tended to be a little on the hard side.

He has spent all his academic life in Birmingham, starting as an undergraduate in physics and returning here after the war during which he was working for the Admiralty and, I think, the Foreign Office. I think it would do him good to get a complete change and from many

[515] H. McManus, 'Classical electrodynamics without singularities', *Proc. Roy. Soc.* **195**, 323–36 (1948).

points of view it would seem a good thing to try and send him to America. I have already warned him that with his present record it will be difficult to find a place for him in any of the departments that are in greatest demand and in writing to you I do not really expect that you will offer him anything at Cornell, though, of course, it would give him a marvellous chance if you did. My object in writing, however, is so that you might bear his name in mind in case you are asked advice by some other place. For a smaller department where there is nobody, or too few people, having a general over-view of modern theory and who can give a rapid judgment of new work and explain it, McManus might be very valuable and, of course, he would be quite prepared to pull his weight on the teaching side. On the whole I think he would be better in postgraduate than in undergraduate teaching. So if you hear of something suitable, will you let me know?

If you want to know more details about McManus, you may remember that Salpeter knows him very well indeed.

Yours sincerely,

Rudi

P.S. I enclose a record of Mc Manus's career.

[104] Rudolf Peierls to Hans Bethe

Birmingham, 30.3.1950

Dear Hans,

I have written about McManus on a separate sheet in case you want to pass that letter on to other people. This is now about everything else. As regards physics, I am rather unhappy about our progress during this winter. Partly, I suppose this was just the result of the department being too large, partly because of the many disturbing factors starting with our trouble over Joanna and finishing with the Fuchs business, neither of which should be remembered during working hours, but neither of which is easy to forget. We have all learnt a great deal from Dyson. It took longer than I thought to learn to speak his language, and here again, all the upsets did not help. However, here is our position for what it's worth. On the field theory side one very good Australian (Dalitz)[516] has applied the Feynman theory to the angular distribution of pairs from the O-O transition in O^{16}. Since everything happens in such a small space, one expects there the radiative corrections to be comparatively large and indeed the terms in the fine structure constant appear multiplied by fairly large factors. Unfortunately these factors don't vary much with angle and the effect is therefore just on the edge of being observable. The whole thing was suggested by preliminary results by Devons[517] who had found something like 10% deviation from simple theory but the radiative correction is in the opposite sense as well as somewhat too small. Incidentally, in the course of his work this man had occasion to check the old Mott formula for the relativistic scattering and found that all the many amendments to this formula in the literature are wrong without exception. He will now publish the correct formula with an explanation of the error in each of the previously published ones. Several people, including a very crazy Pole, are worrying about the two-electron problem with a view to calculating the shift in helium. The first step in this, of course, is to get a consistent derivation of the Breit terms and this now seems to be completely understood. After some struggle over approximations I believe we can now see our way towards getting the second order terms and hence, in principle, the Lamb shift. The evaluation of the matrix element will, of course, involve messing around with wave functions for the excited states of helium and that I expect to be quite troublesome. Another project arising from this work is to write out the two-electron problem up to terms in e^2 only but without regarding the velocity as small. This would cover, for instance, the problem of two electrons in the k-shell of the uranium atom.

[516]In a paper published in 1951, Dalitz investigated the decay mode of the neutral pion. He calculated what has since become known as Dalitz decay. R.H. Dalitz, 'On the alternative decay process for the natural π-meson', *Proc. Phys. Soc.* A**64**, 667–69 (1951).

[517]S. Devons and M.G.N. Hine, 'The angular distribution of γ-radiation from light nuclei. I. Experimental', *Proc. Roy. Soc.* **A199**, 56–73 (1949); S. Devons and M.G.N. Hine, 'The angular distribution of γ-radiation from light nuclei. II. Theoretical', *Proc. Roy. Soc.* **A199**, 73–88 (1949).

On nuclei we have started to look at the problem of the general dispersion formula starting from weakly coupled particles. As long as the coupling is quite weak there will be an appreciable neutron width only for those states which correspond to one excited neutron, all other particles being in their lowest states. In the next approximation there is also a neutron width for those states where two particles are excited and so on. The question to which I would like to know the answer is whether by the time the width of single excitation has been reduced to something of the observed order of magnitude, all the states of 2, 3, 4, ... fold excitation have become indistinguishable or not. It might be that one should not take the neutron width of all states as comparable as is done in the most orthodox form of Bohr theory, but that some of them will have negligible neutron width and therefore are not observed. This would remove certain difficulties in the empirical statistics of resonance levels. We have a method which we think may give the answer to this.

Another promising study deals with the angular distribution of the d-p reaction in light elements. There exist observations of this angular distribution which show that certain proton groups have a very narrow and intense maximum forward, which can only mean an Oppenheimer-Phillips process. We are trying to find a method of calculation which allows first of all to predict the angular distribution of the O-P process and also its relative intensities for different groups.

This is done by the same man (S. Butler[518]) who got some rather nice results on multiple scattering concerning the corrections to be applied in observing large angle scattering with an with an add mixture of multiple small angle scattering. This overlaps with the paper by Snyder and Scott[519] for this purpose gives more accurate results with less effort. We sent a copy of this paper to Snyder with a request to pass it on to you.

---

[518]Stuart Thomas Butler (1926–1982), gained a B.Sc. and M.Sc. from Adelaide before joining Peierls at Birmingham for his Ph.D. in 1949. After research at Cornell, he took up a position at Sydney in 1953, where he stayed until his retirement in 1978.

[519]H.S. Snyder and W.T. Scott, 'Multiple Scattering of Fast Particles', *Phys. Rev.* **76**, 220–225 (1949).

The same man also continued the study which we had started last year of the Svartholm iteration method applied to d-n scattering.[520] The method is workable in principle but needs a lot of computing and we have shelved it until we get hold of a computer. It seems hardly fair to put a research student on to research work of this kind.

We had some fun with the people of the coherent scattering of gamma rays by atoms. Moon has been trying some experiments to observe nuclear resonance scattering and for this purpose one obviously wants to know more about the atomic scattering. This had been done by Franz[521] using the Thomas-Fermi model, but for large angles and short wave length this becomes unreliable because of the singularity in the Thomas-Fermi distribution at the origin which is not real. It was easy to cook up a density distribution from a combination of the Thomas-Fermi function with the correct distribution for the K-shell. At high energies one has, in fact, also to include a kind of Raman scattering in which an electron gets excited so that the gamma ray is slightly softer than the primary but probably indistinguishable in practice.

We have put a good deal of time into the discussion of the analysis of the scattering of tracks in a photographic plate in which Dyson also took a major part. It turns out that the rather high-brow calculations which I reported to you are an unnecessary luxury. It is quite true that they would give, in principle, the most accurate answer, but the loss of accuracy is negligible if one does not count every grain but only a few points along the track and, of course, the effect is very much less. A convenient and fairly good method had already been developed by Fowler at Bristol and we are now trying to put the finishing touches on this and to discuss also an apparent discrepancy in the calibration of this method. We have also resumed an old approach about metals, namely to investigate the validity of the perturbation theory applied to

---

[520] S.T. Butler, 'On Angular Distribution from (d, p) and (d, n) Nuclear Reactions', *Phys. Rev.* **80**, 1095–1096 (1950); S.T. Butler, 'On Angular Distribution from (d, p) and (d, n) Reactions', *Proc. Roy. Soc.* **A208**, 559–581 (1951).

[521] W. Franz, 'Streuung harter Strahlung durch gebundene Elektronen', *Z. Phys.* **95**, 652–68 (1935), W. Franz, 'Rayleigh scattering of hard radiation by heavy atoms', *Z. Phys.* **98**, 314–20 (1936).

the collisions responsible for the ordinary resistance. We have succeeded in proving some rather entertaining theorem, but not yet in developing a practical method of calculation independent of perturbation theory.

We have shelved, for the moment, work on McManus type field equations partly because it seemed that Pais and Uhlenbeck knew so much more about this. I have just received their paper[522] and applied to the kind of equations we always wanted to use, it does not really go beyond what we know. We did, however, discover that classical equations still lead to runaway solutions for certain shapes of the form function and, in fact, we have not succeeded in constructing any form function for which this is not the case.

In general life here is much as usual. The excitement over the Fuchs case has settled down surprisingly rapidly and even more surprising the demand for much more severe checks in security measures has confined itself to the more irresponsible Press, while the man on the street and responsible politicians are extremely sane about it. The exception seems to be a violent speech by Lord Vansittart[523] in the House of Lords who sounds as if [he] had got envious of Senator McCarthy's record.

Thank you for sending me your article on the hydrogen bomb[524] which I found most interesting. We had been thinking somewhat in the same direction but had thought that one could not oppose a claim that hydrogen bombs should be produce[d] for retaliation against attack by the same weapon or some other new form of mass destruction. In any case, the problem is somewhat academic for us as there has been no question (as far as I know) of developing a hydrogen bomb.[525] Nevertheless it seems important to us to think these things out and I have, in fact, just written for our Atomic Scientists' Association to the Federation of American Scientists to see whether it might be a good idea to have a small and informal conference to exchange views about these

[522] A. Pais and G. Uhlenbeck, 'On field theories with non-localized action', *Phys. Rev.* **79**, 145–65 (1950).

[523] See Peierls, *Bird of Passage*, pp. 239–40.

[524] H.A. Bethe, 'The Hydrogen Bomb', *Bull. Atomic Scientists* **6**(4), 106–109 (1950).

[525] The British atomic bomb project had been restarted in 1947, and eventually Britain also developed its first hydrogen bomb which was exploded in 1957. Rudolf Peierls, in 1950 was not aware of the plans to develop the latter.

problems. We feel that this might be of some value, particularly if it is kept informal and without publicity and was not as ambitious a venture as the Jamaican meeting proposed a couple of years ago.[526]
With kindest regards,
Yours sincerely,

Rudi

[105] Hans Bethe to Rudolf Peierls

[Ithaca], 29.5.1950
(carbon copy)

Dear Rudy,
This can only be a short note before I am leaving on my extended and complicated summer travels. I have thought about your letter concerning McManus and find it somewhat difficult to do anything for him. The job situation in this country has become very tight so that I am having considerable difficulties in placing my own students. Perhaps the best chance would still be the Institute in Princeton which I believe has much fewer people this year than it had last year. Whether Oppie would have any money for McManus I am not sure. I shall continue to mention his name and shall write you if anything develops.

In a few weeks I hope to write you a more satisfactory letter.
Yours sincerely

Hans A. Bethe

[526] See letter [84].

[106] Rudolf Peierls to Hans Bethe

[Birmingham], 30.8.1950
(carbon copy)

Dear Hans,
I gather from a letter that Bob Wilson wrote to Moon[527] that you have done some calculations on coherent scattering of gamma rays, correcting the error caused in Franz's calculation because of the Thomas-Fermi distribution.[528] I am writing to let you know that one of my research students has nearly completed a fairly decent theory of this. His method is to construct a density distribution function which over most of the atom is equal to the Thomas-Fermi distribution but at small radii has a correct behaviour and goes over into the exact densities for the K and L shells. This has not yet been prepared for publication because he was checking up on the relativistic corrections which might also modify the formula and was also discussing a kind of Raman effect in which one of the electrons ends up either in a discrete optical level or in a free state of low velocity. This effect will in most cases be rather larger than the actual coherent scattering and there is, of course, a continuous transition from there to the proper Compton effect. In most experiments with really hard gamma rays it may be quite difficult to distinguish this "Raman" effect from the coherent scattering and an estimate of its magnitude is, therefore, relevant.

From Wilson's account it appears that what you have done overlaps a good deal with our calculations here and I thought I should let you know to avoid too much duplication.

---

[527] Philip Moon (1907–1994), graduated from Cambridge in 1928 and continued research there under Oliphant. After another seven years of research at Imperial College, London, he followed Oliphant to Birmingham where, with the exception of the war years, he remained until his retirement in 1974.

[528] See letter [104]. Wilson, Moon and Bethe had been working on related issues in connection with the coherent scattering of γ-rays. See, e.g., R.R. Wilson, 'Elastic Scattering of γ-rays', *Phys. Rev.* **82**, 295 (1951); P.B. Moon, 'The Hard Components of Scattered γ-Rays', *Proc. Phys. Soc.* A**63**, 1189–1196 (1950); P.B. Moon, 'Resonant Nuclear Scattering of γ-Rays: Theory and Preliminary Experiments', *Proc. Phys. Soc.* A**64**, 76–82 (1951).

Another point is that we have recently got interested in some experiments at Bristol on the distribution of energy between the two components of a pair where they find curves which disagree with the Bethe-Heitler theory. One reason for this could be a breakdown of the Born approximation since presumably most of the pairs in a photographic plate are made in silver or bromine for which the Born approximation is rather doubtful. The ideal method to use for this purpose would be the one about which you told me two years ago and on which I think Jack Smith was then working.[529] I have never traced any publication but I then heard a rumour that Wergeland[530] had continued to explore this method and we wrote to him, but he did not seem very enthusiastic about the method though I cannot say whether this refers to the method in general or to the particular application on which he was working. I would, therefore, be very grateful if you could tell me whether you are satisfied with the general principle of this method or how we might go about learning it without re-developing it from scratch.

We have just returned from a very satisfactory holiday to be faced with an outbreak of the British Association at Birmingham. Life is therefore somewhat hectic at the moment and I shall restrict this letter to these two items.

With kindest regards,
Yours sincerely,

[Rudi]

[529] See letter [97].

[530] Harald Wergeland (1912–1987), at the time at the Fysikk Institutt, Trondheim in Norway.

[107] Hans Bethe to Rudolf Peierls

Ithaca, 23.9.1950

Dear Rudy,
I feel very much ashamed that I have not written to you for so long. Your letter of August 30 gives me an opportunity to correct this mistake at least partly.

Your calculations on the coherent scattering of gamma rays seemed indeed to be somewhat similar to mine. To some extent, however, the two complement each other. I have not done anything on modifying the Thomas-Fermi distribution so as to give the correct density near the nucleus. Your student's theory will be very welcome and necessary to treat the transition to somewhat lower energies and scattering angles.

Concerning the K-electrons, I wonder whether your student has taken the relativistic wave functions. I found that the K-electron scattering is increased tremendously by using relativistic rather than Schroedinger wave functions especially at high energies of the gamma ray. At high energies and large angles, of course, only the s-electrons contribute appreciably.

We have now begun to worry about the justification of the basic assumption that the scattering amplitude is given by

$$\int \rho(r) e^{2\underline{q}\cdot\underline{r}} d\tau \qquad (1)$$

where ρ is the density of electrons. The problem can be done very easily by using Feynman's method provided you assume that in the intermediate state the electron may be regarded as free. Since the energy of the intermediate state is about equal to that of the gamma ray this ought to be a good approximation. It seems likely at the moment that the scattering is further reduced as compared with equation (1). This is welcome to us because then the amount to be subtracted from Wilson's experimental scattering is smaller. A good evaluation of the exact theory will take considerable time, and I have asked two of my associates,

Levinger and Rohrlich,[531] to look into this. Is it this problem which your student is now investigating?

I have also looked into the Raman effect and found it to be quite small. So far I have done this by a rather crude method which however ought to be all right for high energies and large scattering angles. I note that all s-state wave functions behave as $r^{\sqrt{1-\alpha^2 z^2}-1}$ near the origin, and that only the region near the origin contributes appreciably to the scattering according to (1). Then I need only know the normalization factor which I find in the books.

However, as you see, a theory for somewhat smaller q and a description of the transition to the ordinary Compton effect using relativistic wave functions, is still necessary.

Rohrlich is making some progress on the calculation (by quantum electrodynamics) of the potential scattering whose measurement is the aim of Wilson's experiments. He will probably have the result for scattering through 0° within two or three weeks, and through 180° a few weeks later. This will be general for any energy. We expect that the 180° scattering will be significant for scattering through smaller angles; we want to assume for the time being that the scattering depends essentially only on q. In addition we have a nice elementary method for treating scattering through small angles which was suggested by Wilson. This small angle potential scattering is of course not measurable.

Wilson's experiments give a definite effect for Th C" gamma rays on lead. We are somewhat concerned that this might be due to the tail of a nuclear resonance since lead is the product of Th C" disintegration. So we shall not believe anything until experiments are completed with different elements. It would be nice if the effect were there.

About the Bristol experiments,[532] I would be inclined to disbelieve deviations from the Bethe-Heitler theory unless they were found with excellent experimental conditions, i.e. very good statistics, exact measurement of the energy of each electron of the pair etc. We may do some such experiments here. Regarding the theory you may have seen

[531] P. Greifinger, J. Levinger and F. Rohrlich, 'Elastic Scattering of Gammas by Bound Electrons', *Phys. Rev.* **87**, 663 (1952).

[532] See letter [106].

a paper by Bess in Phys. Rev. 77, 550 [533]. This theory seems to be all right; it neglects terms which seem negligible for high energies and it neglects screening which also seems justified. It uses standard Coulomb wave functions. The main trouble with it is that it is not well evaluated, and I think it would be worthwhile to exploit it.

The method which I had invented has not prospered very much in the hands of Wergeland. One reason seems to be that he is inclined to make wrong assumptions at some point in the calculation such as that the vector $\underline{q}$ is in the direction of the emitted quantum. Another reason however may be that the largest terms in the matrix element tend to cancel, which of course would make the problem more difficult. I have not had time myself to look at the problem.

The method was to follow the wave function of the incident electron on a straight path through the atom using the WKB method. This gives the phase of the wave function at any point in the atom. The potential is regarded as small compared with the energy of the electron but the phase shift is of course appreciable. The phase is similarly calculated for the wave functions of the final electron whose direction of propagation is assumed to differ slightly from that of the incident. Then the two wave functions are inserted in the normal Dirac matrix element for the radiative transition. I don't know whether this description is of any help.* Also as already mentioned the purely mathematical procedure of Bess is probably more successful. The two seem to be reasonably equivalent because Wergeland was able to obtain from his theory, hypergeometric functions which looked very similar to those of Bess.

We all had a very good vacation in Seattle, Washington and in the Canadian Rockies which are truly beautiful. We are making plans to go to Europe next summer, Uncle Joe permitting. We just ha[d] a visit from Otto Frisch who was very nice. How are things with you?
Yours sincerely,

Hans

*I can't find my notes at the moment. Will send them to you if I do.

[533]L. Bess, 'Bremsstrahlung for Heavy Elements at Extreme Relativistic Energies', *Phys. Rev.* **77**, 550–56 (1950).

[108] Rudolf Peierls to Hans Bethe

Birmingham, 24.9.1950

Dear Hans,
I enclose a second cheque from the American Institute of Physics. Apparently the last batch of work just missed the monthly settlement, so there will be one more similar cheque to come. I am sorry the return of your loan is dragging out in this way. Thanks very much once again, Greetings to all of you.

Rudi

[109] Rudolf Peierls to Hans Bethe

Birmingham, 8.1.1951

Dear Hans,
I understand that H.P. Noyes[534] is applying to you for a fellowship and that you would like to know my opinion of him. I do not think I can do better than send you a copy of the letter I wrote about a month ago to Marshak. I can add little to what I said there, except that by now Dyson has arrived and that Noyes is settling down well to learning the techniques of field theory. He is really a very nice man.

I have just come back from a fascinating trip to India the excuse being a conference in Bhabha's institute.
With kindest regards,
Yours sincerely,

R.E.P.

[534] H. Pierre Noyes (1923–), received his Ph.D. from Berkeley. After positions at Rochester, Berkeley and Cornell, he moved to Stanford, where he has remained since.

[110] Hans Bethe to Rudolf Peierls

Ithaca, 19.2.1951

Dear Rudy,
Thanks for your letter recommending Noyes. I would like to hear more about your trip to India[535] if you have more time to collect your thoughts. Of course I constantly feel that I owe you a long letter about science but there always seems to be too much else in the way.

This letter is to announce the visit of our family to Europe this summer, Uncle Joe permitting. If you are there and are willing to have us, we would like to visit you. As usual in our family we follow Napoleon's principles and travel separately. Rose and the children are planning to go by boat and too get to England about June 1. They would like to come to you some time in early June, and they would like to stay for about a week, if you can accommodate them. I could imagine with the increased family, that space even in your house is becoming limited.

I am planning to come by plane about the middle of June and to spend about two weeks in England. One of these I would like to travel around and spend the other week with you. Again if this fits in with your plans. After that I was planning to go to Uncle Nick's conference[536] which I believe is about July 4 to 10. After that I would like to join the family in Germany.

Please write me what your plans are and how you are fixed up for accommodations. We would all like to see you and to replace by a personal visit what is lacking in correspondence.
With best regards,

Hans

---

[535] Between 14 and 22 December 1950, an international theoretical physics conference ion elementary particles had been held at the newly-opened Tata Institute of Fundamental Research. Rudolf Peierls attended the conference and spent some time afterwards visiting, among others, Raman's institute at Bangalore.

[536] The conference took place between 6 and 10 July 1951, see *Pauli Briefwechsel*, IV/1, p. 339.

[111] Rudolf Peierls Hans Bethe

Birmingham, 28.2.1951
(carbon copy)

Dear Hans,
I am delighted to hear of your impending visit. It will be great fun to have Rose and the children with us and I am sure we can manage to squeeze them into the house somehow. In fact, Genia is planning to take the children to the seaside for a week or two during June and if it should so happen that Rose's visit should fall in that period she might like to join Genia and the small children at a seaside holiday and the best, of course, would be if this were either at the beginning or at the end of that period so that Rose could also have at least a few days at Birmingham.

As regards your own visit, you will, of course, also be most welcome at any time. I am also due to go to the conference in Copenhagen, and our plan is to take the car across via Ostend and to drive up to Copenhagen. I understand that the conference was to start on the 6th. We would probably be able to leave only on the morning of the 1st and cross on the 2nd. It looks possible that we should reach Copenhagen on the evening of the 5th. What would you think about joining us in the trip? I would, of course, hope that this would not curtail the length of your stay in Birmingham since all the other people here, including Dyson, would certainly like to talk to you. However, all these plans are subject to confirmation; it may still be that Genia might prefer to take her holiday elsewhere and I would in that case, go directly to Copenhagen by boat or by plane.

Scientifically the most important result of the Bombay conference for me was that the Bristol people produced convincing evidence in favour of multiple as opposed to plural production of mesons.[537] The point is that for plural production you would expect that for a greater number of mesons there should either be also a larger number of reasonably

---

[537] See *Report of the international conference on elementary particles*, Commercial Print Press, Bombay, 1951.

fast protons ("grey tracks") or else if the energy should be completely dissipated a greater energy in the star formed by low energy tracks. They have looked for correlation of either kind and they are completely absent. On the average, for example, the energy in the star increases only by 55 MeV per meson produced and this is clearly not enough. I also learnt there for the first time about Fermi's theory of meson production which is very attractive and simple as everything else that Fermi does.

Whether it is right is another matter and I think particularly that his explanation of the angular distribution is rather fishy. Of course there was much more to India than physics, but to do justice to the many impressions there would take a pretty long letter and I shall leave it until we meet, I hope, in June.

Yours sincerely,

[Rudi]

[112] Hans Bethe to Rudolf Peierls

Ithaca, 21.4.1951

Dear Rudy,

Thanks a lot for your very nice answer to my letter about the visit to Birmingham. We have intermittently considered the travel problem and have finally made up our collective minds. Rose would greatly prefer to see Genia and the children at Birmingham because they will have something very similar to seashore in Bavaria. She would also prefer to go to London first and to Birmingham afterwards because she says that after the boat trip everybody will be eager for action. Finally, she wants to spend the two weekends during her stay in London with her near relatives, one with her sister near London and one with her brother in Manchester. Both, as you probably know, are now married, and both they and their wives and husbands are busy all through the week.

So now there is the puzzle of fitting all this together with Genia's plans. Not knowing the latter, we have made four alternative schedules which I write you in order of our preference. To start with, Rose will arrive on the de Grasse on May 30 in Plymouth. From thereon the plans divide.

- Plan 1: London from May 30 to June 6 including a weekend visit to Linde, who lives about 1 1/2 hours from the city. Birmingham from June 7 to 13 with a week-end visit to Manchester.
- Plan 2: If Genia is away at the beginning of June but back on June 11 the plan would be the same except that Rose would stay in London through June 8, then go directly to Manchester, then come to Birmingham on June 11 and stay perhaps to the 15th.
- Plan 3: If Genia goes on vacation about June 7 or later, and if she can stand the idea of having our invasion before she leaves, the plan would be reversed. Rose would come directly to Birmingham from the boat and stay through June 6, then go to London; weekends as before.
- Plan 4: If Genia is not planning to be in Birmingham at any time between June 1 and 13, Rose would follow plan 2 up to arrival in Birmingham but stay only one day in Birmingham, and then go for two or three days to see Genia in Devon. This is the least favourable from our point of view.

Please let me know which of these plans, if any, fit in with yours. And can I ask you to do us a very great favor. After you have selected one of our plans, can you make a reservation for Rose and the two children at some hotel in London for the time she would be in London? I understand that it is often quite difficult to get reservations in London and that it is good to make them well in advance. Also, you know much better what hotels are suitable. If available, they would like the Cumberland but it is probably too late to make a reservation there. Something in the neighborhood of Hyde Park would be fine but it does not need to be. They want to visit museums and their relatives who mostly live in Northolt, except for Linde who lives in Essex. Of course it should not be the Savoy, which I am told charged 50 shillings a night already three years ago; in other words it should not be exorbitant but about middle class.

Now for my plans. I have made a reservation on the plane arriving in London Tuesday, June 19, early in the morning. I want to adjust my plans to your preference, namely whether you want me to come to Birmingham first or visit other places first. Thanks a lot for your invitation to join you in driving to Copenhagen but I think I would prefer to go directly by boat or plane, because there will be enough locomotion in my summer as it is. This means that I have the time from June 19 to July 4 or 5 available. I should like to spend about a week at Birmingham during that time.

Rose and I would also be interested in knowing whether the Dyson's will be in Birmingham during our visits.

I hope it all comes true.
Sincerely yours,

Hans

[113] Rudolf Peierls to Hans Bethe

Birmingham, 28.4.1951
(carbon copy)

Dear Hans,
Thank you very much for your letter. Of the various possibilities, we would, I think, best like plan 2, in other words, to have Rose and the children with us on or about the 11th June. This means that they would need hotel accommodation in London from May 30 until June 6th and we shall try and book in a suitable place and report to you later.

About your own plans: It would probably be best, if you could come to Birmingham starting on the 23rd or 24th June, since the following week is comparatively free from meetings. I have to attend, as far as I know two short meetings during that period, but there will be many other people wanting to talk to you. We expect to depart for Copenhagen some time on 1st July, and the 30th June is our degree day, at which amongst others, Chadwick is to receive an honorary degree. He is rather inclined to evade such ceremonies by being unwell at the last minute and if he does that on this occasion, I could cut all proceedings. But if he comes, I ought to be there.

Dyson is at the moment on the continent, but is expected back shortly. I think he intends to stay here approximately until the end of June, so he is likely still to be here when you come. Mrs. Dyson is in Switzerland, and will not return here.

We are all looking forward to seeing you all again in instalments.
With best regards,
Yours sincerely

[Rudi]

[114] Hans Bethe to Rudolf Peierls

Ithaca, 14.5.1951

Dear Rudy,
Thanks very much for your letter. The arrangements you suggest are fine with us. Accordingly, Rose is planning to come to you on June 11 and to stay approximately four days. She will probably come from Manchester and go on to Oxford. She will let you know the details when she is in England.

I shall come as you suggest on June 25 and stay probably through the week. Among other people, I should like to see Rose's sister and brother while I am in England and I understand that this can be done best on week-ends. So I shall probably spend the week-end of the 24th in London and that of July 1 in Manchester.

We have now arranged the trip in such a way that I will see Rose and the family in London the first two days after my arrival.

Preparations for the trip are progressing. Monica is just recovering from the chicken pox, just in time to be fully recovered and no longer contagious on the day of the sailing. Henry had them two weeks ago.
Yours sincerely,

Hans

[115] Hans Bethe to Rudolf Peierls

Ithaca, 26.5.1951

Dear Rudy,
Thanks a lot for arranging the hotel for Rose and the children in London. They left happily last Tuesday and should by now be about in the middle of the ocean. When last seen they intended to follow the plan which you found best for you, that is to come to you on June 11.

My plans are now to spend June 19 and 20 with my family in London, June 21 and 22 in Cambridge with Frisch, June 23 with my sister-in-law Linde, and June 24 with my old friend Werner Sachs. On Monday morning June 25 I probably shall have to go to the American Office of Naval Research in London so I shall take an afternoon train to Birmingham. I shall write you the time of arrival when I am in England. Probably I can stay through Saturday, June 30th. It will be very nice to see you all again.
Yours sincerely,

Hans

[116] Rudolf Peierls to Hans Bethe

Birmingham, 2.1.1952
(carbon copy)

Dear Hans,
I see from your Christmas card, which has just arrived that somehow I have not yet told you that my visa has come through alright and that I hope to arrive in Princeton on 10th January.[538] Genia will not come. I come to Princeton with the resolution to travel as little as possible, but I do want to see you before you disappear to the wilds. Are you

[538] Rudolf Peierls spent a sabbatical at Princeton in early 1952. He had had difficulties in obtaining a visa. See Peierls, *Bird of Passage*, p. 322.

likely to be at Princeton or at least in New York? Are you coming to the Physical Society conference in New York?

Apart from wanting to see you in general I would also like to discuss again the scattering of gamma rays.[539] We have discovered reasons to think that your approximation of treating the intermediate state as free gets worse as the energy of the gamma rays goes up. I would have written about this earlier, but we are still trying to check some points and we have not yet got a formal proof, but it looks very likely.
With kindest regards,
Yours sincerely,

[Rudi]

[117] Hans Bethe to Rudolf Peierls

[Ithaca], 8.1.1952
(carbon copy)

Dear Rudy,
It was good to hear that you are coming so soon. It would be very nice if we could see each other before I depart west. As usual we are in a turmoil before leaving and I don't quite see how I shall get finished in time. Therefore I have decided not to go to the New York meetings of the Physical Society,[540] especially as the most interesting problems in meson and nuclear physics will be discussed at a small meeting in Rochester at the end of this week.[541] Besides one does not find time for quiet discussions at the big meetings anyway, so I think not much will be lost as far as our personal meeting is concerned.

Rose would of course very much like to see you too and so we wondered whether we couldn't persuade you to come here to Ithaca after

[539] See letters [104], [106–107].
[540] The New York Meeting of the APS took place between 31 January and 2 February 1952.
[541] On 11 and 12 January 1952, the Rochester Conference on Meson Physics took place.

all. The best time, in fact just about the only possible time, would be one of the three days preceding the New York meeting that is January 28 to 30. Take your choice. Of course we could pay your travel expenses and of course everybody else in the Laboratory would be happy if you could come.

I should like to hear about elastic scattering of gamma rays. We have not done much about this since I saw you but we may have some other things to tell you.

Hoping that you will come,
Yours sincerely,

Hans A. Bethe

[118] Rudolf Peierls to Hans Bethe

Princeton, 21.1.1952

Dear Hans,
This is about my proposed visit. I am making a reservation to fly from Newark on flight 7 of Robinson Airlines on Jan. 29, which is due at 9.43 p.m. I can then easily make my own way to your house or to wherever else you say. I would then like to go on to New York by train, leaving on the evening of Jan. 30. As it seems a little complicated to make the rail reservation from Princeton, I wonder whether your office could be good enough to arrange that for me.

All this is on the assumption that you still feel up to receiving visitors so shortly before your departure. I expect in any case to see you here when you pass through. My lecture in New York has been fixed for the Saturday afternoon so I shall probably get back rather late on Saturday, but expect you will be visible on Sunday morning.
With kindest regards,
Yours sincerely,

Rudi

[119] Rudolf Peierls to Hans Bethe

Birmingham, August 1952

Dear Hans,
I owe you many letters on different topics, but the specific occasion for this letter is that we have been discussing the possibility of holding a conference on nuclear structure in Birmingham during the second half of July 1953[542] and before finally fixing the date and subject Moon and I are writing to you and one or two other people to ask for comments. We would, of course, very much hope that you would come yourself.

We intend to run the conference on similar lines to the one held here in 1948 when you were present, and it will again last about a week and consist of introductory reviews followed by informal discussion.

We are trying to find funds from which to offer hospitality to some visitors and perhaps even to help with travelling expenses. Do you think if we do not succeed that the lack of such help would be a serious obstacle to people from America?

I do not think I even wrote a bread and butter letter after the visit to Cornell, but I think you will have known anyway how much I enjoyed the visit. I hope your stay on the Hill has been successful and enjoyable, and I imagine you are now making plans for returning to civilisation. I certainly had a most enjoyable time at Princeton, where apart from learning a lot, I made some progress with my non-local theory, but not yet enough to be sure that the whole thing will work. I am still trying to answer some of the questions that are holding things up. We had some discussions in Princeton with Jensen,[543] about nuclear photo-effect at high frequency, where there seemed on the face of it a contradiction between the paper by you and Levinger,[544] and the view by Goldhaber

[542] After the successful Birmingham conference in 1948, Rudolf Peierls and his colleagues at Birmingham organised another conference on nuclear physics in July 1953.

[543] J. Hans D. P. Jensen (1907–1973), Professor of Physics at the University of Heidelberg since 1949, had spent much of the early 1950s in the US as visiting professor at different universities. In 1952 worked at Princeton.

[544] J.S. Levinger and H.A. Bethe, 'Dipole Transitions in the Nuclear Photo-Effect', *Phys. Rev.* **78**, 115–129 (1950).

and Teller[545] which was taken further in a paper by Jensen[546] which you probably know. We came to the conclusion that this was only a superficial contradiction, because when you derive the relation for the mean values of 1/E, the quantity that appears in the first place is the mean square fluctuation in the centre of gravity, this is the same as the mean square fluctuation in the centre of electricity. Now I believe that in spite of the success of the one-particle model that it is a bad approximation to treat the particles as independent and therefore neglect products for this purpose, since the tendency for the nuclear forces is certainly to keep this distribution of neutrons and protons uniform. I think the situation is similar to that in the electron theory of metals where one normally assumes that the electrons are independent, but where this must not be done if one wants to compute the fluctuations in electric density. In fact, as you know, Jensen estimates the tendency towards uniformity from the appropriate term in the empirical mass effect formula and gets quite good agreement about introducing the concept of alpha-particles as sub-units, which you think necessary.

You will have had a copy of the draft paper by Brown and Woodward,[547] on atomic scattering of gamma rays. This has now been tidied up for publication. We had some correspondence with Levinger and Rohrlich and it looks as though the various pieces of work nicely supplement each other, but a lot still remains to be done. In particular for the case of lead, which is probably of specific interest in connection with the Cornell experiments, there is probably nothing for it but to do numerical computations based on exact hydrogen wave functions, even for the continuous spectrum. We are planning to consider how best to conduct such a calculation, but we would probably not have any facilities here to embark on it and I thought that if I could interest Rose and his collaborators at Oak Ridge in this problem, they might be in

[545] M. Goldhaber and E. Teller, 'On Nuclear Vibrations', *Phys. Rev.* **74**, 1046–49, (1948).

[546] Helmut Steinwedel, J.H. Jensen and P. Jensen, 'Nuclear Dipole Vibrations', *Phys. Rev* **79**, 1019 (1950); Helmut Steinwedel and J.H. Jensen, 'Hydrodynamics of nuclear dipole oscillations', *Z. Naturforsch.* **5a**, 413–20 (1950).

[547] G.E. Brown and R.E. Woodward, 'Coherent Scattering of Gamma-Rays by Bound Electrons', *Proc. Phys. Soc. A* **65**, 977–80, (1952).

the best position to undertake this. However, obviously the first thing is to think out what is really involved.
With kindest regards to all of you,
Yours sincerely,

Rudi

[120] Hans Bethe to Rudolf Peierls

Ithaca, 15.12.1952

Dear Rudy,
I have the feeling that I have owed you a letter for quite a while. In particular, you asked me about an International Conference in Birmingham[548] next fall. You may have decided about this long ago. If you haven't I want to point out the existence of an International Conference in Japan in September 1953,[549] which you probably have been invited to. I don't know whether you want to go, but in any case, a considerable number of people possibly will go, and it may therefore not be a good idea to run a competition.* I probably shall not go, but I have not decided yet.

An additional reason for writing you is that a very good graduate student of mine, Stanley Cohen,[550] wants very much to go to England after he finishes his Ph.D. He has applied for a Fulbright but it is of course not clear whether he will get it. If he gets it, he would like to come to you. Some time ago you told me that you might have fellowships or positions of your own available for good Ph.D.'s from Cornell. This one I can recommend wholeheartedly. What are the possibilities for him at Birmingham? Naturally, his going there would be dependent on your

[548] See letter [119].

[549] International Conference on Theoretical Physics, Kyoto and Tokyo, September 1953.

[550] Stanley Cohen completed his Ph.D. under Hans Bethe. He later gained fame among his fellow students and researchers when he made a plaster mould of Hans Bethe's bust and sold it for a dollar a piece to his colleagues. See A.N. Mitra, 'Hans Albrecht Bethe (1906–2005): Personal Glimpses', *Current Science* **88**, 1855 (2005).

reaction to the other letter enclosed with this one, but supposing you were at Birmingham he would be very anxious to come.

There is another question connected with the same man. He is working on a good and accurate theory of the energy levels of electrons in heavy atoms. After he had gone quite a way on this calculation, we heard a rumor that a student of yours was working on the same subject. I think this is not necessarily undesirable because a lot has to be done on this problem and two different people may find different approaches to it. However, I think you should know that this is going on and maybe the two people concerned should exchange their wisdom.

We are making considerable progress on the scattering of mesons by nucleons.[551] It is a straightforward application of pseudo-scalar coupling. The approximation made is that we consider only processes in which one additional meson is produced in the field, or a nucleon pair, or both. We hope this is a good approximation but we have no proof for that. Apart from this approximation the problem is treated exactly, using a method similar to Tamm and Dancoff.[552] We can reproduce the essential features of the phase shifts of the important partial waves as measured by Fermi, et al. The only major discrepancy is for the S state of isotopic spin 1/2. We are only at the beginning of these calculations, so that much of it can still be wrong, and we have neither quantitative results nor a consistent theory. But it is very interesting and exciting.

Hoping to hear from you soon,
Yours sincerely,

Hans

Merry Xmas!

*This letter was held up by a secretarial confusion. In the meantime I have learnt that the conference is going to take place, and in July, so there won't be any interference.

---

[551]See letter [121], note 553.

[552]I. Tamm, 'Relativistic interaction of elementary particles', *J. Phys. (USSR)* **9**, 449–60 (1945); S.M. Dancoff, 'Non-Adiabatic Meson Theory of Nuclear Forces', *Phys. Rev.* **78**, 382–85 (1950).

[121] Hans Bethe to Rudolf Peierls

Ithaca, 15.12.1952

Dear Rudy,
This is to ask you whether it would be possible for you to come here to Cornell as a Visiting Professor for one year, namely the academic year 1953–54.

Unfortunately, we are losing Dyson to the Institute of Advanced Studies. During his first year here, he had the feeling that he did not work very effectively; others did not. Particularly, during the summer he accepted some work which did not satisfy him. This fall he is working with very great efficiency on the scattering of mesons by nucleons; in fact we are all working with him.[553] However, before this creative period started, he had come to the conclusion that a University with all its graduate students, examinations and other diversions would not permit him to devote his full energy to his work. He therefore asked Oppenheimer whether the standing invitation he had to the Institute could be converted into a permanent appointment. Oppie of course took up this suggestion most eagerly and made him an offer which also financially exceeds anything that Cornell University can match. We are very sorry to lose him, but find it impossible to prevent it. We have offered Ed Salpeter an Associate Professorship to take Dyson's place at least as far as the teaching is concerned. Ed has accepted this position in preference to the Chair of Theoretical Physics at Canberra. However, he wants to visit Australia for one year and, for personal reasons, he would prefer this to be the first year of his appointment, that is the next academic year. Oliphant has already offered him to come to Canberra for one year as he wants to, but the exact time has not been settled yet. Salpeter has written to him, asking whether next year will be alright. Cornell University has agreed to give him a leave during his first year he can make his arrangement with Oliphant.

---

[553] A joint paper was published in 1954. F.J. Dyson, M. Ross, E.E. Salpeter, S.S. Schweber, M.K. Sundaresan, W.M. Visscher and H.A. Bethe, 'Meson-Nucleon Scattering in the Tamm-Dancoff Approximation', *Phys. Rev.* **95**, 1644–58 (1954).

This leaves us minus a professor of theoretical physics for this coming year. At the same time, it gives us an opportunity which especially Wilson and I have been seeking for quite a while. We have wanted for a long time to invite you to come here for a considerable period. The University has agreed to my approaching you on this matter. We can offer you for the academic year a salary of §10,000. The University starts on September 21, 1953 and ends on June 8, 1954.

I do not know, if course, whether you can get away from Birmingham for this length of time. However, I thought that perhaps it could be arranged if you started this long ahead of time. If you could come, all of us would be delighted. I am sure you would want to bring your family, and I am sure that suitable living quarters could be found.

You would have to teach presumably one course each semester, 3 hours a week. You could choose among the set of courses ranging from three courses in classical physics through a year-course in quantum mechanics to some more advanced courses like nuclear theory. The other courses will be given by Morrison[554] and myself. As you know, there is a lot of experimental work going on in the Nuclear Laboratory and I am sure you will find the contact with this work interesting. You may have heard that Wilson is building a new synchrotron which is to give 1 to 1.5 Bev with a small amount of iron. We hope this will be working next summer so that the coming academic year may a be particularly interesting time.

In addition, we are doing a lot of theoretical work and we hope, with Dyson's help to bring meson theory into a shape such that it can actually be applied to the calculation of specific processes. This also ought to be in full swing my next fall.

There are always a considerable number of young men who have taken their Ph.D.'s at other Universities and who come here with some kind of fellowship. This year we have three such people from the United States, plus a man from Turkey. In addition, we have three or four Research Associates who are paid by the Laboratory. There is a con-

[554]Philip Morrison (1915–2005), obtained his Ph.D. from Berkeley in 1940. After his wartime work on the Manhattan Project, he worked at San Francisco State University, University of Illinois, Cornell University and M.I.T.

siderable number of graduate students, some of whom are doing their theses in theoretical physics. Altogether, this makes quite a stimulating group. The group is usually quite international; for next year we expect to have a Frenchman, an Egyptian and presumably one or two Englishmen, and possibly an Austrian or a Japanese.

I hope very much that all this will sound attractive enough for you to consider coming here. If you decide to do so, you would make us very happy. Especially I would like very much the idea of repeating the times of Manchester, 1933. Please let me know what you think of it.
Yours sincerely,

Hans

[122] Rudolf Peierls to Hans Bethe

[Birmingham], 16.12.1952
(carbon copy)

Dear Hans,
I am writing once again about a further possible link in the chain between Birmingham and Cornell. This time it is about R.D. Dalitz, who will be writing to you saying that he is applying for a Commonwealth Fund Grant and asking whether you are prepared to have him if he gets the award and also what the possibilities are of a Research Fellowship, or similar post at Cornell.

As regards the first part of the question I shall eat my hat if you do not say that you will be glad to have him, but as regards the second part you might like to know more about him.

Firstly you met Dalitz when you were here last and I think we had some quite lengthy discussions with him, but I do not know how much detail you remember. Then he is, of course, also well known to Dyson and you should be able to find out what Dyson thinks of him.

As far as my own view goes I have an enormous respect for Dalitz. He is good at field theory, for example when he worried about the internal pair creation in the Oxygen 16 decay, the papers by Feynman were quite new, and when he found that those would be useful for him he learnt the drill in a month or two and produced a reasonable answer. He knows a lot about nuclear physics, for example both the shell model experts here like Flowers, and the experimentalists, made a habit of consulting him frequently, much to their benefit, but somewhat to the detriment of Dalitz's own research output. He is good in talking to experimentalists and has constructive ideas, for example he realised that the pairs found at Bristol close to the stars represented the decay of a neutral meson into one pair and one photon, calculated this to show it gives the right magnitude and showed how this can be used to estimate the life-time. He volunteered last year to give a course on nuclear reactions, which was new to him then and he gave a first rate course on this. As part of this he produced the elementary derivation of the Butler formula, which was noticed independently by several other people.[555] It never occurred to him to write this up as a paper, this was just part of giving a good course of lectures. There was later some interest in the polarization of the protons, which on the Butler approximation is zero, and this came out more easily from Dalitz's technique than any other and only then was I able to persuade him to write a paper on the subject. This year he is giving us a similar post-graduate course on meson physics, both for experimentalists and our own group. I had thought of giving this course myself to avoid overloading Dalitz, but he was keen on doing it and I am now glad that I let him, because he is doing a much better job than I could I have done.[556]

He is also a first rate undergraduate teacher, though this will concern you less and one of his functions is to give our somewhat crazy courses in the Physical School, which as you may remember involve doing Maxwell's theory in twelve lectures and wave mechanics in another twelve. The Physics Department have pressed me hard to keep Dalitz on this particular assignment, unless I can do it myself, because nobody else has done it so well.

---

[555] See letter Dick Dalitz to Rudolf Peierls, 19.3.1952, *Peierls Papers*, Ms.Eng.misc.b205, C.75.

[556] Dalitz later published research arising from his work on the τ-meson. R.H. Dalitz, 'The Decay of the τ-Meson', *Proc. Phys. Soc. A* **66**, 710–13 (1953).

He is also running our Seminar and since he has taken it over I am for the first time in the position of not having to worry about it, but just to ask who will be talking next week.

With all these activities and his sense of duty it is not surprising that he has not published as much as he might have, even though his output is quite creditable. Another reason for this is that he has set his heart on meson theory, where it is much harder to publish papers than, for example, on nuclei, where he could have got out a lot more. He has also very high standards and will not undertake a calculation by sloppy methods, perhaps he is even a bit too critical in this respect. One of the objects of trying to spend some time at Cornell would be just the hope of getting a more concrete programme in that direction.

He has made himself so useful here that I am not at all happy at the thought of being without him, but of course it ...
(last page missing)

[123] Rudolf Peierls to Hans Bethe

[Birmingham], 5.1.1953
(carbon copy)

Dear Hans,
Thank you for your letter of 15th December, which arrived a day or two ago and for your invitation. This is most tempting indeed, I would love to stay at Cornell for some time and even more to have a chance of working with you again at leisure. However, I do not think it can be done. My group here still has such a rapid turn over that I am a very important element in its continuity. If I went away for a year I would probably have to start from scratch when I returned. In any case it would not be fair on the other members of staff here, who would have to deputize for me and whose chances of getting some research done would suffer the most because they would loose my help and because they would have to do my work instead.

It would also be unreasonable to leave just when the synchrotron in the Physics Department here is nearing completion (it might in fact be working) so that the experimentalists are at last beginning to think about mesons rather than about engineering problems.[557] On the more formal side I could not expect to get study leave from the university so soon after spending a term at Princeton. I could presumably get leave of absence without pay, provided I could assure the university that the work of the department could continue satisfactorily, but I am not sure this would work financially. However, the major difficulty is that I feel I can't leave the other people here in the lurch.

This is particularly acute next year when presumably Dalitz will not be here, because his presence made a lot of difference during my absence in Princeton. I certainly would not want to bring any pressure on him to postpone his departure since I think he needs a change of scenery.

With all this, much as I felt tempted by the invitation, I think I must not accept.

Of course the situation may change with time and if the possibility of such an invitation might arise again some time later I would be very happy to look at it again.

If at the present you are likely to be somewhat shorthanded I think you might find my suggestion of having Dalitz with you quite useful. He would certainly be quite willing to do some teaching and to help look after research students and the like, provided this is not on such a scale as to prevent his doing physics himself. I think you would also find that he is a very good person to discuss things with.

Once more many thanks and kindest regards,
Yours sincerely,

[Rudi]

---

[557]The proton synchrotron at Birmingham University came into operation in June 1953.

[124] Rudolf Peierls to Hans Bethe

[Birmingham], 5.1.1953
(carbon copy)

Dear Hans,
In reply to your letter about Stanley Cohen,[558] I would of course be very glad to have him if he can get a Fulbright grant. Does he want, at this stage, a formal letter saying so? On the question of a local grant, this is not quite so easy, our Fellowships here are limited in number and I do not know how many vacancies I will have next year, if any, since the plans of one of two people who are holding such Fellowships at the present are not yet decided. I also expect some fairly strong applicants and therefore one would have to make out a fairly good case for Cohen, if he were to have any chance. From what you say it seems as if he has not published anything yet and while this does not rule him out, it would certainly need a very firm supporting statement from you.

When the time comes to advertise our Fellowships, which is likely to be in two or three months, I shall let you know and perhaps you would then be good enough to tell me your opinion about him explicitly.

As regards his problem, we had already heard about this from Salpeter and Gerry Brown, who is looking after this work, and has sent a summary to Salpeter. The numbers of course are still changing a little, because one or two corrections of interest have turned up, at the same time our confidence in the terms already worked out is increasing and the really surprising part of the situation is how simple it all is in the end. However, this work is concerned only with the K-absorption edge and we are not at present proposing to look at other energy levels. The main outstanding uncertainty in the K-absorption edge is the contribution of Lamb-shift terms and we think we have got a method which will make the numerical computation of this term possible and will at the same time give the results for the scattering of gamma rays in the K-shell of a heavy element, which I believe are still of interest.

---

[558] Letter [120].

I am now in touch with Hartree to see if this problem can be done on the Cambridge machine.
Yours sincerely,

[Rudi]

[125] Hans Bethe to Rudolf Peierls

Los Alamos, 16.1.1953

Dear Rudi,
Thanks for your letter. It is, of course, a great disappointment to us that you cannot come, but we more or less expected it.

I see only one point where we possibly could help. You say that it may be difficult financially to come to us if you take leave without pay. If this is a major reason for your inability to come, we could certainly try to see whether we could not raise the offer from Cornell University. Unfortunately it seems to me from your letter that this is only a minor point. It is too bad because it certainly would have been nice.

You will have seen the copy of my letter to Dalitz. I presume that he now will be coming to us and it will be very good of him if he will take some of our teaching load. We are looking forward to having him.
Yours sincerely,

Hans

[126] Hans Bethe to Rudolf Peierls

Ithaca, 29.5.1954

Dear Rudy,
You may know that I am coming to Glasgow[559] and you may have guessed that we are all coming to England. We would very much like to

[559]Hans Bethe intended to attend the International Nuclear Physics Conference in Glasgow in July 1954.

visit you for a few days during the week before the Glasgow Conference. Would it be all right with you if we came from Tuesday evening, July 6th to Saturday morning July 10th?

I must warn you that we are 5; we have acquired an extra child, the 11 year-old boy of one of Rose's friends, for the trip to Europe. If it is difficult to accommodate us all, please just say so; in this case you could find a place for all or some of us to stay in your neighborhood. We shall have a car so that transportation is no problem.

Could you let me know if the dates are convenient and what the arrangements will be? My address will be:

P.O. Box 1663, Los Alamos, New Mexico

where I shall be until June 18th. If possible, could you send a copy to Rose at her sister's, Mrs. L. Davidson, 2 Oliver St. Shenfield, Essex. She expects to arrive there on June 15th.

With best wishes to Genia and hoping to see you soon.
Yours sincerely,

Hans

[127] Rudolf Peierls to Hans Bethe

Birmingham, 4.6.1954
(carbon copy)

Dear Hans,
Thanks you very much for your letter. We had already had a letter from Rose giving very similar information.

We of course are delighted to hear that you will be visiting us. As far as we can see now we can put up all of you, but if there should be any unforeseen domestic complications we shall certainly find some suitable accommodation in the neighbourhood.

We saw yesterday in the papers the report of the panel about Oppenheimer and we are indescribably shocked.[560]
Yours sincerely,

[Rudi]

[128] Rudolf Peierls to Hans Bethe

[Birmingham], 3.7.1954
(carbon copy)

Dear Hans,
Here are the instructions for your trip: I assume you start from Oxford. You leave Oxford by the Woodstock Road which gets you onto the road to Stratford-on-Avon without any difficulty. In Stratford you turn right just after passing the narrow bridge and immediately left again, (All this is well marked) and you are then on the main road to Birmingham. You pass through two villages, Henley-in-Arden and Hockley Heath, which have speed limits and at the next speed limit sign you are in Shirley, a suburb of Birmingham. You still continue on the main road to the City Centre. You will hardly notice passing the city boundary and a little later you will pass a very complicated direction sign for the ring road. You ignore this and still continue for about three quarters of a mile, when the road bends a little to the left and comes to a roundabout (College Arms Pub) immediately after this you take a turning to the left into College Road and this is about the only place which is not obvious. You go to the top of College Road and bear right into Wake Green Road, follow this for about a mile, when the road carrying the main traffic bends left and right again into St. Mary's Row. At the bottom of this you continue straight across some traffic lights (Moseley village)

[560]In 1953, Robert Oppenheimer was suspended from the Atomic Energy Commission as an alleged security risk and he had his security clearance taken away, in part due to criticism from Edward Teller. In October 1954, he was unanimously re-elected director of the Institute for Advanced Study. In 1963, Oppenheimer was awarded the Enrico Fermi Award which was widely seen as a gesture of political rehabilitation.

into Salisbury Road. At the bottom of this you bear left following the main traffic into Edgbaston Road which later becomes Priory Road. After passing two traffic lights, Priory Road goes up hill, at the top of the hill you bear left along a piece of dual carriage way (Church Road), where this ends, at a roundabout, you turn right into Arthur Road and the second turning left is Carpenter Road, with our house, number 18, at the corner.
If you get lost, telephone and we will rescue you.
Yours sincerely,

R.E. Peierls

[129] Rudolf Peierls to Hans Bethe

[Birmingham], 13.8.1954

Dear Hans,
I enclose a reply which we received about the document which you so kindly tried to get for us. As it appears to be out of print for the moment there is probably no need for you to do anything until you get home.

We forwarded to you yesterday a letter from Cambridge which has been sitting here as we were not quite sure of your address.[561] We now got your address from your secretary in Cornell, but my secretary also wrote to you via Frankfurt, but you needn't now reply to this.

We had a very hectic time after Glasgow because there was a dense flux of visitors through Birmingham and all that week I felt like playing one of those multiple chess games, where you go round making one move at each table. However, I am glad all these people decided to come here after Glasgow and not before otherwise we would have got much less out of your visit.

I hope you are having a pleasant time in Switzerland and the weather is better than here and better than we are likely to find in Cornwall,

[561]The Bethe family had been travelling through Europe, eventually reaching their holiday destination in Sils-Maria, Switzerland.

where we are now going off for three weeks. Life is made slightly more complicated by the fact that our car got a crumpled wing as the result of a skid and we hope it will just be ready in time for us to go off, but all the preparations for the trip have to be done by pedestrian means.
With kindest regards to all of you,
Yours sincerely,

Rudi

3rd September

This letter got delayed until it was too late to send it to Switzerland. Meanwhile the document has arrived and we have started reading it. It makes fascinating reading and, since (at least as far as I have got) every bit is interesting, it will take quite some time to get through it.

We are now nearly at the end of our Cornish holiday, which was most successful, and even the weather was reasonably kind, mostly dry, with marvellous surf-riding. We visited the farm where we spent the summer of 38, and where you came to visit us. It was still the same woman there, and she was very pleased to see us.

Meanwhile Gaby got her exam result and, to everybody's surprise, including her own, she got a second, which is very good for her. She still seems to enjoy the work in her looney-bin, and in a week's time she goes off for a holiday in Italy.

Ronnie spent a few weeks as a shift worker in a factory making black-currant juice, and now after some time with us in Cornwall, is recruiting members for the local branches of the Labour Party. Combined with his ordinary work, this certainly makes for variety.

The only new thing in physics is the work of Matthews and Salam,[562] which may mean that — similarly to their earlier paper about fermions in an external boson field — the series expansion of the propagator converges after all, provided one expands the right quantity. However, a lot has to be looked at before one can be sure that it does mean that.

---

[562] P.T. Matthews and Abdus Salam, 'Renormalization', *Phys. Rev.* **94**, 185-191 (1954).

[130] Rudolf Peierls to Rudolf Peierls

Birmingham, 19.3.1956

Dear Hans,
I have just received you letter about Kap[u]r as I am leaving for the Rochester Conference.[563] I shall certainly be glad to consider him but he will have to compete for our research fellowships with some other people. I evidently cannot do much about this until I return, but shall then try to assess the probabilities.

I hope this letter will reach you somewhere and I wish you a successful Easter trip.

I hope you remember your promise to visit Birmingham. I expect you will be here on 30th June anyway, and the week ending on that day would suit us very well, but earlier would do also if you prefer.
Yours sincerely,

pp. R.E. Peierls

[131] Rudolf Peierls to Hans Bethe

Birmingham, 25.4.1956

Dear Hans,
I hope you've had a good time in Italy. I am now writing to ask whether we could settle the time for your visit to Birmingham. We mentioned several weeks in June, and if we had actually agreed on a definite period, I am afraid I have not kept a note of it. I am now of course anxious to arrange my programme accordingly so as to be as free as possible when you are there.
Yours sincerely,

Rudi

[563]The Sixth Annual Rochester Conference on High Energy Physics took place in Rochester from 3 to 7 April 1956.

[132] Hans Bethe to Rudolf Peierls

Cambridge, 11.5.1956

Dear Rudy,

We have discussed the week of 11 June. It looks as if Rose and the family cannot come and bring me to Birmingham. In this case it is just as satisfactory to me to arrive in Birmingham only in the evening of Monday, 11 June. So please don't go to any trouble trying to change your engagement at Aldermaston.

Some more travels have descended on me so I think I will come on Monday evening to Saturday morning, June 16.

I mentioned to you several times Mr. Goldstone[564] who has been working with me on the Brueckner theory. He is unusually good. He has now talked to Hamilton[565] and agreed with him that it would be better for him to go somewhere else for his further research studies. He has been in Cambridge for five years. His first choice of course would be Birmingham. I warned him that you were much overcrowded with students and that this was an awfully late time to make up his mind for next year. However, if you can still fit him in in any way I think he is very much worth while.

Yours sincerely,

Hans

[564] Bethe and Goldstone published a paper together. H.A. Bethe and J. Goldstone, 'Effect of a repulsive core in the theory of complex nuclei, *Proc. Roy. Soc. A* **238**, 551–67 (1957). Their joint work with Brueckner became known as the Brueckner-Bethe-Goldstone theory.

[565] James Hamilton (1918–2000) received his Ph.D. from Manchester in 1948 and became College lecturer in Mathematics and Fellow of Christ's College, Cambridge, before becoming Professor of Physics at Imperial College London (1960–64) and Professor at NORDITA in 1964.

[133] Rudolf Peierls to O. Hood Phillips[566]

[Birmingham], 6.6.1956
(carbon copy)

Dear Hood Phillips,

I am now writing to give you some notes about Bethe.

Firstly about his work. Bethe is a pupil of Arnold Sommerfeld, who was probably the most famous and most successful teacher of theoretical physics in the last generation, and whose work was characterised by a very positive attitude with the emphasis on the solution to practical problems rather than on general principles. Bethe, whose attitude was strongly influenced by Sommerfeld, is now occupying the position in physics once held by his former teacher. The main impression his work makes is one of immense strength. There is hardly any branch of modern theoretical physics in which Bethe has not made contributions of vital importance. For example, some of his work on the theory of solids is now about 25 years old and still standard equipment for theoretical physicists. He developed a powerful method to clarify problems in the structure of alloys, he showed how to calculate realistically the stopping power of various substances for particles. He made important contributions to the theory of nuclear forces and of nuclear reactions. He was among the first to work out the corrections to atomic energy levels, which had led to the development of modern field theory and he has assisted in an important way our understanding of mesons and their interactions with nuclei. This is by no means a complete list of the important things he has done.

In addition he has written important texts, such as books, some as review articles, which are very characteristic in that each not merely reviews published work in a certain field, but adds a new quantitative discussion and completes investigations which had not been done completely or not correctly, so that each such article represents a stage in the development of a branch of physics.

[566] Owen Hood Phillips (1907–86), Professor of Law at Birmingham University.

During the war he was active in several fields, first in radar and simultaneously in the theory of shock waves caused by explosions and later in atomic energy. He was the head of the theoretical physics division of the Atomic Weapons Laboratory at Los Alamos from its formation to the end of the war and to him must go a good share of the credit for the work of that laboratory. Those who have seen him at work get in the first place the impression of immense strength. For him problems are there to be solved and not to be worried about, and there are few problems which have attracted his interest to which he would not find a reasonable solution in a short time. He seems to talk and act slowly and deliberately, by knowing just what to do and where to go he does get there with incredible speed. One is sometimes reminded of the sensation of seeing an elephant run, when one is also struck by the speed of progress of an apparently slow and lumbering mode of progress.

The story goes that he was once consulted by a colleague about a certain calculation and replied that this was well worth doing. "It would take you about a week to do it. It would take me two days". Those who know him would realise that this was not a boast, but a statement of fact.

I know that puns on names are generally frowned upon in such connections, but nevertheless you might like to know the story that Gamow, one of our colleagues who likes to make people laugh, once persuaded Bethe to contribute to some work he was doing jointly with a student by the name of Alpher, mainly for the pleasure of then publishing a joint paper under the authors Alpher, Bethe and Gamow, which he likes to refer to as the α, β, γ paper. I have also heard that in a seminar at a French University a research student referred to a difficult problem which he said was so hard that "même le grand Bête" had been unable to solve it.

In private life, Bethe was famous as a young man both for the quantity of food he could consume and for the slow and deliberate pace at which he went about it. In a way his appetite for food was similar to his appetite for physical problems.

I assume that you have the usual facts about his biography. I might mention specially the fact that he has held his present chair at Cornell University since 1935, which for a physicist is rather unusual continuity.

It may also be on interest that he worked in this country from 1933 to 1935, first at Manchester and then at Bristol, after he had left Germany on Hitler coming to power, and that he is now spending a year at Cambridge where his presence had been a great source of strength through the ideas he has brought along and through the inspiring guidance he has provided for many students and colleagues.

I am passing this letter on to Moon who no doubt will add any further points that I may have omitted.
Yours sincerely,

R.E. Peierls

[134] Rudolf Peierls to Hans Bethe

[Birmingham], 22.6.1956
(carbon copy)

Dear Hans,
About the journey from Cambridge I believe that in the evening it may be quicker to come by Coventry. You go as before to St. Neots, Bedford, Northampton, Weedon and the Daventry By-pass but you then continue along the main road to Coventry. It probably pays to take the Coventry By-pass, in other words follow the Birmingham signposts just outside Coventry and the rest of the way to Birmingham is straightforward. You pass the Birmingham city boundary just by the Birmingham Airport and about 2 1/2 miles further you cross the ring road which you recognise by a black-and-white direction sign with a large number of destinations. You continue for another 2 1/2 miles, when just after passing the Singer Motor Works (a tall continental style factory building) on your left and past a park on your left, you turn left at the traffic lights into Wordsworth Road.

At the next major intersection bear left into Golden Hillock Road and take the first turn right into Walford Road (a rather inconspicuous sign says Edgbaston). This runs into a T-junction with traffic lights. You turn right into Stratford Road, immediately left into Stoney Lane, and right again into Highgate Road (which is in fact a continuation of Walford Road by which you came). Then you go straight on for about 1 1/2 miles, the road changes its name to Belgrave Road, and you finally cross the Bristol Road by the Bristol Cinema. You go on for one more block when you run into another T-junction, turn right into Spring Street, and at the next corner turn sharp left into Gough Road. This ends in another T-junction which is Carpenter Road, you turn right and out house is at the next corner.

If you find very heavy traffic on the road from Weedon to Daventry, it may be better to go through Warwick. I do not know whether you remember the road, but you turn left on the Daventry By-pass following the signpost to Southam and Leamington. You then go on to Warwick and the Birmingham road. After passing under the railway bridge at Olton, you pass the city boundary and immediately bear left onto Boulton Boulevard (wide grass verge) and at the end turn left into Shaftsmoor Lane. This takes you to the Stratford Road. You bear left towards Birmingham and immediately turn left again into College Road, right into Wake Green Road, which if you follow the main traffic leads you to the traffic lights at "Moseley village". You continue straight into Salisbury Road and bear left into Edgbaston Road, which eventually becomes Priory Road. You cross two traffic lights and up a hill and at the top turn left along a piece of dual carriage way (Church Road) and where the double road ends turn right into Arthur Road and at the second intersection you are there.

As it happens, Gaby, who is now moving to Cambridge will want to come home for that weekend and she would be grateful if you could give her a ride. We are telling her to get in touch with you, but you might like to have her phone number in Cambridge which is 55303.
Looking forward to seeing you.
Yours sincerely,

R.E. Peierls

[135] Rudolf Peierls to Hans Bethe

[Birmingham], 22.10.1956
(carbon copy)

Dear Hans,
I have heard a rumour that you are embarking on a computational programme on the Brueckner-Bethe model of the nucleus, and I have therefore asked De Dominicis to write up a note summarising his progress to date.[567] You may also find this of use for orientation. We would of course also appreciate you comments, particularly as regards the question whether one could continue doing these simple calculations with a separable potential, while presumably machine calculations of a more ambitious kind will be starting.

The note includes a very amusing exact solution for a potential which is almost indistinguishable from the hard core. This was discovered too late to be used in the numerical work, and it may not be of great practical value since it is probably impractical to carry out the required integration without expanding the integrand by something very like your method.

I have just received an invitation to the meeting at the Stevens Institute, to which you are also going. Since I propose to come to the Rochester Conference[568] I doubt very much whether I can make a second trip so shortly before the other, but I shall reply suggesting that they should send an invitation to de Dominicis. He is returning to France at Christmas, but it may be that if he has an invitation the French authorities may pay his fare.
Yours sincerely,

[Rudi]

---

[567]C.T. deDominicis, Ph.D. student at Birmingham who obtained his doctorate in 1957. He published his results in C. deDominics and P.C. Martin, 'Energy of Interacting Fermi Systems', *Phys. Rev.* **105**, 1417–19, and C. deDominics and P.C. Martin, 'Saturation of Nuclear Forces', *Phys. Rev.* **105**, 1419–20.

[568]The Seventh Annual Rochester Conference on High Energy Physics took place between 15 and 19 April 1957.

[136] Rudolf Peierls to Hans Bethe

[Birmingham], 25.11.1956

Dear Hans,
The American Institute of Physics have now sent you the last cheque. The delay was due to the fact that one batch of translations was lost by the Mexican Post Office — fortunately I had kept carbon copies. My computing department tells me that the three cheques add up to $ 339. If this is confirmed would you deduct interest and service charge and if then any balance remains, I know of a young man not too far from you who might be persuaded to accept it.[569] Thank you very much again for the whole transaction.

De Dominicis has just returned from Copenhagen after two weeks work with Martin there. They have a lot of further quantitative results including estimated of the "W" terms and the effects of the reduced mass with and without hard core and with and without Pauli principle. I shall not however attempt to summarize these.

With kindest regards to all of you.
Yours,

Rudi

[137] Rudolf Peierls to Hans Bethe

Birmingham, 9.2.1957

Dear Hans,
I am about to propose to Gerry Brown for promotion to grade 1 (senior lectureship) and it usually helps on such occasions to have outside opinions about the merits of his research work. I believe you are familiar with at least some aspects of his work and I wonder whether you would

[569]The 'young man' referred to in the letter was Ronnie Peierls, Rudolf Peierls' son, who, after completing his undergraduate studies at Cambridge, had moved to Cornell to study for his Ph.D. with Hans Bethe.

be willing to write such a letter which I would have to have by 25th February if it is to be used for the appropriate meeting. You will not be surprised to hear that I am thinking of paying a visit to Cornell either before or after the Rochester conference. My travelling dates depend somewhat on MATS authorities by whose courtesy I am expected to get to Rochester this time, but with luck I should get to the U.S.A. on the 11th April. It would then seem most convenient to visit Ithaca before the conference, but I could make it afterwards if that is more convenient for you, or I may be forced to make it afterwards if I don't get on a plane at the right time.
With kindest regards,
Yours sincerely,

Rudi

[138] Hans Bethe to Rudolf Peierls

Ithara, 15.2.1957

Dear Rudy,
Excellent! We are all looking forward to your visit either before or after the Rochester meeting. As far as I know, there is not much reason for us to prefer one or the other time. It is possible that after the conference the Russian physicists will visit Cornell for a day or so, and in this case we would of course be very grateful if you would do some extra work as an interpreter. At present we expect the Russians to come but, of course, such things may change in the course of two months.

Ronnie is doing fine.[570] He is starting some calculations on the production of K-mesons by gamma rays, in connection with the experiments which are just beginning on our synchrotron. He is a most pleasant companion to have at home.
With best regards,
Yours sincerely,

Hans

[570]See letter [138], note 531.

[139] Hans Bethe to Rudolf Peierls

Ithara, 15.2.1957

Dear Rudy,
I am very happy that you are considering Dr. Gerry Brown for promotion to a senior lectureship.

I have known Dr. Brown for many years, and have always been impressed by his knowledge of physics and the high quality of his work. Several years ago he worked on a new derivation of the interaction between the spins of two electrons in the helium atom, a problem which had been left in a rather unsatisfactory state after the early investigations by Breit. The derivation by Brown is simple and convincing. Salpeter and I have made much use of it when presenting the subject in the new edition of the Handbuch der Physik.

Later on Brown cleared up the problem of the elastic scattering of high energy gamma rays by the electrons of a heavy atom. I had used a rather naïve theory of this effect, on which several calculations had been done by some of my collaborators. In a paper with you, Brown gave the first correct theory, and he has pursued this theory in very effective numerical calculations.

During my last visit I was especially pleased to talk to Gerry Brown at some length on the polarization of protons of 1 Bev when scattered by complex nuclei or by other protons. He showed a very complete mastery of the subject; he knew all the work that had been done by others, and he understood the problem in a simple and physical manner. He made interesting suggestions for experiments, and he knew exactly how to calculate the most relevant quantities for the planning and the interpretation of the experiments.
Yours sincerely,

Hans A. Bethe

[140] Rudolf Peierls to Hans Bethe

Birmingham, 21.2.1957

Dear Hans,
I believe Elliot H. Lieb,[571] now in Kyoto, has written to you to ask if you have a job for him for the next year, and I am therefore writing to you to tell you what I know of him.

He came to us directly from his first degree at M.I.T. and we did not expect that he would be ready to start on research straight away. However, he turned out to have rather more knowledge and understanding than this qualification normally implies, and he settled down quite quickly to doing useful work (while of course continuing to broaden his background).

His first piece of work was a calculation about magnetic moment corrections in atoms, suggested to him by Brown and published in the Philosophical Magazine, 46, 311 (1955).[572] He then spent the remaining two years working with Edwards on a problem in field theory. The problem was to apply the Feynman integral over history's method to the case of a meson field interacting only with itself acting through a [??] coupling using the philosophy developed by Edwards, which is similar to Feynman's treatment of the polaron problem. He got quite a lot of interesting results which formed his Ph.D. thesis. A paper on this has been submitted to the Royal Society, but because of his departure for Japan there has been some delay in getting this into an acceptable form.[573]

He is now on a Fulbright grant in Kyoto largely because of a very strong personal interest in Japanese culture. I gather that this is not

---

[571] Elliott H. Lieb (1932–) had completed his Ph.D. at Birmingham in 1956. After research positions at Kyoto, Illinois, Cornell and IBM, he took up an Associate Professorship at Yeshiva University, before becoming Professor of Physics at Northeastern University and later at MIT. Since 1975 he has been Professor of Physics at Princeton.

[572] E.H. Lieb, 'Second Order Radiative Corrections to the Magnetic Moment of a Bound Electron', *Phil. Mag.* **46**, 311–16 (1955).

[573] E.H. Lieb, 'A Non-Perturbation Method for Non-Linear Field Theories', *Proc. Roy. Soc. A* **241**, 339–363 (1957).

the ideal set-up to gather inspiration, but he manages to keep going on some further investigation of the polaron problem, mainly with a view to learning what relation, if any, the Feynman approach has to a representation in terms of conventional wave functions.

Lieb is a very charming young man with a lively mind and a great thirst for knowledge. He was a bit of a problem child when he came here (I believe Weisskopf sent him here for personal education as much as for research training), but he developed very satisfactorily. He is still a little immature in that he is not too critical towards his own work and he tends to assume rather easily that when he gets an answer differing from what he found in the literature that he must be right and everybody else has made a mistake. But I am sure he will grow out of this and it would be very useful for him now to get to a problem related directly to experiments and involving, if possible, features of real practical physics. He told me that amongst other things he was thinking of going to the Institute in Princeton, but I discouraged that idea strongly. This is not what he needs at present whereas Cornell, I think, would be just right. He is a good person to argue with and he is the sort of man who goes looking for arguments and does not go to sleep if left alone. If you could find him a grant he would be sure to earn his keep.

I might perhaps add that in my opinion while he is a much less mature person than Downs, he has a greater ultimate ability of the two.
Yours sincerely,

Rudi

(handwritten addition)
Thank you very much for your letter about Brown, which was just what I hoped you would say.

[141] Rudolf Peierls to Hans Bethe

Birmingham, 1.4.1957

Dear Hans,
I have at last the full information about my journey. I am now scheduled to arrive at McGuire Air Base. N.J. on the morning of Wednesday 10th April, a day earlier than I planned. I shall then stay a day in New York and, if it is convenient to you, fly to Ithaca some time on Thursday, probably Thursday afternoon and then proceed to Rochester some time on Sunday. All this of course can be changed at the last minute.
Yours sincerely,

Rudi

(handwritten addition)
If you want me to give a seminar, I could talk on "A variational approach to collective motion". But that is as you like, I have no strong urge to perform.

[142] Rudolf Peierls to Hans Bethe

[Birmingham] 1.5.1957

Dear Bethes,
Having returned home safely and only seven hours behind schedule, and having made a little hole in the pile of waiting correspondence, I now want to write and thank you again for your manifold hospitality. I tremendously enjoyed the days in Ithaca, with the unique combination of talking physics, being amongst old friends and seeing Ronnie. He seems to be getting really a lot out of work and life there — not that even without personal inspection there was much doubt that he would.

I had a most instructive time in Brookhaven, and rather liked the place though I can see that living there for a long time must have its disadvantages. In New York I stayed at the "New Yorker", which at Easter time when all the local high school children, as well as the girl scouts of New York are holding their conventions there it is more than usually depressing. I did my shopping in no time, and had some time to visit Columbia, although the people there had by that time had their fill of visitors.

Then I spent the best part of a day at McGuire field, having by chance the company of Caianiello.[574] Our combined stock of stories lasted well over the time that we both had to wait for our planes.

Returning here I found that Genia had yet painted another room, apart from various domestic complications. Now, however, she has taken the children off to Bournemouth for a week. This was in the first place for the benefit of the children, but at the same time Genia will get some well-deserved rest, too.

My first job here was to read six draft papers — there were no more, because he has not yet got round to writing up the rest yet! Incidentally there seem to be new good experiments by Mann in Philadelphia on gamma-ray scattering.[575] At 1.3 MeV these now agree with our theory, at 2.5 MeV they exceed the (extrapolated) theory considerably. According to Brown this can be understood as Delbrück scattering, which should be predominantly imaginary. The imaginary part, which is related to real pair production, should even be reasonably calculable.

Before slipping too deeply into physics in what started as a general letter, I had better stop.
Yours sincerely,

Rudi

---

[574]Eduardo Renato Caianiello (1921-1993) obtained his first degree from Naples in 1944 and his Ph.D. in physics from Rochester in 1950. After posts at Turin, Rome and Princeton, he was appointed to the Chair in Theoretical Physics at the University of Naples. Later he focussed on cybernetics, founding the Laboratory of Cybernetics at Arco Felice. Later he moved to Salerno.

[575]A.M. Bernstein and A.K. Mann, 'Scattering of Gamma Rays by a Static Electric Field', *Phys. Rev.* **110**, 805–814 (1958).

[143] Rudolf Peierls to Hans Bethe

Birmingham, 8.1.1958

Dear Hans,
I am writing to mention the case of a very bright American, James S. Langer,[576] who is expected to get a Ph.D. here this summer, and who is going to return to the United States after that.

He is a graduate of Carnegie Tec'. and came here immediately from his Bachelor's degree in 1955. We do not usually expect students who have completed an undergraduate course in an American institution to be ready for research straight away; however, Langer was unusually well trained and unusually mature and we found it reasonable treat him on the same basis as one of our graduates.

He produced, under the guidance of Edwards and Matthews, a paper on the Effect of Virtual Pairs of Strange Particles in the S-Wave Scattering of Mesons by Nucleons, which, as you may remember, Matthews summarised at the last Rochester Conference, and which is published, (*Nuovo Cimento* **6**, 674, 1957.[577]) Since then, he has changed to more conventional nuclear theory, and has been working under the supervision of Gerry Brown. He has already made some very nice contributions to the derivation of the optical potential from the dispersion theory, following the work of Brown and de Dominicis, and I have every confidence that he will produce a good Ph.D. thesis before he leaves here.

He is a very charming person, speaks well and clearly, and has been successful in what little part-time teaching we asked him to do. He would like to continue for a while in nuclear theory and would be quite interested in a job which would involve some teaching, provided of course it left him with adequate time to continue research.

---

[576] James Langer (1934–), received his BS at Carnegie Institute of Technology in 1955 and his Ph.D. from Birmingham in 1958. He returned to Carnegie Mellon in 1958. He became Professor of Physics and a permanent member of the Institute of Theoretical Physics at University of California, Santa Barbara in 1982.

[577] J.S. Langer, 'Strange-Particle Effects in S-Wave Pion-Nucleon Scattering', *Nuovo Cimento* **6**, 674–81 (1957).

In considering his plans for next year, his first choice turned out to be M.I.T. because of the connection of what he is doing with the main line of interest of some people there, and his second choice is Cornell. We have, of course, made an approach also to M.I.T., but the object of this letter is to find out what the prospects would be in your department.
Yours sincerely,

Rudi

[144] Rudolf Peierls to Hans Bethe

Birmingham, 8.1.1958

Dear Hans,
I had a letter from a Mr. B.H. Brandow,[578] about the possibility of his coming here for a year, and I would therefore be grateful, if you could tell me a little bit about him. From his letter it sounds to me as if he has not yet reached the stage of starting a research problem, and may still have to spend a good deal of next year attending graduate courses. If so I am a little doubtful whether a change at this stage would be useful for him. In any event we are again likely to be under pressure for places, though it is too early yet to assess this quantitatively and he may have to compete with others. I would therefore be grateful if you could give me an opinion of his promise and of the stage he has reached in his training.

I might mention in this connection that we are happy to have Kabir here. He is not only a charming person, but also very bright and enthusiastic and has already made himself very useful. He still seems to show some reluctance to come down to brass tacks with algebra, and with numbers, but I am keeping some pressure on him in that respect and that seems to be working all right.

I expect he will stay here also for next year. I shall then, however, still have a Research Fellowship vacant on a similar level arising from

[578] See letters [145–146].

the departure of Nina Byers, another American, so if you happen to hear of some good person with a Ph.D., who would like to spend a year or more in this country, I would be glad to know. We have now a new arrangement by which the D.S.I.R. provide Senior Visiting Fellowships which can be on a level up to professorial and can provide travelling expenses, so we could consider even a fairly senior person, though, of course, such a Fellowship would still not be comparable to the senior levels of the Fulbright, or similar grants.

I am looking forward to hearing your reaction to Brown's latest remarks about the doubtful factor in the strength of the spin orbit potential and nucleon-nucleon scattering at high energies. It seems to me from a superficial look at the problem that he had quite a strong argument.

Yours sincerely,

Rudi

[145] Hans Bethe to Rudolf Peierls

[Ithaca], 16.1.1958
(carbon copy)

Dear Rudi:

This is about your inquiry concerning Mr. Brandow. I entirely agree with you that it does not make very much sense for him to come to you at such an early stage in his career. He is just a beginning graduate student, taking this semester an introductory course into quantum mechanics, and he will need a lot more formal course work during the next academic year. He may be ready to start research in the Fall of 1959. When he told me of his plans to go to England next year, I tried to dissuade him but did not succeed. Perhaps a letter from you expressing the same opinion independently would do the trick.

Brandow's work in Introductory Quantum Mechanics is about average according to Ed Salpeter. Average is pretty good, since the class is well selected. I think it would be better for him to stay here at least another year, and perhaps come to you for his thesis if he wants to. This, however, would probably mean that he should get his Ph.D. from you since Cornell University requires that the last year before the degree is taken here [...][579] Silly regulation but they insist on it.

I am very happy that you are enjoying Kabir, and that you are putting pressure on him to come down to brass tacks. He is really very bright and worth a lot of education.

We are similarly happy with Ronnie who is working very hard and getting good and interesting results on production of K mesons by gamma rays. He is becoming one of the experts in the field, and it is a pleasure to have him around.

It is good to know that you have a Research Fellowship vacant. There are usually a number of people who would like to go to you for a year after their Ph.D. I don't have any specific person in mind at the moment, but I will remember it when the occasion arises.

I am terribly busy these days. Perhaps the main cause is the Scientific Advisory Committee to the President which is extremely active and has a lot of subcommittees similarly active. This work, however, is quite satisfactory because it seems that we do have some success and some influence. In addition to this, I have gotten involved in one of the missiles projects to the extent of one week out of three, and in view of all this I have taken a half-time leave from Cornell University. The consolation is that there is hope that all this is temporary.

With best regards,

Yours sincerely,

Hans A. Bethe

[579] Part of sentence missing in manuscript.

[146] Rudolf Peierls to Hans Bethe

Birmingham, 3.3.1958

Dear Hans,
I am writing mainly about the Birmingham-Cornell pipeline. Firstly, to say that Jenkins about whom I wrote to you recently, has now had and accepted a firm offer from the University of Pennsylvania. Since he plans to go largely into the non-nuclear applications of the many body problem and plasma theory, this would seem to be a sensible arrangement.

Jim Langer, about whom we corresponded before, has now a firm offer from Carnegie Tec' which he is inclined to accept, except for the fact that is still not know whether Nick will be there next year, and this is clearly an important consideration. He still has an application in to M.I.T. and Weisskopf, when he was here, was favourably impressed by him, but was not yet sure that the job would be available for him. In other words, he may still be in the market, but the position is uncertain.

I would like to mention two other names very different from each other in case you may be interested in either or both of them. One is Dr. J. Dabrowski,[580] a quite senior member of the staff (aged about 31) of the University of Warsaw, who has been here since March '57 on a Polish government grant. He is a pupil of Rubinowitz and has published a number of papers of which the latest are concerned with nuclear theory. Here in Birmingham he has done calculations on the Jastrow method and has calculated with this the binding energy of O^{16}.[581] This is, in my opinion, a very nice piece of work showing firstly that this comes out extremely simply and secondly that the results are encouraging, though the approximations used and the forces assumed are not yet sufficiently realistic to look for a quantitative check.

---

[580]Janusz Dabrowski returned to Poland and, with the exception of some research leave abroad has remained there since, as professor of physics at the Institute for Theoretical Physics, Warsaw.

[581]J. Dabrowski, 'On the Binding Energy of the ^{16}O Nucleus', *Proc. Phys. Soc.* **71**, 658–667 (1958); J. Dabrowski, 'On the Binding Energy of the ^{16}O Nucleus. II. Higher Clusters', *Proc. Phys. Soc.* **72**, 499–504 (1958).

He now finds that he could get a further year's leave of absence, and would like, if possible, to spend a year in the United States, but could not expect to get from Poland a grant on a sufficiently large scale to make this possible. He has already an invitation to spend a year with Tobocman at the Rice Institute, but this is not necessarily the best place for him, both because of the size of the group there and because of its geographical position. He would be extremely happy if there was still a chance of spending a year in Cornell, but because it is so late I am writing about him to Princeton.

The other name is G.H. Burckhardt,[582] a graduate from Oxford, who completed his Ph.D. here last autumn after two years research work (which we regard as quite exceptional). His thesis was concerned with the deduction of the ionisation in the beginning of fast electron pair tracks because of the small distance between electron and positron. You may remember that I mentioned this work when I was last at Cornell. He has made a very sensible analysis of the quantum theory of this phenomenon and the result is the deviations from the classical picture are nowhere of practical importance, but I do not think that this detracts from the value of his work. Apart from this he has mainly been working with Dalitz, when the latter was here, first on the analysis of the tau decay and later on the problems of pion physics, and he has done a good deal of work in collaboration with Cassells at Liverpool, clearing up points relating to the theory and interpretation of experiments. He is therefore practised in using a practical approach and in talking with experimentalists. I can recommend him very highly. He is married; his wife is a mathematics teacher and could no doubt earn her keep.

Dalitz has already put his name to Lee[583] to suggest that he might spend a year at Columbia, but so far there is nothing definite yet from there.

[582]G.H. Burkhardt took up a research position at Columbia University and later moved to CalTech. He later moved on to become the director of the Shell Center for Mathematical Education at the University of Nottingham.

[583]Tsung-Dao Lee (1926–) received his Ph.D. at the University of Chicago in 1950. After appointments at the University of California and at Princeton, he moved to the University of Columbia where he eventually became Enrico Fermi Professor of Physics.

As regards movement in the opposite direction, you will remember our correspondence about Mr. Brandow.[584] This problem has been complicated a little by the fact that we have had the papers about him from the Fulbright Commission, and, from past experience this indicates that he is on their short list and will in fact get the grant subject to the approval from Washington. However, we are not allowed to tell him this at the present stage. This put me in some difficulty because it seems wrong to refuse to accept a man who holds a Fulbright Award, and we have never yet done so in the past. I have therefore replied to the Fulbright people that we would be willing to have him, but have written to Brandow himself explaining the various reasons why next year might be wrong time to come and suggesting that he considers seriously whether to come the following year. I do not know whether if he gets the Fulbright Award it would be reasonable for him to try and get this postponed (I do not think he would get a promise of a grant for the following year, but he might get at least an indication that an application for the following year would be considered with sympathy).

I also had correspondence with another man now at Cornell: this is Levinson a graduate student who came to you from Glasgow and is now apparently under Salpeter's supervision. He seems inclined to go off into abstract philosophy (Neumann's theory of Hilbert space) and apparently was told by Salpeter that this approach was not regarded as being fruitful at this particular time at Cornell. He therefore wrote to me and asked if he could do things like that here, but I replied, of course, that in such matters our attitude was precisely the same as that of your department. I pressed him strongly to listen to Salpeter's advice, but if he was determined to do these abstract high-brow things he would have to go elsewhere.

Finally, a question about physics. I heard from Weisskopf that you have some results about the imaginary part of the nucleon-nucleus potential in an infinite nucleus and that this comes out smaller than usually assumed, indicating presumably a much stronger absorption in the surface regions. I was about to start a research student on substantially

[584] Baird H. Brandow, worked at Cornell and Copenhagen, and later at Los Alamos.

the same problem and would therefore be grateful if you could confirm that the rumour is correct and would in due course send more details about your work when this gets written up.
With kindest regards,
Yours sincerely

Rudi

[147] Hans Bethe to Rudolf Peierls

Ithaca, 20.3.1958

Dear Rudi,
Thanks a lot for your letter of March 3.
I am glad to hear that Jenkins and Langer have satisfactory offers. We are in the meantime were saturated, actually somewhat more than saturated, because it is possible that two people may stay for part of the time whom I had expected to leave. So I am afraid I cannot make any offer to Dabrowski or Burkhardt.

Concerning your physics questions, we have indeed some results about the imaginary part of the nucleon-nucleus potential. If we assume that the effective mass of a nucleon is about .7 of its actual mass, the imaginary potential comes out to be about 1/2 Mev for slow neutrons, about 4 Mev at 14 Mev neutron energy. Both values are about half of the observed, and perhaps even somewhat less. So one has to assume a very large contribution of the nuclear surface, as you say in your letter.

These results were obtained by Mr. Gordon Shaw.[585] He first considered the Born approximation with an exponential potential, similar to the Italians, but doing the integrals more accurately. More recently he has taken the numerical results on the *G* Matrix of Brueckner and

[585] Gordon L. Shaw, obtained his Ph.D. at Cornell in 1959; he continued his research at Indiana University, Bloomington, at the University of California, San Diego, at Stanford and at the University of California, Irvine where he became professor in 1968.

Gammel[586] and deduced the imaginary potential from these.[587] The result is only 20% higher than from the conventional potential with Born approximation. The 20% may be attributed to the fact that G is larger than v for an attractive potential, and also partly to the non-central forces. It is not clear to us, however, how strongly the result would be influenced by singular potentials (a repulsive core was included by Brueckner and Gammel). For different conventional potentials with the same effective range and Blatt strength parameter, the imaginary potential comes out to be about the same.

One possible modification may come from the effective mass. Thouless[588] has developed a theory of the actual one-particle energies in which he shows that these energies are considerably larger than the model energies used in the Brueckner theory.[589] The correction comes mainly from the Pauli principle. Since Thouless' correction is greater for low levels, it tends to increase the effective mass for real nucleons. This seems to be contradicted by experiments on the optical potential but, if it were true at least near the Fermi energy, it would lead to an increase in the imaginary part of the optical potential which is proportional to about the square of the effective mass.

With best regards,
Yours sincerely,

Hans

[586] K.A. Brueckner and J. Gammel, 'Solution of the Nuclear Many-Body Problem', *Phys. Rev.* **105**, 1679–1681 (1957); K.A. Brueckner and J. Gammel, 'Properties of Nuclear Matter', *Phys. Rev.* **109**, 1023–1039 (1958).

[587] G.L. Shaw, 'Imaginary part of the optical model potential for neutron interactions with nuclei', *Ann. Phys.* **8**, 509–550 (1959).

[588] David J. Thouless (1934–), obtained his Ph.D. from Cornell; in 1959 he joined Peierls' department at Birmingham; after further research at Cambridge he became professor at Birmingham (1965–78). He also did research at Kingston, Yale, and Seattle before taking up a professorship at Washington University in 1980.

[589] D.J. Thouless, 'Application of Perturbation Methods to the Theory of Nuclear Matter', *Phys. Rev.* **112**, 906–22 (1958). See also David J. Thouless, 'Single-Particle Energies in the Many-Fermion System', *Phys. Rev.* **114**, 1383–90 (1959).

[148] Hans Bethe to Rudolf Peierls

[Ithaca], 23.5.1958

Dear Rudi,
Thanks for your letter, I had just spent an evening with Maximon[590] last week, hearing about his plans. It seems that he is not really very anxious to spend another year in Europe, especially since he had three good and unsolicited offers from various places in this country since he came here. He seems somewhat inclined to earn some money which is more easily done in this country than in Europe.

One important factor in his decision will be whether you will be at Birmingham all year. He had heard some rumor that you might be in this country for one semester.

As to his qualifications, I don't think he is correctly being described as lazy. In fact, I think he works quite hard. He does enjoy other things besides physics, and it was most enjoyable to hear his accounts of his travels through Europe of which he has made the best possible use.

Maximon does tend to be an expert in a very narrow field. He has clearly approached the ideal of a scientist in the old joke, namely he knows everything about nothing. He is not broad in his knowledge, education and research. He wants to change his main field of interest and he is very intelligent so that he will make good use of any opportunities such as being with you. I believe he would know enough about nuclear physics to give a good lecture course on it, and he is certainly very competent in classical physics and quantum mechanics.

So I can recommend him as a lecturer and as a pleasant person to have around. You will not get very exciting research from him at the frontiers of physics but if and when he digs into a field seriously he will get very significant results and do a thorough job.
Yours sincerely,

Hans A. Bethe

---

[590]Leonard C. Maximon, had been Bethe's graduate student at Cornell. After work at Brown University, Rhode Island, Trondheim, Manchester, he joined the National Bureau of Standards and became Research Professor at George Washington University.

[149] Rudolf Peierls to Hans Bethe

Birmingham, 29.12.1958

Dear Hans,

I am writing once more about a possible item for the Birmingham-Cornell pipeline. This is Graham McCauley[591] who is expected to get his Ph.D. here next autumn. He is a Northern Irishman, who studied in Cambridge and took his Part III there at the same time as Ronnie. I cannot remember whether you actually met him in Cambridge. He is extremely bright, quick, reliable and very powerful in mathematics. He has so far completed some work with Gerry Brown about high energy nucleon scattering and polarization effects, where his particular job was to investigate the effect of inelastic scattering with low excitation, which experimentally is hard to distinguish from elastic scattering. This was a nice piece of work for which he could probably already get a Ph.D., but at the moment he is changing over to dispersion relations and, at my suggestion, is following up the idea of Bohr and Mottelson to derive the dispersion relations from macroscopic causality, and this is making quite good progress. If you could find him some kind of fellowship or associateship, I think you would like him and he would undoubtedly benefit greatly from a year or so at Cornell.

With kindest regards,

Yours sincerely,

Rudi

[591]Graham McCauley received his Ph.D. from Birmingham before taking up a fellowship at Cornell. In 1963 he returned to Birmingham.

[150] Hans Bethe to Rudolf Peierls

[Ithaca], 28.1.1959
(carbon copy)

Dear Rudi:
Thanks a lot for your letter recommending McCauley. He sounds very good, and there is at least a 50% chance that we can offer him a job. We have an offer out to a student of Wheeler's,[592] but it is more likely than not that he will prefer to stay with Wheeler and continue work on general relativity. If he doesn't come and with a slight possibility even if he does, we can make an offer to McCauley. We have always fared very well with students from you.

A proposal in reverse. I have a good student who is finishing, Mr Alan H. Cromer.[593] His thesis was on polarization of nucleons so it would be particularly appropriate for him to work with Gerry Brown for a year.

H.A. Bethe

[151] Hans Bethe to Rudolf Peierls

[Ithaca], 12.2.1959
(carbon copy)

Dear Rudy,
Welcome to this country:
As you see from the enclosed we are offering a job to your student, Graham McCauley. I hope he will come.

---

[592] J.A. Wheeler (1911–), received his doctorate from John Hopkins University in 1933; he was Professor at Princeton between 1938 and 1976 before taking up a Professorship at the University of Texas at Austin.

[593] Alan H. Cromer received his Ph.D from Cornell and took up a position at Northeastern University. Later he published widely on topics of physics, philsophy and science education.

I also hope very much that you will be able to come to Cornell very soon; if possible, for a week or so.[594] We have some funds from which we can pay you. I hope you can give us a colloquium as well as a theory seminar, but I want to find out about our other engagements before writing you concerning dates.
With best regards,
Yours sincerely,

Hans A. Bethe

[152] Rudolf Peierls to Hans Bethe

New York, 18.2.1959

Dear Hans,
Thank you very much for your letter. I am very pleased that you have sent an invitation to Graham McCauley. He will now be suffering from an embarrassment de richesse. Because he has also an invitation to come to Columbia. I have written to him to tell him all I [k]now about the attractions of both places, and he will now have to make up his mind.

Thank you also for your invitation to spend some time with you. I do want to come, of course, but life is difficult. Easter week, which you suggest, would be perfect, because there are no lectures here that week, but unfortunately I am already committed to spending the first three days of that week at McMaster. I now regret I let myself in for that; I had some reasons, but they were probably foolish. Otherwise I lecture here every week on Mondays and Wednesdays until some time in the middle of May, which is getting a bit late. I am rather reluctant to cancel lectures since this is about all I do for a living here.

This effectively blocks my talking to your colloquium or seminar, at least on the regular days, or coming for a whole week. I could come any Wednesday night (if there is a late plane) or Thursday morning and stay

[594] Rudolf Peierls had spent a semester at Columbia University in early 1959.

for the rest of the week. This includes Easter week when I could come from McMaster. But perhaps the second half of that week would not be so suitable for talking shop from the Cornell point of view? There is a strong possibility that Genia might be here then — and probably will join me for three or four weeks from about March 20 or so, and even if the second half of Easter week is not suitable to do business, we might still come for a merely social call.

I am sorry this is so complicated, but to sum up, if you pick the second half of any week not earlier than March 4, I could make it as far as I now know. March 18 would not be so good, because I only get back that morning from the Tallahassee conference, and my preference would probably [be] to come after McMaster, i.e. late on 25 March, or early next morning.

I have nothing much to talk about for a colloquium type meeting. I go around talking on such occasions about "Nuclear Models" which means a bit of Brueckner &c, and Brown et al. about the optical model. But to do this in Cornell would be taking coals to Newcastle (or neutrons to Oak Ridge). To a theoretical seminar I could report on the problem of centre-of-mass motion in the shell model which I seem to have solved at last after a couple of years' struggle.

I would appreciate it, if you could let me know your preference as soon as convenient in case I get trapped into something else.
With kind regards,
Yours sincerely,

Rudi

[153] Hans Bethe to Rudolf Peierls

Ithaca, 24.2.1959
(carbon copy)

Dear Rudi,
Thanks for your letter of February 18. It's too bad you have such a busy schedule. But we shall be happy to take you in the second half of any week if you can't come in the first half.

March 26, unfortunately, is out for me because I have to go to Detroit, and the following week is out because Cornell is on vacation. Rose and I would, of course, like you to come when Genia is here, too. So I would suggest April 9–12 (Thursday through Sunday) or the same period a week later. How would this be?

We would like you to give a theoretical seminar on Thursday afternoon at 4:30. On Friday we have a Nuclear Physics Journal Club, and it would be very nice if you could tell us quite informally about the work going on at Birmingham including experiment.

Please let me know when you can come.

Yours sincerely,

Hans A. Bethe

[154] Rudolf Peierls to Hans Bethe

New York, 12.3.1959

Dear Hans,

I find I never replied to your letter of Feb. 24. and although we already agreed on dates by phone, perhaps I should keep the record straight.

As I told you April 9–12 is fine for us. I think we would arrive at the first reasonable plane on Thursday (which means, I believe, 11.35) but we could also get the evening plane the night before and arrive at 10:14 p.m. on Wednesday if that is an advantage.

As a subject for my seminar talk I would like to take the work which I am doing at the moment which concerns the use of analytical behaviour on a Riemann surface to derive a new and simple form of the nuclear dispersion formula (expressing Kapur-Peierls-Wigner-Eisenbud in terms of physical complex eigenstates) but which I mainly hope to extend to high-energy problems. The latter part is still mainly conjecture, but I may have something more definite in time. A suitable title would be "The use of complex eigenvalues in scattering problems". I shall also be glad to say something in the Journal Club about current work at Birmingham, but I don't think I have enough here to fill a whole meeting.

Yours sincerely,

Rudi

[155] Hans Bethe to Rudolf Peierls

[Ithaca], 18.3.1959
(carbon copy)

Dear Rudi,
Thanks for your letter of March 12.

Concerning transportation, you may want to take the overnight train leaving Penn Station about 11.50 p.m. and arriving here about 8 a.m. It now has roomettes.* However, if you prefer the plane, the one arriving here at 11:35 a.m. on Thursday would be fine. There isn't terribly much point in arriving at 10:14 p.m. the previous night since there isn't enough time to start talking that evening, and since the planes are quite often late. Please let me know your choice.

I assume you will be in the doublet state, and we hope of course that you will stay with us.

Your topic for the theoretical seminar sounds very interesting, and is just the thing I would like to learn. We shall also schedule you for half the meeting of the Journal Club on Friday.
It will be fun seeing you.
Yours sincerely,

Hans A. Bethe

*also double bedrooms.

[156] Hans Bethe to Rudolf Peierls

[Ithaca], 6.4.1959
(carbon copy)

Dear Rudi,
Of course we want you to have breakfast at our house. Whether we shall meet you at the station at the crack of dawn I do not know yet, but if we don't please take a taxi to our house (209 White Park Road).
See you on Thursday.
Yours sincerely,

Hans A. Bethe

[157] Rudolf Peierls to Hans Bethe

Birmingham, 13.8.1959

Dear Hans,
I am writing to ask whether you would be willing to act for us as external examiner for Graham McCauley. The position is that his thesis is complete, but he is now on a visit to Copenhagen and between the time he returns and his departure to Cornell, there will be no opportunity of giving him an oral examination. One possibility would be therefore if you were good enough to act as examiner which, as you may remember from the case of Schaefer,[595] in which you kindly helped us, involves firstly reading his thesis and saying whether you regard it as adequate (no long report is needed) and secondly giving him a short oral examination, partly about any points in the thesis and for the rest about physics in general, which is not as searching as the corresponding American test, but I hope will be more successful than in the case of Schaefer. We would not, of course, observe the usual custom of the internal examiner also being present at the oral, but that is all right. The recommendation should be received by some date in the middle of November.

If you regard this as inconvenient, please do not feel under any obligation to accept, because we have other ways of dealing with the situation, e.g. we are allowed to dispense with the oral, if we can appoint two external examiners to read the thesis instead of one, but if you are willing to do it, it would probably be the most attractive solution.

I should have mentioned that the title of the thesis is "Semi-Classical Theory of High Energy Nuclear Scattering". It applied the method used by Brown, Glauber and others to inelastic scattering with low excitation in order to see in particular whether the experimentally unavoidable admixture of inelastic scattering will distort the results about polarisation.

---

[595] Glen Willard Schaefer (1930–1986), a graduate of Toronto University, had received his Ph.D. from Birmingham in 1955. After working at the English Electric Atomic Power Dvision, he joined the University of Loughborough before returning to Canada.

While I am writing I realise that I don't seem yet to have written to you after my visit to Cornell and I should have long before now told you how much I enjoyed the visit, both personally and scientifically, and incidentally I appreciated very much the exceedingly generous treatment which your department provided in the way of fees and the like. This would also seem to be the right point to say how grateful I am for all you have done for Ronnie, who is now getting launched in the world after having been given a wonderful start.

I should have said all these things long ago, but I have never yet stopped running round in circles, first to complete what I wanted to do at Columbia, then to catch up with the arrears here, and then to get away to the Kiev Conference,[596] which was not very exciting, as you will have heard, mainly because nobody has made any exciting discoveries during the year, an omission for which one could hardly blame the organisers. As you will have heard, there were also a variety of minor handicaps for which one probably can blame them.
With kindest regards,

Rudi

[158] Hans Bethe to Rudolf Peierls

[La Jolla], 31.8.1959
(carbon copy)

Dear Rudi:
Thanks a lot for your letter of August 13. I shall be glad to act as external examiner for McCauley. I am interested in the subject of his thesis, and it is a good idea to know his thesis before he starts work at Cornell.

It was awfully nice to have you at Cornell, and I hope you will soon com again, if possible for a longer time.

[596] The Ninth International Annual Conference on High Energy Physics was held in Kiev in 1959.

Wilson and one or two others have told me about the Kiev conference, much in the same vein as you, only more strongly. Wilson was terribly disappointed, especially in comparison with the 1956 meeting.

We are just back from a very enjoyable vacation in the Canadian Rockies, with quite a lot of trials and glaciers. La Jolle is almost a continuation of the vacation.

With regards, also from Rose,

Yours sincerely,

Hans A. Bethe

[159] Hans Bethe to Rudolf Peierls

Ithaca, 18.11.1959
(carbon copy)

Dear Rudi,

Enclosed I am sending you my report on McCauley's thesis and examination. I hope it is all right to send it to you, and you will deal further with it. I mean everything I say in the report: I like his thesis very much and he did very well in his examination, quite differently from Mr. Schaefer.

McCauley is now working quite hard on dispersion relations a la Mandelstam, and will give us a seminar about this subject today. He thinks he can apply these relations to the photoproduction of K mesons which would be very nice.

I am afraid I shall be further removed from physics since I have to go once more to the Geneva Conference, leaving here on Friday. I hope it will do some good. If you happen to come to Geneva during the next three or four weeks, it would be awfully nice to see you.

With best regards. Also to Ronnie and Genia.

Yours sincerely,

Hans. A. Bethe

[160] Rudolf Peierls to Hans Bethe

Birmingham, 19.1.1960

Dear Hans,

Another case for the Birmingham-Cornell pipeline, but this time probably more the Department than for the Laboratory.

Robin R. Stinchcombe[597] is one of our students who easily obtained First class Honours in Mathematical Physics in 1957, and should complete his Ph.D. here next summer. He has been working on solid state problems and particularly on the effect of magnetic fields on electrons in metals.

In his first year he wrote, following our usual practice, a thesis on the Masters level, and I enclose a summery of this to give you an idea of what it is about. Since then he has been looking at this and associated problems on a more rigorous basis using the new methods of van Hove. It did not seem to us obvious how to extend the "Diagonal singularity property" of Van Hove to the case of a magnetic field, but I suggested a way to do this, and this seems to work.

As a result of this work he is not a specialist in magneto-resistance, but has a thorough understanding of the statistical mechanics of transport problems. He also has an excellent general knowledge of solid state theory and a good deal of knowledge of theoretical physics in general.

Stinchcombe is a man of excellent ability with initiative and drive, and would, I think, benefit greatly from a year or so spent with your experts in similar fields, he is quite able to look after himself and is a very pleasant person with no nonsense about him. He is single so that any of the usual fellowships are associateships would be fine for him.

He has really worked more with G.V. Chester[598] than with me and if desired Chester could well give a second opinion.

[597]Robin B. Stinchcombe obtained his doctorate from Birmingham in 1960. After research at Cornell, he joined the physics department at Oxford, where he stayed for the remainder of his career.

[598]Geoffrey V. Chester obtained his Ph.D. from Kings College, London. Afterwork at the Radar Research Establishment and research at Yale and Chicago, he joined Peierls' department as a Research Associate and later as Lecturer and Reader, before taking up an appointment as Professor of Physics at Cornell.

I have also written on his behalf to Kittel in Berkeley,[599] but it is not obvious to me whether Cornell or Berkeley would be better for him, if he had the choice.
With kindest regards,
Yours sincerely,

Rudi

[161] Rudolf Peierls to Hans Bethe

[Birmingham], 6.1.1961
(carbon copy)

Dear Hans,
I am writing once again about a possible candidate for a post doctoral fellowship of some sort at Cornell. This is a bright young Canadian, D. W. L. Sprung,[600] who in expected to complete a Ph.D here next summer.

He came to us in 1957 from Toronto with an excellent academic record and he struck us immediately as very bright and lively. His knowledge, however, was still deficient in some ways, (this is not the first time we have noticed this with a Toronto graduate. You will remember the problem of Schaefer, whom you examined here), and we therefore made sure that he filled in background while starting on research. This is the reason why he is taking four years over a Ph.D. His first piece of work, on which he wrote a qualifying thesis, was done for Gerry Brown on an attempt to use WKB for the scattering of neutrons by a heavy nucleus at rather lower energies than was normally done. There were some reasons for hoping that this would prove possible but

[599] Charles Kittel had studied at Cambridge and obtained his Ph.D. from University of Wisconsin in 1941. After research at MIT and Bell Labs he became Professor of Physics at Berkeley in 1951.

[600] Don Sprung (1934–), received his Ph.D. from Birmingham in 1961. After research at Cornell, he joined the faculty of McMaster in 1962 where, with the exception of several sabbaticals, he has remained since.

Sprung found that one ended up with small differences between large quantities so that the method was not practical. This work was intelligently and competently done and he should not be blamed for the negative result.

He then turned at my suggestion to the idea of Brown by which the nuclear spin orbit force could be understood as the relativistic correction to the repulsive core, an idea which I believe I mentioned to you some time ago and which I find very plausible. He did some rough and ready calculations on this, which are published in a paper by Sprung and Willis of which I am sending a reprint separately.[601] Sprung has been entirely responsible for the theory in this paper, (under Brown's supervision), Willis merely helped to organise the computing side. It was always clear that the approximations used in this paper are of doubtful validity and that it can be used only for general orientation, and Sprung is attempting to make a proper theory of this problem. Since it is essential to put in projectional operators to eliminate negative energy states, one comes out with an integral differential equation of the structure not unlike the Bethe-Goldstone equation. Sprung is exploiting the analogy to construct a solution. He is making good progress with this and I am confident that he will complete a respectable Ph.D thesis on this subject by the summer. It also looks as if this procedure will come out with a spin orbit force which at least is qualitatively of the right kind.

As a by-product of his numerical studies he discovered some errors in other people's results and he wrote a little note about this of which I am also sending you a preprint and which is going to come out in the Physical Review.[602] This of course is not deep but it does show that he knows his way around the numerical side.

As a sideline he has done some teaching for us, giving a short course intended to introduce biologists to the mysteries of calculus and he has done this very successfully.

Sprung is very intelligent, hard working and lively, he is also a very pleasant and businesslike person. He is not as powerful a mathematician

[601] D.W.L. Sprung and J.B. Willis, 'Spin-orbit potential in proton-proton scattering', *Proc. Phys. Soc.* **76**, 539–44 (1960).

[602] D.W.L. Sprung, 'Relativistic Formula for the Spin Correlation Coefficient C_{KP}', *Phys. Rev.* **121**, 925–26 (1961).

as McCauley, but he knows what he is about and in my opinion he is a good man to have around.

He plans ultimately to return to Canada, but he feels rightly that he would benefit from another year or two in the United States before settling down. I think a period at Cornell would be a wonderful opportunity for him. He would I think also be very good at talking with experimentalists. He has already done some of this here in connection with proton-proton scattering and polarisation experiments.

He has a wife and a child of just over a year.

Please let me know what you think.

I am sorry you were not able to come to the Pugwash Meeting at which according to rumours we were hoping to see you. You would have heard that on the whole this was quite useful and informative. We did not expect to solve any problems there, but on the whole I came back from there somewhat more optimistic about the chances of a reasonable agreement than I was before.

With kindest regards to all of you,
Yours sincerely,

[Rudi]

[162] Rudolf Peierls to Hans Bethe

[Birmingham],13.11.1961
(carbon copy)

Dear Hans,
I am writing in the first place to correct a wrong statement I made to you at the Solvay Conference.[603] You will remember that I referred to the result by Elliott Lieb and a collaborator on the problem of ferromagnetism in one dimension and said that we had found a flaw in his argument. On further consideration, however, it turned out that we were wrong and as far as we could see, he was right.

[603]The 12th Solvay Conference of Field Theory took place in Brussels between 9th and 14th October 1961.

His result goes somewhat against the grain, or rather against the impression one has from other methods of calculation, in particular the Heitler-London method, but there is no obvious contradiction, unless one can construct a model in which the Heitler-London approximation is positive. We are now trying to see whether this can be done or whether one can prove that it is impossible. The main thing is that I slandered Lieb and wanted to get this right.

I enclose my article about the book by Kahn[604] which you have not seen. This was followed by further correspondence in the "Spectator" but I have no copy of that.

Yours sincerely,

[Rudi]

[163] Rudolf Peierls to Hans Bethe

[Birmingham], 24.2.1962
(carbon copy)

Dear Hans,

I am writing this letter to kill two birds with one stone. The first bird is to ask you if you could be good enough to write me a short letter giving me your opinion of Graham McCauley, who seems to be willing to come here next year. We know him, of course, but your opinion would be of help to the appointing committee and we need in any case some information about how his work has progressed since he went to Cornell.

I also would like to say that I have read your famous Cornell lecture[605] about disarmament etc. and I found it extremely interesting and a wonderfully sane and balanced presentation of the subject. One gathers that you have come in for some nasty attacks from various quarters

[604] Peierls had reviewed Herman Kahn's *On Thermonuclear War*, Princeton University Press and OUP, Oxford, 1961. The review appeared in *Spectator*, 28 April 1961 under the title: 'Agonising Misappraisal'.

[605] In 1958 Hans Bethe had headed the presidential study of nuclear disarmament and was an advisor to the United States at the Geneva nuclear test ban talks.

as a result and I hope they were not too much of a headache. Someone sent me an article from an unidentified magazine (it looks like "Life") with a nasty and vicious attack. I hope this reaction is not general.

I was interested in one remark you make, namely, that one motive for supporting the idea of a test ban treaty was that it would preserve the technical advantage of the United States.[606] I had not previously realised that this was a strong factor and it is certainly worth making the point clear to the American public that such a treaty would have been a good thing, but I am a little concerned over the effect on Russian readers of your lecture and therefore, also indirectly on people talking with the Russians, which I shall probably have to do this summer at a Pugwash meeting, since it provides a good reason why the Russians felt justified in their reluctance to accept the test ban treaty, at any rate until their own tests have made further progress. However, if this was a factor in the situation it has to be admitted.

Are you coming to the Pugwash Conference in Cambridge?[607]
With kindest regards,
Yours sincerely,

[Rudi]

[164] Hans Bethe to Rudolf Peierls

[Ithaca], 2.3.1962

Dear Rudi,
Thanks for your letter of 24 February. Graham McCauley had just told me that he would like to return to Birmingham.

We were very happy with McCauley. He knows a great deal about dispersion theory and about physics in general. He participates very intensively in our seminars both the Theoretical Seminar and the Nuclear Physics Journal Club. For some time he was in charge of the

---

[606] Similar ideas had already been expressed earlier by Bethe. H.A. Bethe, 'A Case for Ending Nuclear Tests', *Atlantic Monthly*, August 1960, pp. 43–51.

[607] In 1962, British scientists hosted two Pugwash conferences, one in Cambridge and one in London.

Journal Club and did a very good job. He is always well informed and his questions and remarks are very much to the point.

In addition to this, he has worked very hard all the time. Therefore, it is somewhat surprising that he has not published anything. In some cases when letters in The Physical Review have appeared, McCauley remarked that he had found the same result a few months earlier but had considered it too trivial to publish. Probably I should have talked to him more often and should have encouraged him to publish his minor results.

His main trouble has been that he got involved in some numerical calculations of the pion-pion interaction. He thought he could calculate the double spectral function from the unitary relation, and gradually get up to higher energies in this manner. This was a noble thought but it proved futile and got him involved in a lot of numerical calculations on our computer. The failure of these calculations has depressed him somewhat, but he is now his old cheerful self again and realizes that she should change his subject of work.

In addition to his research, McCauley is assisting Morrison this year in giving a new course on the ideas of physics to non-physicists. Morrison says he is doing well.

McCauley has also been very helpful in research to the graduate students and some of the research associates here. Once he gets out of the $\pi - \pi$ problem which was simply too involved, I am sure he will be productive and very good.

Yours sincerely,

Hans A. Bethe

[165] Hans Bethe to Rudolf Peierls

[Ithaca], 2.3.1962
(carbon copy)

Dear Rudi:

I thought I should write the letter on McCauley separately because you may want to show it to others. In this letter let me first tell you that it was a great pleasure to have a visit from Ronnie a few weeks ago. Both he and his wife were flourishing and a pleasure to have. We have invited him to come here as a Visiting Assistant Professor next year, and we are all looking forward to this.

I am glad you liked my famous Cornell lecture about disarmament.[608] It may by now be published by the Saturday Evening Post, either completely or in part, after the AEC has expurgated it.

Your remark about the possible effect of one of my statements is certainly pertinent. I have never understood why the Russians wanted the test ban in 1958 when it was clear that they were considerably behind. But they took the initiative all along, first proposing it in 1955, then pressing it, then unilaterally stopping their tests, then at the Experts Conference pressing for an immediate ban. So they can hardly blame us that we were trying to lead them into a treaty unfavourable to them. By the way, I already said in my article in the Atlantic Monthly in 1960[609] that a test ban treaty would be to our military advantage, so it is not really a new statement.

Obviously there will be problems at the Pugwash meeting this coming summer but I think it should not be too difficult to explain the Western position in this matter. I don't expect to come to that summer Pugwash meeting[610] but I am planning to come to the Rochester Conference at CERN.[611]

With best regards to all of you,
Your sincerely,

[Hans]

---

[608]See letter [163].

[609]H.A. Bethe, 'The Case for Ending Nuclear Tests', *Atlantic Monthly*, August 1960, pp. 43–51.

[610]The Ninth Pugwash Conference was held in Cambridge between 25th and 30th August 1962, the Tenth Pugwash Conference was held in London between 3rd and 7th September 1962.

[611]The first Rochester Conference at CERN took place between 4 and 11 July 1962.

[166] Rudolf Peierls to Hans Bethe

Birmingham, 2.5.1962

Dear Hans,

This letter is probably quite academic, but one never knows.

It all arises because I have been asked to move to Oxford to succeed Lamb,[612] as you will no doubt have heard by rumour. The rumour will have further told you that I am going to accept, which is more than I know, because there are some problems to be settled, but it is quite possible that the rumour might be right. If I do go, it would be in October, 1963.

It is, therefore, pretty clear that there will be no professor there during the next academic year, and this brings me to the point of my letter. At the Solvay Conference[613] you mentioned that you were going to take a sabbatical year. You were then thinking of specific good places for specific good reasons, but just in case these plans have not completely worked out for some reason, maybe you should know that there is a fairly attractive place which should spare professorial salary for a year, without any particular chores.

I have not mentioned any of this to anyone in Oxford, because it seems a very long shot, but if by any chance you were interested, I should be most surprised (and no doubt you would) if they did not jump at the idea.

I have no great illusions about the prospects, but they do seem worth a one-and-threepenny stamp.

Yours sincerely,

Rudi

---

[612] Willis E. Lamb (1913–), received his Ph.D. in Physics from the University of California in 1938. He moved to Columbia University where he became Professor of Physics in 1945. He was Professor at Stanford (1951–56, Oxford (1956–62), Yale (1962–74) before taking up his Professorship at the University of Arizona where he has remained since.

[613] The 12th Solvay Conference had taken place in October 1961.

[167] Hans Bethe to Rudolf Peierls

[Ithaca], 16.5.1962
(carbon copy)

Dear Rudi,
Thanks a lot for thinking of me. However, as you already believed, your letter is academic. My sabbatical will be 1963–1964, and in addition, I don't want to tie myself down to one place. But thanks anyway.

I have already heard rumors that you had accepted Oxford, and I am glad to hear how matters actually stand — including the fact that the rumor may be right. There are also rumors about your successors in Birmingham.

I know how difficult it is to make such a decision; it is a pleasant agony.
With best wishes,
Yours sincerely,

[Hans]

4. Physics: Not a Young Person's Pasttime

Having been offered the Wykeham Chair of Physics in 1961, Rudolf Peierls, after lengthy negotiations, decided to accept it in early 1962 and he took up his appointment in the autumn of 1963. When asked about the reasons for his decision to move from Birmingham to Oxford, he would later refer to the need for change after quarter of a century at the same university. But it was more than simply the desire of change. Peierls liked the challenge. After successfully building a school of theoretical physics at Birmingham, he wanted to achieve something similar in Oxford. If Peierls' role as a senior academic in the United Kingdom had already undergone some changes towards the end of his time at Birmingham, this change became even more pronounced during the last decade of his university career at Oxford. This is visible in his correspondence which mostly deals with private, political, administrative or general scientific questions. Peierls' publications in the 1960s and 1970s mirror the impression conveyed by his correspondence. Among his publications one finds those of general conceptional or explanatory kind, such as his writings on the 'laws of physics',[614] the development of quantum theory,[615] nuclear matter,[616] or Fermi-Dirac statistics;[617] those of political content such as contributions in the Pugwash

---

[614] R.E. Peierls, 'Wo stehen wir in der Kenntnis der Naturgesetze?' *Phys. Bl.* **19**, 533–39 (1963).

[615] R.E. Peierls, 'The Development of Quantum Field Theory', *The Physicist's Conception of Nature* (ed. J. Mehra), Reidel, Boston, 1973, pp. 370–79; R.E. Peierls, 'An Examination of the quantum jumps in the growth of quantum mechanics', *Sci. Am.* **216**(1), 137–140 (1967).

[616] R.E. Peierls, 'Nuclear Matter', *Endeavour* **22**, 146–150 (1963).

[617] R.E. Peierls, 'Fermi-Dirac Statistics', *Aspects of Quantum Theory* (A. Salam and E.P. Wigner, eds.), CUP, Cambridge, 1972, pp. 117–127.

context,[618] works on the relationship between science and politics,[619] or indeed, biographical contributions.[620] Evidently, Peierls' research had undergone a gradual change which had already been evident in his final years at Birmingham and which accelerated during his time at Oxford. He had been among the outstanding figures of the last generation of universalists in physics, and unlike many of his colleagues of his generation, he refused to choose one narrow field as a focal point of his attention and instead tried to keep his interests broad. The increasingly rapid pace of developments in subject areas such as particle physics made it difficult to keep up with the trends in the discipline for anybody keen on dividing his attention between different specialisations. In addition, Peierls felt that his age was beginning to make itself shown by the speed with which he was capable of picking up and utilising other people's ideas and concepts.[621] He still exchanged ideas with colleagues and former students and facilitated information transfer, but he did not feel able to contribute anything original to the subdiscipline.

The same could not be said for Hans Bethe. In 1967, he was awarded the Nobel Prize in Physics for the discovery of the stellar energy production mechanism. The prize, awarded for work published just before the outbreak of the Second World War, was the first Nobel Prize ever awarded in astrophysics. Apart from some brief astrophysical

---

[618]R.E. Peierls, 'Problems of Collective Security', *Proc. 13th Pugwash Conference*, Taylor & Francis, London, 1964, pp. 259–264; R.E. Peierls, A possible use of "black boxes" in connection with a comprehensive test ban', *Proc. 16th Pugwash Conference*, Taylor & Francis, London, 1966, p. 272; R.E. Peierls and C.F. Powell, 'Pugwash and the World', *Proc. 18th Pugwash Conference*, Taylor & Francis, London, 1968, pp. 209–212; R.E. Peierls, 'Britain in the atomic age', *Bull. Atomic Sci.* 26 June 1970, 40–46; R.E. Peierls, 'A Ban on the Development of New Weapons?' *Proc. 19th Pugwash Conference*, Taylor & Francis, London, 1970, pp. 280–283; R.E. Peierls, 'International Aspects of the Pollution Problem' *Proc. 21st Pugwash Conference*, Taylor & Francis, London, 1971, pp. 294–303.

[619]R.E. Peierls, 'The scientist in public affairs: between ivory tower and the arena', *Bull. Atomic. Sci.* 25 November 1969, 28–30; R.E. Peierls, 'Physicists and politics', *The Times Lit. Suppl.*, 18.12.1969, 1442.

[620]R.E. Peierls, 'Lev. D. Landau', *Phys. Bull.* **19**, 147 (1968); R.E. Peierls, 'J. Robert Oppenheimer', *Dict. Sci. Bio.* (Sr.Ed.C.C.Gilliespie), **10**, 213–218, Scribner, New York, 1974.

[621]Peierls, *Bird of Passage*, p. 304.

excursions,[622] Bethe had not done any significant work in the field since then, and, in fact, did not return to it in a focused way until 1978, a few years after first being approached by his colleague and friend Gerry Brown about a possible collaboration. While at Copenhagen, where Brown was professor at NORDITA, Bethe read through the existing literature on the core collapse of massive stars, and he quickly realised that all supernovae calculations contained an error. The consensus was that the core collapse ended when the core density reached about 10 % of the nuclear density, whereas in fact it continued to densities well in excess of nuclear density. Together with Jim Applegate and Jim Lattimer, Bethe and Brown wrote the paper, nicknamed 'babble' by William Fowler,[623] and generally referred to as such in the astrophysics community. It derived the equation of state in stellar collapse from simple considerations, notably Bethe's early insight that the entropy per nucleon remains small in the order of 1 (in units of k), during the entire collapse.[624] Babble turned out to be the beginning of a long and fruitful astrophysical collaboration between Bethe and Brown. For nineteen years the two scientists spent the month of January together on the West Coast, at Santa Barbara, Santa Cruz, or Caltech, producing more than twenty joint papers in the process.[625]

Rudolf Peierls and Hans Bethe had been at the cutting edge of physics for several decades and had contributed to the development of theoretical physics since the mid 1920s. Their knowledge of 20th century physics and in particular their assessment of the development of nuclear weapons was in great demand in scientific publications,[626] and increasingly, the subject was discovered by the media. Radio

---

[622] E.g. G. Baym, H.A. Bethe and C.J. Pethick, 'Neutron Star Matter', *Nucl. Phys. A* **175**, 225–71 (1971).

[623] William Alfred Fowler (1911–1995), graduated from Ohio State University and obtained his Ph.D. from CalTech in 1936. He joined the CalTech faculty as a research fellow, then as professor and eventually Institute Professor.

[624] H.A. Bethe, G.E. Brown, J. Applegate and J.M. Lattimer, 'Equation of State in the Gravitational Collapse of Stars', *Nucl. Phys. A* **324**, 487–533.

[625] See letters [193], [198].

[626] See letter e.g. letters [191], [193].

programmes were broadcast, and tv documentaries and discussions,[627] and their advice was thought on numerous of these.[628] Among the more controversial productions was the BBC's programme *The Building of the Bomb*, for which Peierls had recorded an interview. Although his interview was not used when the programme was screened in 1965, Peierls was disturbed by the inaccuracies of the programme as a whole, and he took up the issue with the Director-General of the BBC, Sir Hugh Carleton Greene.[629] Moreover, when the BBC and their producer R. Reid, planned a similar programme on the history of the H-bomb, Peierls warned Hans Bethe and thereby led to a more cautious co-operation of American colleagues with the BBC.[630]

Rudolf and Genia Peierls had always enjoyed leading a nomadic existence, and during their Oxford years, both before and during retirement, they continued travelling large parts of the world. The longest period away was a sabbatical in the US in 1967. Peierls revived his contacts with the University of Washington in Seattle, which he had visited for a summer workshop in 1962 and now spent six months there as a visiting professor. While on the other side of the Atlantic, Peierls not only visited some of his children who had by then settled in the US,[631] but he also attended the New York meeting of the American Physical Society, where he met many of his old friends and colleagues, not least Hans Bethe.[632] The trip to Seattle also gave Peierls the opportunity to revisit a problem on which he had worked on in the late 1930s. At a symposium on dislocations at Seattle, he was discovered, much to his surprise, that a paper in which he had discussed a simple model

[627]Letter [203].

[628]Peierls helped with several programmes including BBC TV's *Series on Quantum Physics* (1963), *Ethical Responsibility* (1964), BBC Radio's *The Bomb* (1962), *Growing Points in Physics* (1966–67), *A Fable in his Life Time — Paul Dirac* (1972), NBC's *The Decision to Drop the Atomic Bomb* (1964), ABC's *The Struggle for Peace* (1966), CBC's *On Klaus Fuchs* (1973).

[629]See letters exchanged between Peierls and Greene in *Peierls Papers*, Ms.Eng.misc.b219, D.23.

[630]Letters [170–71].

[631]The three older Peierls children, Gaby, Ronnie and Kitty had settled in North America.

[632]Letters [175–77].

suggested by Egon Orowan of the structure of an edge dislocation[633] had become a classic in this research field. Peierls' student, Frank Nabarro later extended the argument and corrected a major algebraic error in Peierls original paper,[634] and the phenomenon became widely known as the Peierls-Nabarro force. The Peierls couple continued to visit Seattle regularly, and Rudolf Peierls gladly accepted the offer of a three-year part-time professorship at the University of Washington when he retired from Oxford in 1974.

Hans Bethe and Rudolf Peierls continued to take an interest in questions of nuclear energy, nuclear weapons and arms control. Peierls' work on nuclear weapons had ceased in 1945, but his concern with the problems of nuclear armament remained with him all his life. On 9th July 1955, Bertrand Russell and Albert Einstein had issued their famous manifesto calling on scientists of all countries to save the world from nuclear war. This resulted in a series of conferences on 'science and world affairs', the first of which was held at Pugwash, Nova Scotia, and gave the movement its name: Pugwash Conferences. Under the first president Cecil Powell and Secretary General Joseph Rotblat, the movement grew with increasing numbers of scientists getting involved in the annual conferences and regular meetings and workshops. Peierls played an active part in the Pugwash movement, serving on its continuing committee from 1963 to 1974 and as its chairman between 1969 and 1974. He had always given high priority to the Pugwash Conferences since attending his first such conference in Moscow in 1960, and in retirement, he still tried to attend and contribute whenever possible.[635]

---

[633] R.E. Peierls, 'The size of a dislocation', *Proc. Phys. Soc.* **52**, 34–37 (1940).

[634] F.R.N. Nabarro, 'Dislocations in a simple cubic lattice', *Proc. Phys. Soc.* **59**, 256–72 (1947).

[635] See publications: R.E. Peierls, 'How deep is deep enough?' *Proc. 37th Pugwash Conference, Sept. 1987*, Taylor and Francis, London, 1987, pp. 414–16; R.E. Peierls, 'Test Ban and Verification'? *Proc. 39th Pugwash Conference, July 1989*, Taylor and Francis, London, 1989, pp. 311–12; R.E. Peierls, 'Energy Conservation?' *Proc. 40th Pugwash Conference, Sept. 1990*, Taylor and Francis, London, 1987, pp. 530–31; R.E. Peierls, 'Why are nuclear weapons tests claimed to be necessary?' *Proc. 41th Pugwash Conference, Sept. 1991*, Taylor and Francis, London, 1991, p. 372. See also R.E. Peierls, 'Strategic Defence Initiative', *Phys. Bull.* **37**, 217 (1986); R.E. Peierls, 'Is there any logic in the Arms Race?' *The Challenge of Nuclear Armaments*, (eds.

And even in the last years of his life, when he was health was declining and it was becoming increasingly difficult for him to engage in travel and writing, he kept up his determination to contribute to the nuclear debates with his last publication being devoted to these issues.[636]

Hans Bethe, similarly, worked tirelessly to create effective tools for verifying and validating negotiated agreements to slow down the arms race between the superpowers. He argued against President Johnson's so-called 'thin missile defense', and even more forcefully against President Reagan's Strategic Defense Initiative. And even after the end of the Cold War, on the occasion of the 50th anniversary of the explosion of the first atomic weapon over Hiroshima, he called on his colleagues to 'cease and desist from work creating, developing, improving and manufacturing further nuclear weapons'.[637]

With regard to scientific research Hans Bethe, at the age of 71 — as he himself would later describe it — apprenticed himself to his friend and junior colleague Gerry Brown, when he set out to make significant contributions to an aspect of astrophysics that he had not previously considered in depth.

In contrast, Rudolf Peierls during the last years of his life, concerned himself with broader conceptual questions. He thought about two scientific issues in particular — questions of measurement, which he had been in dispute over with John Bell[638] for some time,[639] and the discussion of symmetries and broken symmetries. Peierls had exchanged views with P.W. Anderson[640] about this issue in an open controversy

---

A. Boserup, L. Christensen, O. Nathan), 379–80 (1986); R.E. Peierls, 'The case for defence', *Nature* **328**, 583 (1987).

[636]R.E. Peierls, C.R. Hill, R.S. Pease and J. Rotblat, *Does Britain need nuclear weapons?* London: British Pugwash Group, 1995. See also letters [208], [210–11].

[637]H.A. Bethe, 'Cease and Desist', *Bull. Atomic Sci.* **51 (6)**, 3 (1995).

[638]John Stewart Bell (1928–1990), received his Ph.D from Birmingham in 1956. After work at the British Atomic Energy Agency in Malvern and at Harwell he moved to CERN.

[639]R.E. Peierls, 'In defence of measurement', *Physics World* **4**, (1), 19–20 (1991); this was a reply to John Bell's 'Against "measurement" ', *Physics World* **3** (8), 33–40 (1990).

[640]Philip Warren Anderson (1923–) received his doctorate from Harvard in 1949; from 1949 he worked at Bell Labs; from 1967 he was Professor of Theoretical

largely through the medium of academic journals,[641] and it was on this topic that Peierls delivered his last published and public lecture, the Dirac lecture in 1992.[642]

During the last years of his life, Rudolf Peierls was troubled by a number of health problems, and he suffered a deterioration of his eyesight which restricted his reading and made correspondence more difficult. Despite all this, he continued to lead an active and independent life well into his eighties. In the summer of 1994 however, after suffering a combination of heart, lung and kidney problems, he decided to move into a residential home close to Oxford. It was another crucial stage at the dawn of Rudolf Peierls' life, and another step which his close friend from student days, Hans Bethe, shared in thought. On hearing of his friend's illness, Bethe sent an encouraging letter of reminiscences[643] — to which Peierls, in visibly frail hand writing, replied with similar recollections.[644]

Having been independent since leaving home well over sixty years earlier, Peierls nevertheless settled well into his new environment, one of the few residents at Oakenholt who would word-process circular letters to friends and family and read scientific papers in enlarged script on a computer screen! But his health deteriorated further throughout 1995. It was evident to Hans Bethe, in early September 1995, that the friend he had made almost seven decades earlier at Sommerfeld's seminar in Munich, would be taken away from him. Bethe sent his farewell, again recalling their interconnected lives and their long-lasting friendship and he summed up their shared experiences: 'It is good to remember all those days ··· You had a full and good life, and I thank you for letting me participate in it.'[645]

Physics at Cambridge; since 1975 he has been Professor (since 1997 emeritus) at Princeton.

[641]See P.W. Anderson, 'Broken Symmetry can't compare with ferromagnets', *Physics Today*, (5), 117 (1990); R.E. Peierls, 'Spontaneously Broken Symmetries', *J. Phys. A* **24**, 5273–79 (1991).

[642]R.E. Peierls, 'Broken Symmetries', *Contemporary Phys.* **33**, 221–26 (1992). See also letter [205].

[643]Letter [212].

[644]Letter [213].

[645]Letter [214].

Hans Bethe defied the notion that physics is young persons' pastime. He published significant papers in every decade from the 1920s into the 21st century. Bethe and Peierls shared a love of physics and a commitment to advances in the sciences. Physics had been central to their lives from an early age and it would remain so until the very end of their lives, if in different manifestations of scientific activity. They both had 'full and good lives', which were further enriched by their mutual friendship that lasted for almost seven eventful decades.

[168] Rudolf Peierls to Hans Bethe

Oxford, 26.2.1964

Dear Hans,
Thank you for your letter. You need not apologize because I knew my request was rather an imposition and I did not count on your being able to help.[646]

As it happened the problem was in any case somewhat academic, because it turned out that with the competition from other branches of pure and applied mathematics I was not able this year to get any of the theoreticians in; it is conceivable that this depended somewhat on the choice we did make about who should head the theoretical physics list.

As a result the same problem will come up again in a year's time and therefore if you have any comments to make they would still be very welcome either now or at the end of the year when I can bring the list of candidates up to date.

While on the subject of unwritten letters it seems that Kabir[647] gave your name as a reference in applying for a Fellowship at New College here, and that the Warden (Sir William Hayter[648]) did not have a reply from you. Possibly some letter went astray.

This again is not serious because there are other letters about Kabir, and of course I know him well. However, if possible it would be quite nice to have a letter from you. The probable date of the committee meeting is 9th March, so there is just a chance that a letter might get here in time.

---

[646] Refers to a request for advice on a senior appointment at the physics department at Oxford.

[647] Prahaban K. Kabir, had worked at Princeton, Cornell, Calcutta and Lyon, before taking up a fellowship at Oxford in 1965. He later joined the physics department at the University of Virginia, Charlottesville.

[648] Sir William Hayter (1919–2007), British Ambassador to Moscow 1953–1957, Warden of New College Oxford 1957–1976.

I am delighted to hear about your plans. I am sure one way or another our paths will cross, and I do hope you will manage to include Oxford in your diary.
With kindest regards,

Rudi

[169] Rudolf Peierls to Hans Bethe

Oxford, 15.5.1964
(carbon copy)

Dear Hans,
Thanks you for your letter. I am delighted you will be visiting here. The appropriate consultation between your targets has taken place and it seems that the first week in June suits everybody and the suggested order is 1 Oxford, 2 Cambridge, 3 Imperial College.

As far as Oxford is concerned, we hope that you will come as early as you can during the weekend and stay perhaps till Tuesday night. You will then have Wednesday and Thursday in Cambridge and you would have Friday and Saturday at Imperial College. This divides the week roughly equally. I imagine that if for any reason you want to spend more time in any one place and less in another this could still be adjusted.

Assuming this timetable is satisfactory we should hope that you might be willing to give us a talk, either on Monday or Tuesday. Since these are not our regular meeting days I should like to leave the day and time open until I have been able to consult with other people but presumably you need not know the time in advance. There is a good direct train to Cambridge at 6.48 p.m. (arriving at 8.56) and a reasonable one at 7.55 arriving 10.22 p.m. A Tuesday afternoon seminar would therefore be compatible with your going to Cambridge the same evening. There are also good trains in the morning.

I hope you will stay with us. By the sound of your letter you are not bringing Rose, but is she was able to come she would be most welcome, of course, even at short notice.
Yours sincerely

[Rudi]

[170] Rudolf Peierls to Hans Bethe

Oxford, 24.9.1965

Dear Hans,
I have had a conversation with a B.B.C. producer, Robert Reid, which I think I should report to you.[649]

He recorded last year a programme about the history of the atom bomb, particularly about the decision to use it. It started off with some old background and this contained a lot of inaccuracies. One instance was that it showed a picture of the usual conference at Copenhagen in a context which suggested that it was a Göttingen meeting, and on that picture appeared the face of Oppenheimer who at the relevant time had been in neither Göttingen or Copenhagen. This face had obviously been touched up. I assumed all this was just carelessness and that perhaps they had assumed some suitable face to be that of Oppenheimer. I wrote to the man to complain and he now came to see me about it. It emerged that all this had been done quite deliberately to make a nicer presentation. I did not like this attitude in a documentary programme but there is no deep significance.

However, in the course of the conversation something very different came out. The programme had included an interview with Oppenheimer in which he appeared to be answering the question whose precise wording I do not, of course, remember, but which had to do with the arguments on the committee about whether or not the bomb should be used, and the attention, if any, they gave to the Franck Memorandum.[650]

---

[649] Robert Reid was the producer responsible for the programme 'The Building of the Bomb', screened in March 1965. Peierls was disturbed by inaccuracies and distortions in the final version of the programme and took the matter up with the producer and with Sir Hugh Carleton Greene, the Director General of the BBC. See *Peierls Papers*, Ms.Eng.Misc.b202, C.19 and Ms.Eng.Misc.b219, D.23.

[650] James Franck's memorandum to Wallace and his report to the Secretary of State of War Stimson warned about the possible arms race with the Soviet Union and urged that bombs only be used in extreme circumstances. See Alice K. Smith, *A Peril and a Hope*, Chicago University Press, Chicago, 1965, pp. 31–32.

Oppenheimer's answer to this was that he could not remember. When I saw this programme I was quite staggered by this and so were many other people. It now transpires that in the editing process the question got changed and that the original question which Oppenheimer in fact answered was about what Arthur Compton[651] had said in the Committee on a particular date. I think I need not elaborate on my reaction to being told that a programme sponsored by the B.B.C. could let this happen.

I am now writing because the same producer is planning another programme to deal with the H-Bomb controversy, and I know he is shortly going to the United States to get, if possible, interviews with you as well as Bradbury, Smith and Ulam[652] My first reaction was to write to you and the others to suggest that you have nothing to do with the new programme. However, I also know that Edward Teller and some others have agreed to be interviewed and that the producer himself feels that he might get a very one-sided story. For this reason it would seem very desirable that some sensible people with a different point of view should be heard though obviously you must make up your minds yourselves.

In view of the past experience, I would, however, suggest that if you agree to be interviewed you make a firm condition (a) that any editing of your remarks be submitted to you for approval, and that your words be used in full unless you have authorised any specific extracts or cuts (this was a precaution taken by Edward Teller on the first programme and was observed in spite of some technical inconvenience to the producer). (b) That the wording of the questions to which you reply be shown on the programme in the exact wording put to you unless, of course, you

---

[651]Arthur H. Compton (1892–1962), Head of Department of Physics at Washington University in St. Louis, where he became Chancellor in 1946. From 1941, along with Vannevar Bush, he was Head of Office of Scientific Research and Development.

[652]Norris Bradbury (1909–1997), replaced Robert Oppenheimer as director of Los Alamos Scientific (later National) Laboratory; Henry de Wolf Smyth (1898–1986) Professor at Princeton 1936–1966, member of the Atomic Energy Commission 1949–54, US representative to the International Energy Agency 1961–1970; Stanislaw M. Ulam (1909–1984), Polish-born American mathematician who proposed the Teller-Ulam design of thermo-nuclear weapons.

have approved any alterations in advance. I may yet raise this matter at high level in the B.B.C. but I have not yet made my mind up about this.
Yours sincerely,

Rudi

[171] Hans Bethe to Rudolf Peierls

Ithaca, 5.11.1965

Dear Rudi,
Thank you very much indeed for your letter of 24 September concerning the BBC program about the history of the atom bomb, and their plans for a similar history of the H-bomb. I am very grateful to you for putting me on guard about my interview with Mr. Robert Reid of the BBC. Normally, I am used to giving such interviews quite innocently. After receiving your letter, I made him promise to submit to me any material that he may quote on the basis of your interview.

I believe your protest has had its effect on Mr. Reid. In order to avoid similar troubles, he is going to eliminate all personal quotations, and all personal appearances of the characters in the drama. Instead, he is writing a script much of which he has showed me for my comment. It seemed on the whole fairly good, but I was able to correct [two] or three mistakes in fact. He took these corrections very gracefully. At many points I would have given different emphasis, but I did not find much that troubled me greatly.

In the meantime, Mr. Nath will have appeared in Oxford. It took him much longer to finish than I had thought. Also, I was rather disappointed by his general knowledge of physics which came out in his final examination. He is quite ignorant of the more elementary parts of physics, not only classical but also elementary quantum mechanics on the level of Schiff.[653] On the other hand, he did a good job on his

---

[653] Leonard Schiff, *Quantum Mechanics*, 3rd edn., McGraw-Hill, 1968.

thesis. It will do him good to teach, and I hope he will have to learn some of the more elementary parts of physics in the process. I am sorry I recommended to you a man who turns out to be less well trained than I thought he was.

Otherwise, life is busy but happy. Henry has started studying at the University of Wisconsin,[654] and seems to be managing quite well. He gets good grades, is interested in his work, and seems to have lots of friends.

With best regards to you and Genia,

Yours sincerely,

Hans

[172] Rudolf Peierls to Hans Bethe

[Oxford], 14.12.1965
(carbon copy)

Dear Hans,

It is ages since I received your letter of 5th November. You need not feel apologetic about Nath.[655] He seems an extremely nice person and is making himself useful though he still has to spend a good deal of time on finishing his thesis. His teaching will not start till next term and I was glad to have your warnings as to know what possible difficulty to watch for.

I talked recently with him about his plans. His original intention was to return to India next year, but he has had second thoughts about this in the light of various reports, and he now feels that he has to mature some more before it would be wise to go back. In the light of this his first choice would be to try and return to the Unites States next year. One of the reasons being financial; he would like to send some money home eventually, and on an English salary that is not so easy. His first

[654] Henry Bethe, Hans Bethe's son.

[655] See letter [171].

choice would be Cornell, and he told me he was about to write to you to enquire about the possibilities.

Alternatively, he might stay here, but it is not yet clear whether there will be funds to extend his appointment here which was made for only one year in the first place. This doubt does not reflect any lack of interest in having him here for longer but only shortage of money in general together with the need for supporting some other people who are already here and looking after people whom we should like to have for special reasons.

I thought it would be useful to make these remarks to explain the situation since otherwise the fact that he already is thinking about leaving might create the impression that we do not like him or that he is unable to settle down. Both these impressions would be quite misleading.

Yours sincerely,

[Rudi]

[173] Rudolf Peierls Hans Bethe

Seattle, 11.1.1967

Dear Rudi,

Thank you very much for your letter of 4 Jan, which caught up with me here. I hope you have meanwhile received my letter from Oxford which explains that much as I would like to accept your invitation to visit Cornell, it does not look possible this time. I do hope, however, to see you at the New York meetings.[656]

Yours sincerely,

Rudi

---

[656] Between 30 January and 2nd February 1967 a meeting of the American Physical Society took place in New York.

[174] Hans Bethe to Rudolf Peierls

Ithaca, 18.1.1967

Dear Rudi,
Rose and I would like it very much if you and Genia could have dinner with us on Wednesday evening, February 1. We would like to go early to dinner, about 6 o'clock, partly because the New Yorkers have the habit of eating late and partly because we would like to be finished by about 9 so that we can catch a plane to Ithaca at 10.15. We are only inviting you so that we can have a leisurely talk during dinner. Please write me whether you can come.

It would probably be best, if we met at the end of our joint session on Wednesday afternoon. Rose and I are staying at Loew's Midtown Motel which I believe is close to the Hilton.
Yours sincerely,

Hans

[175] Hans Bethe to Rudolf Peierls

Ithaca, 20.1.1967

Dear Rudi,
I thought it might be useful to co-ordinate our talks at the New York meeting to some extent. In the main I am planning to give a speech very similar to the one at Gatlinburg of which I enclose a copy.[657] However, I want to put a little more emphasis on the statistical theory of the finite nucleus. Mrs Nemeth[658] has made some progress about solving the integral equation for the density which I want to report. In addition,

[657]Hans Bethe gave a paper at the International Conference on Nuclear Physics, Gatlinburg, Tennessee, 1967. See *Proceedings of the International Conference on Nuclear Physics*, Gatlinburg, Tennessee, 1967, Academic Press, New York, 1967.

[658]Judit Nemeth had worked with Hans Bethe at Cornell before taking up a post in Budapest in 1968. See J. Nemeth and Hans Bethe, 'A simple Thomas-Fermi calculation for semi-finite nuclei', *Nucl. Phys.* A**116**, 241–55 (1968).

I want to talk for about three minutes on the use and justification of the local density approximation in the theory of the finite nuclei, by which I mean both a Hartree-Fock and a Thomas-Fermi theory. I use the local density approximation essentially as defined by C.W. Wong,[659] not Brueckner, Gammel and Weitzner.[660]

I hope all this does not interfere with your talk; if it does, please write me and I will change my talk accordingly.

Yours sincerely,

Hans

[176] Rudolf Peierls to Hans Bethe

[Seattle], 21.1.1967
(carbon copy)

Dear Hans,

Your dinner invitation sounds wonderful, and I am almost sure we can make it. The "almost" arises from the fact that Genia is still in England, and will probably arrive in Boston on Wednesday, where I shall make contact with her on Saturday. It is just conceivable, though most unlikely, that she will already have made some kind of firm commitment. In such a contingency I would of course let you know as soon as possible.

Yours sincerely

[Rudi]

---

[659] Chun Wa Wong obtained his Ph.D. from Harvard in 1965 before taking up a post at University of California, Los Angeles. See letter [177], note 661.

[660] K.A. Brueckner, J.L. Gammel and H. Weitzner, 'Theory of Finite Nuclei', *Phys. Rev.* **110**, 431–45 (1958).

[177] Rudolf Peierls to Hans Bethe

[Seattle], 24.1.1967
(carbon copy)

Dear Hans,
Thank you very much for the advance information about your talk in New York. I see no reason to ask for any change in your selection of topics (a) because there is not much overlap, and where there is it will probably lead to constructive discussion (b) because I had in any case kept my plans flexible in order to fit in with whatever you are going to say, and (c) above all because all you say is interesting and worth saying and I would not be able to say it.

I do not have a previous manuscript with which to return your compliment, and of course I don't have my New York talk written down. I am still sorting out things to some extent, but roughly speaking what I want to do is the following:

First I want to make some general remarks about why one wants to do this kind of thing. One hears remarkably often the view that all this is a waste of time because once field theory has been put in order, one should just solve the many-nucleon problem in interaction with pions and everything else, and all the answers will come out. Of course, the exponents of such views will be at other sessions, but one may give a little moral support to the people who may be there.

Then I want to make some brief remarks about the forces, and there will be a little overlap, because I want to comment both on the question of Many-body forces, and on non-local and other non-conventional forces. Here I shall not disagree with you, but place the emphasis a little differently, so that I am quite happy about your remarks preceding mine.

After that I want to turn really to finite nuclei, say around oxygen, and the approximations one wants to make there are really complementary to your Thomas-Fermi philosophy (which is not saying one might not even get useful information about oxygen via TF) except that your reference to the use of the local density approximation by C.W. Wong & Co. presumably gets close to my subject. I shall not try to review all the existing calculations, and give numbers — the real reason is of course that I am too lazy to collect all the detailed information, but I rationalize this by the argument that it is a little too early for such a review because the answers have not yet settled down sufficiently.

I do want, however, to discuss methods a little. It seems to me that many of the eminent authorities are too hypnotized by the Brueckner expansion. For example, the Brueckner reaction matrix for a finite nucleus is a horrible animal to define and to write down if known let alone to calculate. The separation or reference spectrum methods, or a judicious combination of them allow one to calculate this beast approximately, but if one goes beyond the crudest approximation, all the difficulties and complications of the Brueckner matrix are still there. If one uses physical intuition to approximate to some of the harder quantities it may be very hard to say how one should correct for the errors involved, or at least estimate them, because it is not clear what has been left out and what the next term should be.

I believe it is useful to have a formalism in which it is quite clear what has been left out and in which one can see what terms in the expansion would have to be taken into account to deal with this. This view is also taken strongly by C.W. Wong.[661] Of course it is also important that the approximation be reasonably convergent — a series of which you have to calculate fifteen terms is not much use. My impression is, for example, that Wong tends to go too far in writing too many many-body effects into the definition of the one-particle states which causes all sorts of trouble (e.g. lack of orthogonality of the states) without evidence that this really pays in terms of more rapid convergence. One can see that this reduces the size of some higher-order terms, but since one cannot have a series without higher order terms it does not matter to increase them somewhat, as long as they do not become larger in order of magnitude.

[661]See Chuan Wa Wong, 'Application of the Reaction Matrix Theory (1). The Calculation of Nuclear Reaction Matrix Elements', *Nucl. Phys.* **91**, 399–32 (1967). See also Chun Wa Wong, 'Spin-Orbit Splitting in the Conventional Spherical Hartree-Fock Theory', *Nucl. Phys. A* **108**, 481–92 (1967).

A similar question relates to the self-consistency. One is tempted to require full self-consistency in the Hartree-Fock spirit. This can be stated by requiring the matrix elements from the ground state to one-particle excitations to be zero. One can use a much simpler model potential (e.g. harmonic oscillator for a light nucleus like O 16) and keep the difference amongst one's correction terms. The matrix elements for one-particle excitations are then non-zero, but they are usually very small, and it does not look as if they would make higher-order terms appreciably larger. What one gains is a much greater ease in evaluating, for example, second-order terms.

You will see that this is largely a commercial for the approach we are developing in Oxford. However, we have so far only very partial results available, and this is more a programme than a calculation. However, I think my plea for flexibility is more general than that.

I hope this will give you some idea of what I propose to talk about and I think you will agree that there will be no serious demarcation dispute, as the trade unions would say.

I am not too happy about the attempts to push this kind of discussion into the newspapers, because nothing sensible or intelligible will result. I refused to prepare a press release about my paper, but I felt I could not refuse to take part in the press conference on the Wednesday morning where I imagine I shall see you, presumably with not much greater enthusiasm.
Yours sincerely,

[Rudi]

[178] Rudolf Peierls to Hans Bethe

[location unspecified], 28.2.1967
(carbon copy)

Dear Hans,
It was a great pleasure to see you and Rose at length in New York and in such an attractive setting.

I am writing now to ask whether it would be possible to send to David Brink[662] in Oxford a copy of your Gatlinburg talk[663] (or, if it is written up, the text of your talk in New York). The point is that Brink and I are joint contributors to the new journal "Comments" which Weneser and Lederman have started,[664] and your stuff belongs very much to the topic on which we are supposed to comment.

I need hardly say that both Brink and I would be very interested in any further more detailed preprints on the subject.

Yours sincerely,

[Rudi]

[179] Rudolf Peierls to Hans Bethe

Oxford, 21.1.1970

Dear Hans,

Dr. J.M. Irvine[665] has applied for a vacant University Lectureship in this Department, which is earmarked for work in the theory of nuclear structure and/or nuclear reactions. He has given your name as a reference, and my colleagues and I would be grateful, if you could let me have your confidential assessment of Dr Irvine, including his achievements and potentiality in research, his effectiveness in interacting with others, and, if possible, his teaching ability.

Yours sincerely,

Rudi

(handwritten addition)

Hope this will catch up with you somewhere! We look forward to your turning up here at some time!

---

[662]David Maurice Brink, tutor of physics at Balliol College, 1958–1993, Reader of Physics at Oxford 1988–1993, Professor of History of Physics Trento, 1993–1998.

[663]See letter [175].

[664]Leon Lederman and Joseph Weneser (eds.), *Comments on Nuclear Physics*, Gordon & Breach, New York.

[665]John Maxwell Irvine (1939–), at the time Lecturer at Manchester, 1983–91 Professor of Theoretical Physics at Manchester, 1991–96 Principal and Vice-Chancellor of Aberdeen University, 1996–2001 Vice Chancellor of the University of Birmingham.

[180] Rudolf Peierls to Hans Bethe

Oxford, 7.2.1970

Dear Hans,
I see from the correspondence with Heisenberg that your are, or were last week, at Rehovoth, so I might exploit this knowledge to write to you. You said a long time ago that you might be turning up here later on your trip,[666] perhaps in June, and we are hoping you will come. I know that you are visiting many places, but we would be happy, if you could reserve for Oxford as long a time as is possible. As usual, of course, you and Rose, if she is with you, are most welcome to stay with us as long as you have time for. If the time is long enough to make you feel you would rather be independent, something else could, no doubt, be arranged.

Rumours said you might conceivably turn up at the conference in Surrey in April. If this means you may be in England for longer, or earlier than already planned, there is of course nothing special about June; this was mentioned only because you suggested it, and almost any other time would be equally good here.

I wrote to you recently asking for an opinion of J.M.Irvine, who has applied to us for a lectureship.[667] My letter was sent to Cornell and probably by now caught up with you; but, if not, I would very much value your opinion about him.
With kindest regards,
Yours sincerely,

Rudi

(handwritten addition)
P.S. After writing this I received your itinerary from your secretary, from which it seems you will in England only for a short time in April. Can we see you then? Or if you can make a sidetrip from Copenhagen, we would of course be glad to help you with the fare!

---

[666] Hans and Rose Bethe were traveling around the world. As part of this trip, Hans Bethe spent some time at the Weizmann Institute of Sciences in Rehovot.
[667] See letter [179].

[181] Hans Bethe to Rudolf Peierls

Rehovot, 17.2.1970

Dear Rudi,
Thanks for your letter. Yes, of course we want to come to Oxford. We can come for two days and are very much looking forward to this.

The exact time is not quite clear yet because we have not thought about the details of our English visit but we have tentatively fixed the dates for the visit, namely April 22–May 1, which is just two weeks. I am sorry that it cannot be longer this time; as you know we have been travelling around the world which is quite big, after all. We are anxious to settle down for a reasonably long time in Copenhagen.

We shall be happy to stay with you while at Oxford. With best regards to Genia,
Yours sincerely,

Hans

[182] Hans Bethe to Rudolf Peierls

[Rehovot], 17.2.1970
(carbon copy)

Dear Rudi,
I am very happy to recommend Dr. J.M. Irvine for a University Lectureship in Theoretical Physics.

Irvine was at Cornell for two years as a Research Associate and Instructor. He was most satisfactory both in research and teaching. In his research he tried to develop a Hartree Fock theory with effective forces depending on density. He had some success, and published a paper about it.[668] However, he had two difficulties: the major one was that our I.B.M computer did not function properly while he was with us, so that every simple computation took a terribly long time. The

[668] J.M. Irvine, 'The theory of nuclear energy levels', *Rep. Prog. Phys.* **31**, 1–59 (1968).

second was that he tried to do the problem on an oscillator basis which is natural but not very suitable for a density dependent force. The problem was finally solved by Negele in co-ordinate space.[669]

During his work, Irvine demonstrated that he has great mathematical power. Hardly any mathematical problem was too difficult for him. He is completely familiar with all tricks of Nuclear Physics, from Nuclear Matter to the Shell Model to collective motions etc. He is a really well-informed nuclear theorist. In addition he has good physical intuition, and is a pleasure to work with.

Both he and his wife are most agreeable personally. I am sure you would find him an excellent addition to your staff.

Yours sincerely,

Hans A. Bethe

[183] Hans Bethe Rudolf Peierls to

Rehovoth, 22.2.1970

Dear Rudi,

We have now made our plans for the English visit. If it is all right with you, we should like to come to Oxford at the very beginning. We expect to arrive in London on April 22, about noon. Then we want to visit two old aunts of Rose's who live near the Airport. After dinner we should like to take a taxi from there to your house so that we should get there by about 9:30 p.m. or 10:00 p.m. on April 22. Would this be all right with you?

We should then stay until Saturday, April 25. Then we want to go to Werner Sachs. He may come to collect us in Oxford, or else we shall take the train.

Since both the taxi and Werner do not know the geography, would it be possible for you to draw a map how to reach your house from some clear landmark in Oxford?

[669] J.W. Negele, 'Structure of Finite Nuclei in the Local-Density Approximation', *Phys. Rev. C* 1, 1260–1321 (1970).

In your recent letter you kindly offered to pay some travel expenses for me. I should be grateful, if you could. This would involve taxi, and if possible all, or part of the airfare from Germany to England and back. This trip is not included in our fare around the world.

Of course I should be glad to give a talk on Nuclear Matter if you are interested.

Looking forward to seeing you and Genia.

Yours sincerely,

Hans

[184] Rudolf Peierls to Hans Bethe

[Oxford], 28.2.1970
(carbon copy)

Dear Hans,

Thanks you for your letter. The dates you suggest are fine. We wish the visit could be longer, but I can see the problem.

On the 22nd April I think it would be easiest if we came to collect you, presumably around 8 or 8.30 p.m. if you let us know the address. It is really quite near, and that way we could have another hour or so of your company. In case something intervenes which prevents us coming, I enclose a map showing where we are. In that case you should tell the driver to come through Henley, which brings you on to our map by the red arrow. You then follow along the Oxford ring road the signs to Abingdon and Newbury, and from there the map should be adequate.

We would be very happy, if you could give a talk on "Nuclear Matter", probably on the Thursday afternoon. This could be done on one of three levels: either to our theoretical seminar, which is for all theoreticians, including elementary-particle and solid-state people as well as people working on nuclear theory; alternatively, it could be in our nuclear theory discussion group, where you could really take your hair down; or it could also be a meeting of the physics colloquium which normally includes all the experimentalists, both nuclear and otherwise. Please let me know which of these settings you prefer. I guess that the theoretical physics seminar might be the most appropriate. It is conceivable that we might arrange a meeting for the Friday afternoon rather than Thursday, because David Brink is returning from a visit to Orsay about that time and he would regret having to miss your talk.

One minor complication is that Genia and I are supposed to go to a dinner in college on Friday evening. This is an annual event, which it might be awkward to get out of, and as it happens, this applies to Cooke[670] as well. However, Oxford is full of people who will be delighted to be your hosts for that evening, so there is no problem.

We can certainly take care of your fare from Germany etc.

Looking forward to the occasion.

Yours sincerely,

[Rudi]

[185] Rudolf Peierls to Hans Bethe

[Oxford], 20.5.1971
(carbon copy)

Dear Hans,

I am a member of the Scientific Advisory Committee of the French Institute of Advanced Study at Bures, of which you will know.[671] The Institute is active in mathematics and theoretical physics. On the physics side its permanent members, as you may know, are L. Michel[672] and D. Ruelle.[673]

---

[670] Arthur Hafford Cooke (1912–87), physicist, Warden of New College (1976–1985) and Pro-Vice Chancellor of University of Oxford. His wife was a school friend of Hans Bethe's.

[671] The *Institute des Hautes Études Scientifiques* had been founded by Léon Motchane in 1958 as a European counterpart to the Institute for Advanced Study at Princeton. In 1963 Peierls was invited to join the Scientific Advisory Committee on which he served until 1974. See *Peierls Papers*, Ms.Eng.misc.b224, F.12–15.

[672] L. Michel (1923–99), theoretical physicist who held research and teaching posts at Manchester, Copenhagen, Lille and Paris before moving to the Institute des Hautes Études Scientifiques at Bures-sur-Yvette.

[673] David Ruelle (1935–), civil engineer and physicist, studied at Mons, Brussels and Zurich, and held positions at Zurich and Princeton before taking up a professorship at the Institute des Hautes Études Scientifiques at Bures-sur-Yvette.

Negotiations are now going on to secure international support for the Institute and, as a result, it may well be possible in the near future to appoint two more physicists on a permanent basis. I am trying to collect suggestions of suitable names.

So far the orientation of the work has been somewhat abstract, partly because the present and former members were somewhat inclined that way, partly because the presence of first-rate pure mathematicians tends to attract people with formal interests, for example Wightman[674] is a regular visitor there, but the visitors cover a wide spectrum. New appointments might well be people fitting in with the present image of the place, but alternatively it would be an interesting idea to broaden the range by using the two new places, if they can be confirmed, for bringing in people with slightly different, and perhaps more practical emphasis.

The posts are attractive; they correspond in status to full professorships. The salaries are higher than those of say, English chairs, but not as high as American ones.

One knows that at this time there may be people in established positions, for example in the United States, who nevertheless want to change, but it is very hard to know where they are, and for this reason I have been encouraged to use the grapevine to ensure that people like yourself are aware of the possible existence of these posts, and might let me (or Michel, or Ruelle) know if you happen to hear of someone who might be interested.
Yours sincerely,

[Rudi]

[674] Arthur S. Wightman (1922–) obtained his Ph.D from Princeton in 1948, where he eventually became Professor of Physics.

[186] Rudolf Peierls to Hans Bethe

Oxford, 24.11.1971

Dear Hans,
M.B. Johnson[675] has applied for a temporary postdoctoral research position here. It is not yet clear that we have such a position here next year, but we might have. If so, it would be under a research contract relating to nuclear theory. Although we do not have the position yet, a list of candidates is building up.

Johnson has given your name as a referee, and I would be grateful for your opinion of him.
With kindest regards,
Yours sincerely,

Rudi

[187] Hans Bethe to Rudolf Peierls

[Ithaca], 2.12.1971
(carbon copy)

Dear Rudi:
This is to recommend to you Dr. Mikkel Johnson who has applied for a position in your department. He came to me from MIT where he took his Ph.D. with Baranger, and is now spending his second year here as a research associate. I find him extremely satisfactory.

Johnson has worked on several parts of nuclear theory, mostly of a fundamental nature. In his thesis he developed a new method in nuclear matter theory, the use of folded diagrams. This will probably be useful for calculations of finite nuclei in nuclear matter theory, especially if higher accuracy is desired. As a first application, however, he worked on the two-pion exchange contribution to nuclear forces. His results

[675]Mikkel B. Johnson had obtained his Ph.D. from Carnegie Mellon University in 1970. After research at Cornell, he took up a post at the Los Alamos National Laboratory in 1972, first as a staff member and since 1991 as a fellow.

seem to be in approximate agreement with a new calculation by Gerry Brown and Durso[676] Johnson's method is less fundamental but simpler than Brown and Durso.

His main work here at Cornell has been on the equation of state of matter at very high density, particularly for application to neutron stars. This work is being done in close collaboration with me, and the results seem to be very satisfactory.

Most recently he has worked with Molinari[677] from Torino on the simple theory of the interaction of two nucleons outside closed shells. This was stimulated by a phenomenological analysis by Schiffer[678] of the Aragonne National Laboratory, and it looks as if Schiffer's results could be understood in a very pretty and persuasive way. Earlier he worked on spin orbit interaction in complex nuclei where there is a considerable paradox. This work had no results so far, and was interrupted in favor of the neutron stars.

You will see from this that Johnson has worked on very diverse subjects. Mikkel is a delightful person and a first-class collaborator. He visibly enjoys his work, his enthusiasm shows both in his face and in the long hours he puts into his work. He has physical intuition as well as mathematical skill to a high degree. His wife is also exceptionally nice. I can wholeheartedly recommend Mikkel Johnson to you.
Yours sincerely,

Hans A. Bethe

[676]G.E. Brown and J.W. Durso, 'Soft pioneering determination of the intermediate range nucleon-nucleon interaction', *Phys. Lett. B* **35**, 120–124 (1971).

[677]Alfredo Molinari, at the time research associate later Professor of Theoretical Physics at the University of Turin.

[678]John P. Schiffer, completed his Ph.D. at Yale, before taking up a research appointment at the Rice Institute Houston (1954–56); in 1956 he joined the Argonne National Laboratory, where he became director of the Physics Division in 1979. In 1969 he was also appointed Professor of Physics at the University of Chicago.

[188] Rudolf Peierls to Hans Bethe

[Oxford], 8.2.1972
(carbon copy)

Dear Hans,
I believe Dick Dalitz is writing to you to explore the possibility of your coming to Oxford in 1973–74 as a Visiting Fellow of All Souls College. I am writing to add my own voice to his, and saying how enthusiastic we would all be about such a proposal, if indeed it needs suggesting.

I think you know what is being done in the department, and what kind of people are around, though I would be happy to go into details if you wish.

On the practical side, apart from housing which no doubt will be looked after by the college, we would, of course, make arrangements of financial support. The most obvious way of doing so is through a "Senior Visiting Fellowship" granted by the Science Research Council. I have, in fact, at my disposal a research grant which goes on until 1973–74, the subject of which is "Nuclear Dynamics". If you were interested in taking some interest in nuclear problems while you were here (which would, of course, not stop you from being interested in other matters as well!) we would confirm the arrangement forthwith from an existing budget, but if this is not in line with what you might like to do, I have no doubt we could get a separate grant for the purpose, or change the terms of reference of the existing contract accordingly.

The fellowship could carry a full professorial salary (currently £5,472 p.a.) if you are not receiving a salary from other sources in respect of the same period. If you receive some of your Cornell salary, they would pay only the difference which is not likely to amount to much unless any partial support from Cornell could be arranged to relate to part of the time only, in which case we could pay for the rest of the period. Assuming you would receive a salary which would preclude us from paying you a salary from this end, we would still be able to pay your fare, and a subsistence allowance meant to cover extra costs of living away from home, which has to be computed in each case, but would come to something like £1,000 for a year, and proportionately more for a shorter period. This subsistence would not be taxable; any salary we can pay might or might not be subject to U.K. tax depending on whether your appointment here can be classed as a teaching post, which it might be, if you are willing to take some interest in our graduate students. In that case, it would come under the bilateral convention on such matters. Whether it is liable to U.S. tax, of course, I would not know.

I should apologize for this mercenary sounding exposition; it is in any case clear that if you came here, it would not be to make a profit, but it is just as well to be clear about what is involved.

I was grateful to you for bringing Mikkel Johnson to my attention.[679] I was about to send him an invitation when he withdrew because of an offer from elsewhere, but such is life.

Yours sincerely,

[Rudi]

[189] Hans Bethe to Rudolf Peierls

Ithaca, 15.2.1972

Dear Rudi,

Thank you very much for your letter of 8 February. It is a very nice offer, and it would be nice to spend some leisurely time with you. However, rather than being too early, it is already too late. I have already made arrangements for the three coming academic years. In the fall terms I shall still be teaching at Cornell, and for the spring terms I have accepted invitations to Seattle, Copenhagen, and Munich respectively. How I will feel after that, I cannot tell at present, but I have the feeling that I will by that time be somewhat tired of travelling, and will like to stay at home for most of the time.

So thank you very much again. You have worked it out so beautifully, but I am afraid I shall not be able to come. However, I hope to come for a short visit during my semester in Copenhagen.

With best wishes.

Yours sincerely,

Hans

[679] See letter [187].

[190] Hans Bethe to Rudolf Peierls

Ithaca, 28.1.1974
(carbon copy)

Dear Rudi,
This is to recommend to you Dr. Geoffrey West[680] who is applying for a position in your department.

Dr. West was here at Cornell as a research associate from September 1966 through June 1968. He was very satisfactory and highly productive. His time was divided between fundamental particle theory and more phenomenological theories related to the experimental work on our synchrotron.

Last spring I had contact with Dr. West again when he was giving a seminar at the University of Washington; after a seminar we had a long private discussion. The paper he gave at the seminar impressed me very much. He pointed out the similarities between deep inelastic scattering of electrons in the SLAC experiments, and ordinary inelastic scattering of electrons by atoms at energies of the order of 1 keV. He also pointed out differences which he ascribed mainly to the relativistic kinematics of the electrons and the particles produced in the collision in deep inelastic scattering. The theory seemed to me very simple and quite convincing. It is also quite original, being very different from the many current theories. He has developed it with great skill.

Dr. West has a very pleasant personality. I can recommend him wholeheartedly for a position at Oxford.
With best regards.
Yours sincerely,

Hans A. Bethe

[680]Geoffrey West (1943–) graduated from Cambridge University and received his doctorate from Stanford University in 1966. He was the founder and leader of the high energy physics group at Los Alamos and is one of the senior fellows at LANL. In 2003 he joined the Santa Fe Institute as Distinguished Professor.

[191] Rudolf Peierls to Hans Bethe

Seattle, 16.4.1977

Dear Hans,
Thank you for your letter of April 8, which arrived yesterday. My talk at Minnesota,[681] is supposed to deal with the way our ideas of nuclear forces developed, and one point I want to make a kind of leitmotiv is how one started making the simplest kind of assumptions, starting with a static, monotonic, central, "ordinary", zero-range, & c. force, and how one-by-one all these simplifications were proved wrong, so that one has by now to use forces which are about as complicated as they could be — except that the one rule which was first postulated on the flimsiest possible evidence, viz. charge independence, has stood up well.

To convey this picture, I must of course refer, at least in principle, to the various empirical and theoretical arguments which provided information about the forces, and there I am in danger of overlapping with your talk, and in fact with practically every other talk at he conference. For this reason I propose not to write out my talk in advance, but just collect the relevant material and references. Any points already mentioned by others can then be omitted or condensed. (I have the organizers' dispensation from submitting a manuscript at the beginning, on the understanding that I shall tape my talk and then edit it within two weeks of the end of the conferences). You should therefore not worry too much about overlap.

I think all the points on the list in your letter belong to my story, and I shall include them if they are not already adequately covered by you or someone else. It may of course be necessary to summarise the arguments, for continuity, but that type of remark is less overlap than a cross reference, and quite desirable.

---

[681] This refers to a talk to be given at the Symposium on the History of Nuclear Physics, organised by Roger Stuewer ay the University of Minnesota later that year. Conference Proceedings were published: R.H. Stuewer (ed.), *Nuclear Physics in Retrospect: Proceedings of a symposium on the 1930s.* Minneapolis: University of Minnesota Press, 1979.

I was worried when the circular from Stuewer mentioned the proceedings of previous conferences on the subject and requested copies. These have just arrived, and I have not fully absorbed them yet, but my impression is that they are very vague and woolly, and can mostly be ignored. On the other hand I found an article by Wilkinson on the history of isospin, in a book on isospin which he edited and which contains a lot of useful references,[682] though I don't quite agree with his conclusions.

You and Rose seem to have had a very successful trip. Thank you for your postcard, which also arrived recently. Ronda is a place we have so far missed out on our trips.

We had a pleasant drive here from L.A. after a week in the New York area. Here the weather could be better, particularly at weekends, but we have been able to get around a little.

See you in Minn.

Yours,

Rudi

[192] Rudolf Peierls to Hans Bethe

Seattle, 29.4.1977

Dear Hans,

Thank you for your letter with the abstract of your talk, which should be very interesting. I do not think the overlap is anything to worry about, I shall simply take things as read which you have already explained, unless I want to emphasise special points.

I shall try to hand out a list of references (not a complete set, but only the papers I want to mention) so this may turn out to contain papers which in the end I shall not mention because you have done so. I don't think that does any harm.

I should withdraw the remark in my previous letter about the AIP meeting on the history of nuclear physics. When I wrote I had not

---

[682] Denys Wilkinson, *Isospin in Nuclear Physics* (North-Holland, Amsterdam, 1969).

spotted that the volume contained the record of two meetings, and that the second was much more substantial than the first. I enjoyed reading the record of the second meeting and learned some things from it, though the professional historians tried hard to confuse the issues. Of course you were a participant at the second meeting but I do not know whether this is the only reason why it was better than the first.
Yours sincerely,

Rudi

[193] Hans Bethe to Rudolf Peierls

Ithaca, 2.1.1980

Dear Rudi,
Thanks for your letter of 12 December. Unfortunately, I cannot solve the mystery of Niels Bohr's idea about the compound nucleus.[683] I am pretty sure that I did not send a copy of my paper on the one-body theory of capture of neutrons to Copenhagen. However, I gave a talk about it at the American Physical Society meeting in February 1935 which was heard my many people, and I subsequently discussed it quite a lot with other physicists here in the U.S. It could well be that some account of my paper could have travelled to Copenhagen through one of the many physicists with whom I discussed it.

As far as I remember, I did not participate in the conference in the autumn at Copenhagen. I was there during the summer for one month. It may well have been September, but I do not recall the exact time. I remember several seminars in which Niels Bohr participated, but I don't recall any remark by Bohr which indicated to me the germ of the

---

[683] Letter could not be located. The query must have been in connection with Rudolf Peierls' work on the 9th volume of the Collected Works of Niels Bohr. He spent some time in Copenhagen in the winter of 1979/80 in order to collect material for the volume, and he tried to clarify the exact chronology of Niels Bohr's thinking on the compound nucleus and its dissemination among the physics community. This features in the lengthy introductory essay of this volume of the Collected Works. See R.E. Peierls (ed.), *Niels Bohr. Collected Works. Vol. 9 Nuclear Physics (1929–1952)*, North-Holland, Amsterdam, 1986.

compound nucleus idea. Very likely I might not have noticed it even if he gave some indications.

About Contemporary Physics:[684] Unfortunately I have far more writing in the near future that I have time to do. On the supernova work with Gerry Brown, I promised to write a more popular account for the volume to celebrate Willy Fowler's 70th birthday[685] so I can't very well publish once more in Contemporary Physics. On nuclear matter, I have not done anything in recent years, so there are much more competent people to write about it, such as John Negele[686] and Pandharipande.[687] So, I am afraid, for the time being I cannot oblige.

It was good to hear that you are enjoying Copenhagen, and that you were for some time living in paradise, in the Nationbank's guest flats in Nyhavn. Both Rose and I are wishing Genia and you the very best for the New Year. The enclosed circular letter tells you about our experiences in the year just passed.

Yours sincerely,

Hans

(handwritten addition)

In the meantime, your account of the last 5 years arrived, and we enjoyed it very much — but what a travelling life you have!

A happy and perhaps more sedentary 1980 to you both! Where shall we meet again? Denmark '81? Brazil '80?

---

[684]Work with and for the journal *Contemporary Physics* was Rudolf Peierls' longest-running editorial commitment. He contributed several articles himself, but his main aim was to recruit new authors and have new topics investigated through his overseas contacts in Europe, the USA and Russia.

[685]William Fowler (1911–1994). The article referred to is Hans A. Bethe, 'Supernova Theory', in C.A. Barnes, D.D. Clayton and D.N. Schramm (eds.), *Essays in Nuclear Astrophysics*, Cambridge Univ. Press, Cambridge, 1982, pp. 439–66.

[686]John W. Negele, after research at Cornell and Copenhagen, Negele joined the physics department at MIT, where he has been William A. Coolidge Professor of Physics since 1979.

[687]Vijay R. Pandharipande (1940–2006), completed his doctorate at the Tata Institute of Fundamental Research in Bombay. After research at Copenhagen and Cornell, he joined the Department of Physics at the University of Illinois where he became full professor in 1977.

[194] Rudolf Peierls to Hans Bethe

[location unspecified], 15.1.1980
(carbon copy)

Dear Hans,
Thank you for your letter. I think I have resolved the main mystery by the brilliant idea that there were two seminars, one as described by Wheeler in April '35, and another as described by Frisch later in the year, the latter being the one at which the report was about your paper.[688] This removes all contradictions, except for some details about which Wheeler's recollections may well be inaccurate. I have put this solution to Wheeler, and am waiting to see if he is willing to buy it. The other part of the mystery must remain unsolved.

Our travel plans for the immediate future involve travelling South from Vancouver, stopping at Stanford from 18 to 29 February, then heading to Sao Paulo, Brazil, where we are due on 6 March. We plan to call at Los Angeles, but there seems to be time to drop in also in Santa Barbara, where I gather you and Gerry Brown will be at that time. I have just written to Gerry asking him to look at the schedules and phone me in Vancouver when he — and you have got organized in Santa Barbara.

Too bad about Contemporary Physics. Perhaps we can try some other time.

Thank you also for your Annual Report. We did not know about your operation, and it was good to read that all went well, and that it has been worth while.
Love to Rose, also from Genia,
Yours sincerely,

[Rudi]

---

[688] See letter [193].

[195] Hans Bethe to Rudolf Peierls

Ithaca, 1.7.1980

Dear Rudi,
Thank you ever so much for your two letters (and also Genia's) from Sao Paulo. The advice on living conditions was most useful, and especially your advice on energy.

Following your advice I will stress the development of hydroelectric power, and state that this is likely to be cheaper than nuclear. But I am supposed to talk about nuclear power, and give my usual talk about safety, waste disposal, and weapons proliferation. On the latter point, I suppose the Brazilians will be somewhat touchy.

We have had a good spring although I had an awful lot of brief travels between April and now. We are leaving a week from today for Estes Park, Colorado, then on the 27th for Peru where we shall be joined by Monica,[689] and on August 4 to Rio. We expect to be back in Ithaca on August 17.

Thanks once more for your very nice letters, and we remember happily your visit on March 1 in the rain. Next spring we shall again be in Copenhagen, and will be one week in England, probably some time in May. We hope to see you.
All the best, and love to Genia,
Yours sincerely,

Hans

[196] Hans Bethe to Genia and Rudolf Peierls

[location unspecified], 21.2.1981

Dear Rudi and Genia,
How are you, and where? The grapevine says that you, Genia, have recovered from the mysterious disease you had contracted in Sao Paolo.[690]

---

[689] Monica Bethe, Hans and Rose Bethe's daughter, who lived in Japan.

[690] In 1980, Rudi and Genia Peierls had spent three months in Brazil, where Genia contracted a periodic virus infection. See Peierls, *Bird of Passage*, pp. 334–35.

I hope this is true, and you are back to your normal, ebullient self! The people in Sao Paolo said while the disease was rare, it was not unknown to them.

We are making plans for spring and summer. In particular, we'll be in Cambridge at a meeting on supernova astrophysics and physics which I believe will last until 3rd July, as far as I am concerned. We would love to visit you on the 4th if you are in Oxford at that time and if you are willing to have us. Are you still sufficiently connected with America that we can celebrate the 4th of July together?

I hope this works out, we'd love to see you again!

Yours,

Hans

[197] Hans Bethe to Rudolf Peierls

Ithaca, 13.1.1984

Dear Rudi:

Accidentally, I just received copies of my letters to Sommerfeld, including some from 1930–1934.[691] In the early letters I address him as „Lieber und sehr verehrter Herr Professor". In later letters, I just address him „Lieber Herr Professor".

Good luck for your historical researches!

Yours sincerely,

Hans

[691] Hans Bethe was sent the letters in connection with the Sommerfeld Projekt, an edition of a selection of Sommerfeld's correspondence and the presentation of material in a database. See http:lrz-muenchen.de/Sommerfeld/WWW/AS_DFG.html.

[198] Rudolf Peierls to Hans Bethe

Santa Barbara, 2.2.1984

Dear Hans,
Thank you very much for your note with the authoritative answer to my query. This agrees with my recollection, and also with the published letters from Pauli to Sommerfeld. Viki's[692] recollection must be wrong, and I have written to him to say so.

We have now arrived here after working our way down the Coast in improving weather, and find this place as beautiful and the Institute as lively as expected.

We hope our paths may intersect again in the summer. With best wishes,
Yours,

Rudi

[199] Rose Bethe to Rudolf Peierls

[Ithaca], 27.10.1986

Dear Rudi,
"Als wär's ein Stück von mir", the old song goes through my head as I think of Genia. Without much overt evidence of response from me she was a companion of my life since I first met you both in Cambridge in 1939 on orders from Hans to tell you that we were after all getting married. She was very strict then — not to spoil what she had accomplished in socialising Hans and making him into a "good" husband, by which she quite specifically meant that he would help in the house. She and you — you and she, differently yet inseparably, have been Hans' dearest friends since your common Manchester time more than half a century ago. It is as yet impossible to think of you without her, but,

[692] Refers to Viktor Weisskopf.

she would say firmly, that is a matter of time. If you are continuing your nomadic life, please come and stay with us for a while.

After that very brief meeting in Cambridge, I next saw Genia in New York, in 1944. By then I had heard many stories and quotes of admonition. I was 8 months pregnant and she lived up to her reputation, as she instructed me what to do should the baby come on the train back to Los Alamos. But chiefly we talked about England and the war, and her happiness at having Gaby and Ronnie with her again. Later, in Los Alamos, she had other companions and I have wondered since whether she, too, dropped into that strange limbo in which purpose was centered on winning the war, and yet the war itself was a tale, a theory, a background to a daily life that was comfortable, routine, normal or did it retain for her the reality which it had been in England? In any case, her boundless energy brought variety and joy to life on the mesa. We certainly were the beneficiaries of this, especially when Hans needed a vacation from work as well as fatherhood and she took care of Henry for two weeks. He was then 7 months old and of course "completely" spoiled by his mother. I had a vivid picture of how she had projected herself into him when she came to visit the first time after Henry was back with me, and he looked at her calmly, without fear and said in a loud decisive voice "NO". She was delighted. It was his first word and remained his only one for a long time.

You will be inundated with stories now, of Genia's impact on old and young, of her generosity, of her vitality, sagacity, love of life and people. I am very glad that she was part of Hans' and my life. Isn't it good that you were here last spring and we could visit you in June!

My mother asks to tell you that she thinks of you with profound empathy and sympathy. She is basically well, but most of the time very tired so that she feels no longer up to writing.
Please give my love to Jo.
Yours,

Rose

[200] Hans Bethe to Rudolf Peierls

[Ithaca], 26.10.1986

Dear Rudi,
We are both very sad that Genia died. She was one of the greatest personalities I have known.

Genia had a great influence on me, during our year together at Manchester. Early on, she told me "Sie sind doch sehr nett". Surely she and you have long forgotten that. But it gave me more self-confidence in a personal sense, as did my whole life with the two of you. It was the best preparation for my life in America. She taught me that difficulties in external life are problems to be thought about and solved not to be worried about.

There were so many good times afterwards: The vacation in Devon 1936 (or 1935?); I still remember Tintagel. (What did you do with Gaby and Ronnie that summer?) Many happy visits to Birmingham, without and with our children. Life together at Los Alamos, and later in Seattle where you were almost part of the department while we were just frequent visitors.

One of the greatest things Genia did was the education of Jo. As soon as she knew that Jo was a dwarf, she centered all her life on Jo. She studied what one can do to strengthen the body of a person of that type, how to make her use her limbs, and generally prepare her for a normal life. She did all those things for Jo and she succeeded.

Sometimes it was annoying when Genia wanted to arrange everybody's life. I felt this on occasion when Henry had such deep psychological trouble. By then, however, I was old enough so I could ignore it. You were always marvellously tolerant.

Genia lived a full life, and a good life. She helped dozens of people, in addition to her family. How could she be so wise already when she was so very young? How old was she in Manchester? 25?

Rose called Betty Brown, Gerry's wife, but Ronnie had already done so. She also called Ruth Uehling[693] who will tell it to others in the department in Seattle.

Rose and I are happy that we saw you once more, and had such a good time together, end of June last year. Genia seemed so happy at that time, being restored after her disabilities a year ago.[694] I am glad she had that year. Another good year after a very full life.
We will long think of her.
Yours,

Hans

[201] Hans Bethe to Rudolf Peierls

Ithaca, 27.5.1987

We are making preparations for our trip to Europe, especially our visit to Oxford.[695] I seem to remember that you will leave immediately after the celebration on Saturday, 27 June.[696] But in case you don't leave quite so immediately, we have decided to stay in Oxford on Sunday until 15.00, and would love to see you if you are there. We are staying at the Linton Lodge which is very close to you.

It would be very nice, if you have time to do so, if you would send me a note saying whether you will be there on the 28th, c/o my sister,
c/o Doris Overbeck
Rebenweg 8
5450 Neuwied 23
West Germany
Yours sincerely,

Hans

[693]The Uehlings were friends of the Peierls couple at the University of Washington, Seattle, where Rudolf Peierls, after his retirement from Oxford, spent three years as a part-time professor.

[694]The previous year, Genia Peierls had had a benign brain tumour removed which had incapacitated her for some time.

[695]Rose and Hans Bethe were planning to come to Oxford for the symposium on the occasion of Peierls' 80th birthday in June 1987.

[696]On Monday after the weekend celebrations, Rudolf Peierls flew to Turin for a conference at the Centre for Scientific Interchange.

[202] Rose Bethe to Rudolf Peierls

Zurich, without date

Dear Rudi,
I am so very sorry that I cocked out on Sunday![697] A rare opportunity missed. I am doubly glad that I was placed next to you at the banquet. I only hope this proximity did not pass the flu on to you. You'll know by Thursday. (I presumably caught it from my brother who presumably picked it up in Israel — so it's a good Jewish flu.)

It was lovely of Chris to drive us.[698] I wrote him the enclosed card and then realized that I had forgotten his last name — and so has Hans. Would you be so kind as to write an address on the card and send it on? Thanks!

Hans gives his lecture tonight, tomorrow and Thursday, and then we are off to Zermatt and two weeks of walking. Happiness for both of us.

Thank you for providing the occasion to come to Oxford to see you, and to share in what must be in the range of the 400.000th meal in the dining hall of New College.
With warmest love,

Rose

30.6.[1987]

It was lovely to be in Oxford. I enjoyed seeing many people from the past, and seeing how many different subjects you are connected with (of course I knew that).[699] The best was the lunch on Sunday in the circle of your family. Rose has recovered from her cold, Zurich is sunny and hot, and we are enjoying it.

Thank you once more and good luck for the TV debate![700]

Hans

---

[697] Rose Bethe had fallen ill with the flu on the Sunday after the symposium.

[698] Rudolf Peierls' son-in-law, Chris Coppin, took the Bethes to the airport, as they had to travel on to Switzerland from Oxford.

[699] See R.H. Dalitz and R.B. Stinchcombe (eds.), *A Breadth of Physics: Proceedings of the Peierls 80th Birthday Symposium*, World Scientific, 1988.

[700] This refers to a TV debate, in which Peierls took part a few weeks later. See letter [203].

[203] Rudolf Peierls to Hans Bethe

Oxford, 24.7.1987
(carbon copy)

Dear Hans,
Thanks you for your letter and card from Switzerland.[701] I know you are back safely, because I have just been with the Marshaks at the Bristol Conference, who came straight from the encounter at Kennedy.[702]

The TV debate went more or less as predicted. The group consisted, besides Teller and myself, of Sergei Kapitza,[703] Enoch Powell,[704] a woman writer (very CND) and a catholic nun. I cannot remember the name of the chairman, who did not interfere much, except when several people wanted to speak at the same time.

Before we went on air, Edward started a violent quarrel with the chairman, who read out a few sentences about each of us, which he was going to use as introduction. About Teller one of these sentences was "... who is sometimes called the Father of the Hydrogen Bomb, although he does not like this title." Edward blew up and said it was discourtesy to quote that title when it was known that he disliked it. After a long and violent argument, which almost delayed the programme, the chairman agreed to drop the sentence.

It turned out that the title of the programme was not, as I had been told, Star Wars, but "Peace in our Time", so the discussion was on two levels, one consisting of the two ladies saying how dreadful were nuclear weapons and (in the case of the nun) how immoral it was to use force in any circumstances. This was very effectively, if kindly, dealt with by Enoch Powell, who was, for a politician, very restrained, and was very sensible and intelligent.

---

[701] Card [202].

[702] Robert Marshak and his wife were old friends of Hans Bethe's, and they had met them at Kennedy Airport/New York on their way to the conference in the UK.

[703] Sergei Kapitza (1928–), son of Pjotr L. Kapitza, was professor of physics at the Kapitza Institute for Physical Problems at the Russian Academy of Sciences in Moscow.

[704] Enoch Powell (1912–1998), Conservative MP between 1950 and 1974 and Ulster Unionist MP between 1974 and 1987.

Edward was quite predictable. When he told us how far ahead the Russians were with sophisticated defence, I said "This information could only come from intelligence reports and I understand that people who have access to intelligence reports do not have that impression". He did not comment. Sergei Kapitza told me afterwards he was very grateful for this, as it would have been difficult for him to say this.

I also raised the question how S.D.I. affected Europe, was it expected to defend us as well? To which Edward said it was Reagan's policy to be open and share information with everybody who would participate, even the Russians. I reminded him that so far the "sharing" amounted to the European being allowed to work on some small problems whose answer would assist the S.D.I., but they were not told the overall design and there certainly was no indication of hopes to protect Europe. In view of the short flight times, this seemed even harder than to protect the US.

Incidentally he did not claim that a complete shield was possible, nothing could ever be perfect, but it was possible to bring down a large fraction of the missiles. The obvious answer to this came from Powell.

I cannot of course remember all the details of the debate, I have a videotape, but have not yet looked at it, as I do not have a video machine. I want to look at it, because they ran some commercials during our discussion, so that some passages got lost (without telling us) and I am interested to see what got lost.

I did not of course expect to convince Teller any more that I expected he would convince me, but I think some of the viewers will have seen the weakness of his position. In fact someone said it was not really necessary for me to say anything, it was enough to listen to Teller!

I was very grateful for the briefing you gave me, which was a great help.

Nature accepted my review without any substantial change.[705] I checked the question of City College vs University College by asking

---

[705]Rudold Peierls had reviewed Edward Teller's *Better a Shield Than a Sword*, Free Press, New York, 1987, a collection of Teller's essays. See R.E. Peierls, 'The Case for Defence', *Nature* **328**, 583 (1987).

Teller, and he said "O dear, did I say City College?" So I could let the sentence stand.[706]
Best wishes to Rose and her mother.
Yours,

[Rudi]

[204] Hans and Rose Bethe to Rudolf Peierls

Mt Desert Island, 7.7.1988
(post card)

Dear Rudi,
We just reread your Xmas letter of Nov. '88 and were deeply impressed by your energy. Ours is much reduced and so we decided to vacation at sea level, but still where we could walk up rather than flat — hence Mt. Desert Island, where the "mountains" go up to 12–1300′ yet are fun to climb. Did you succeed in reducing your travels? I hope very much that our paths will cross soon.
Love,
Rose,

I am indeed deeply impressed by your energy, between physics and arms control. I am also impressed by Gorbachev who made the whole world look more peaceful. Hope to see you some time soon.
Love,

Hans

[706] This passage relates to a mistake in one of Teller's essays where he refers to University College London as 'City College'.

[205] Rudolf Peierls to Hans and Rose Bethe

Oxford, 25.3.1991
(carbon copy)

Dear Rose and Hans,
I am writing to confirm my conversation with Rose about my threatened appearance in Ithaca. I would arrive from Toronto on 21 May, probably in the afternoon, but I have yet to study the plane connection, and, if I may, stay until the morning of the 24th, to go on to Long Island.

I shall be at a conference at McMaster University on 17 May, address: c/o R.K. Badhuri, Dept. of Physics, McMaster Un. Hamilton Ont. L8S 4N1 phone (416)525-9140 ext. 3185. On the Thursday and Friday night I shall be staying at the visitors' inn, in Hamilton. From Saturday 18 to Tuesday 21 May I shall stay with Mrs. I. Jephcott, 323 Rosemary Toard Toronto, Ont. M5P 3E4; phone (416)483-6777.

If people are interested, I could give a seminar on "Spontaneously Broken Symmetries". After my controversy with Phil Anderson, which you may have seen in *PHYSICS TODAY*,[707] I have made some more progress, and have, I think, a new way of looking at the phenomenon.
Love,

[Rudi]

[206] Hans Bethe to Rudolf Peierls

Pasadena, 8.1.1992

Dear Rudi,
I was terribly sorry to hear about your medical problems: first your intestines and now phlebitis. Please take it easy! Stay in bed as much as you can, and let your friends and others help you with shopping, cleaning and everything else. I hope your troubles will soon clear up!

[707] R.E. Peierls, 'Reflections on broken symmetries', *Physics Today* **44**(2), 13–14 (1991).

We enjoyed your end-of-year letter, as we do always. It is amazing how much you do, travelling going to meetings, and visiting friends and relatives. I get tired just reading your letter. And you go to such difficult countries, e.g. China. I have given up travelling anywhere except the U.S. and occasionally Europe. But I hope I can go to Europe in June–July 1993: the Royal Society has invited me to give the Bakerian Lecture in June.[708] This, I think, is a great honor, thinking of my predecessors, and I'll have some trouble making it sufficiently interesting. Who, by the way, was Mr. Baker?

At present, I am again in Pasadena, having just arrived yesterday, Gerry Brown is here also we share an appartment, and he not only cooks but also provides the ideas for the physics we are doing. Our office is right next to Willy Fowler's. As usual, there are lots of interesting people around. We are trying to work on the transition from hadrons to quarks and gluons which seems to take place at a temperature of 150 MeV, as a second-order transition.

I am tremendously interested in the great changes in Russia. Today's N.Y. Times had a picture of the Moscow Patriarch, in full robes, shaking the hand of Yeltsin.[709] So Christmas has returned to Russia and Leningrad has reverted to St. Petersburg. But the hard part is yet to come, getting the economy going again.

Will the peasants deliver enough food? Can they get the factories which are now (or have been) to 60% on war time production, to produce civilian goods, and goods of sufficient quality? It is a tremendous problem. Soluble?

Once more, I hope you get well soon. Gerry Brown joins me, and so does Rose.

All the best!

Hans

[708]The Bakerian Lecture, originated in 1775 through a bequest by Henry Baker FRS of £100 for an oration to be spoken or read annually by one of the Fellows of the Society is the Royal Society's premier lecture in the physical sciences.

[709]Boris Yeltsin (1931–2007), had emerged as the first elected President of Russia in 1991. Initially he followed a reconciliatory vis-à-vis the Russian Orthodox Church.

[207] Rudolf Peierls to Hans Bethe

Oxford, 26.1.1992[710]
(carbon copy)

Dear Hans,
Thank you for your letter and for your injunction to take it easy. I am being very sensible and spend much of my time with my legs up, with intervals of walking around, as instructed by my doctor. So far I have had an army of friends doing my shopping, but next week I shall venture into town.

I am glad to hear you will give the 1993 Bakerian Lecture, and I hope you will find time to visit Oxford. You ask about Mr. Baker. He was an 18th century fellow, whose only distinction seems to have been that he left the society a sum of £100 to establish an annual lecture.

As you say, the developments in Russia are very interesting. The picture is very confusing. In a week's time two friends from St. Petersburg, distant relatives of Genia, will come for a visit, and I am looking forward to their impressions.

As for the economy, I think people are terribly naive to think that as soon as you make rules for a market economy, things will immediately start to work. After all, our market economies in the West (such as they are, not really text-book market rules, but much government intervention) have developed over generations, and involve many traditional mechanisms and conventions, which will not grow overnight.

The fragmentation into smaller and smaller units, as in the USSR and Yugoslavia is worrying, particularly because of its effect on the further proliferation of nuclear weapons. The news today is of a recruiting drive by Iran for Soviet nuclear weapons experts. Evidently, the IAEA regime is not strong enough. They would never have discovered what Iraq was doing let alone stopped them, if Iraq had not lost the war.
Greeting to Gerry and other colleagues.
Yours,

[Rudi]

---

[710] The copy is dated 26.1.1002, which is evidently a typing error.

[208] Rudolf Peierls to Hans Bethe

Oxford, 23.9.1992
(carbon copy)

Dear Hans,
I suppose you have seen the article in the September Bulletin by Goldberg and Powers, claiming that Heisenberg kept information about atom bomb design hidden from the German authorities and even from his colleagues, and thereby prevented a German bomb being built.[711] The New Scientist of 5 September refers to the article under this headline "Heisenberg principles kept bomb from Nazis".[712] I am about to write to the New Scientist in protest, but I do not think it is my job to write to the Bulletin. Do you know whether someone is picking this up, or are you?

I hear you are not coming to the Chicago meeting of the US and European Nuclear Societies,[713] but you are one of the "Pioneers" and therefore no doubt have been sent the "consensus statement". I told them I am not going to sign this, and gave my reasons — copy enclosed. I would be interested in your comments.

I do not remember whether I adequately congratulated you on the Einstein Prize[714] — I heard about it from Gerry so long before the news broke officially. Joe Rotblat is giving his prize money to the Pugwash Foundation and asked me whether you might be inclined to do the same. Any chance?

I was twice in Germany, once in July at a meeting in Heidelberg, when I went for a weekend to Nürnberg, to see my school friend Heinz Rudolph, whom I had not seen for 65 years! Then a week at a Pugwash

[711] Stanley Goldberg and Thomas Powers, 'Declassified Files Reopen "Nazi Bomb" Debate', *Bull. Atomic Sci.* **48**, Sept. 1992, pp. 32–40.

[712] Dan Charles, 'Heisenberg's principles kept bomb from Nazis', *New Scientist*, 5 Sept. 1992, magazine issue 1837.

[713] The autumn meeting of the International Nuclear Societies Council took place in Chicago on 15 November 1992.

[714] Hans Bethe and Joseph Rotblat had been awarded the Einstein Peace Prize for 1992.

conference in (East) Berlin and saw a little of West Berlin where I had not been since 1934.

Love to Rose, and to Linde, if she is still there.

Yours,

[Rudi]

[209] Hans Bethe to Rudolf Peierls

Ithaca, 22.10.1992

Dear Rudy,

In your recent letter you mentioned the article by Goldberg and Powers in the September *Bulletin*[715] and indicated that you do not agree with it. I am now beginning to write a short article for the *Bulletin* on the same subject.[716] I enclose the first few pages of my draft, plus some pages of notes for a talk I'll give at Cornell on November 19.[717] You will probably disagree with me also, so please let me know.

It is nice to know that you are in this country. How long will you stay? It would be nice to see you some time during your visit to the U.S.

All the best,

Yours sincerely,

Hans

[715]See letter [208].

[716]Hans A. Bethe, 'Bethe on the German Bomb Project', *Bull. Atomic Sci.* **49** (1), 53–54 (1993).

[717]Hans Bethe gave numerous talks on this subject. Digital recordings of two of these (An Evening with Hans Bethe: The German Atomic Bomb Project, 9.11.1993; and Hans Bethe discusses the Manhattan Project with introduction by Carl Sagan, 6.4.1994) are accessible on DVD. *Remembering Hans Bethe*, The Internet-First University Press, Cornell University.

[210] Rudolf Peierls to Hans Bethe

Oxford, 7.2.1993
(carbon copy)

Dear Hans,
I do not know how long you are staying in Pasadena,[718] so I am sending this letter in two copies.

You ask about the level of the Bakerian lecture. As far as I know, the audience will be scientists, predominantly physicists, but only a minority of theoreticians, and few experts on astrophysics or whatever your subject will be. So I would say it should be a serious lecture, with substantial content, intelligible as far as possible to the non-expert, but not excluding some passages of recent results that would be best appreciated by the experts. Is that a reasonable picture?

Talking of the Bakerian lecture, I would be interested to know about plans for your visit. The point is that this year's Pugwash Conference in Sweden is from 10 to 15 June. I am certainly going to be at your lecture (in fact I am supposed to take the chair as Atiyah[719] has to be somewhere else). I hope you will come to Oxford during your visit, and I would not like to miss that. So if you plan to come here after the Bakerian lecture, I would skip the Pugwash Conference completely. If you plan to come here before the lecture, I might attend the second half. I expect you will stay with Linde, but if for any reason this will not work, you (and Rose, if she is with you) would be welcome to stay here.

I have another question: I am reviewing Power's book for the New York Review of Books.[720] I have read about 3/4 of the book in proof. I find it has many interesting bits of information I did not know. I also

[718]Hans Bethe spent the first month of the year at a university on the West Coast, often at CalTech, to pursue research with Gerry Brown. See above letter [206].

[719]Michael Francis Atiyah (1929–), at the time Director of the Newton Institute for Mathematical Studies in Cambridge and President of the Royal Society.

[720]Peierls reviewed Thomas Powers, *Heisenberg's War: The Secret History of the German Bomb Project*, (Knopf, New York, 1993). R.E. Peierls, 'The Bomb that Never Was', *New York Review of Books*, 22.4.1993, 6–9.

find that he is very sloppy, and practically every item I can check from my own knowledge is somehow garbled. So I am suspicious of what I cannot check. Where he quotes from documents I assume he quotes correctly, but I do not necessarily accept when he quotes from such secondary sources such as Robert Jungk[721] or David Irving.[722]

One episode which I found surprising is when Bohr arrived at Los Alamos with the famous sketch by Heisenberg and claimed that the Germans might have a weapon using slow neutrons, and that you and others had to argue to convince him that this made no sense. I find this surprising since Bohr was the first to point out that you could not get a powerful explosion with thermal neutrons. But Powers cites documents which seem to leave no doubt. No doubt you remember the episode: it is right that Bohr was temporarily so misguided?
Greetings to Gerry Brown and/or Rose,
Yours,

[Rudi]

[211] Hans Bethe to Rudolf Peierls

[location unspecified], 19.2.1993

Dear Rudi,
Thanks a lot for your prompt letter, and thanks for your advice about the Bakerian Lecture. It turns out that the public lecture which I gave last week at Berkeley, as a rehearsal fits your description quite well. It was a success.

Our plans are to go to Oxford after the lecture. We probably want to look at a museum in London on Friday (11 June) in the morning

[721]The Austrian writer and journalist Robert Jungk (1913–1994) had written a controversial account about the Manhattan Project: *Brighter Than a Thousand Suns* (New York: Harcourt Brace, 1958).

[722]David Irving (1938–), British historian and writer, who wrote about military history and World War II. His status as a historian has been widely discredited when an English court found him to be an active holocaust denier.

and then take a train to Oxford some time in the afternoon. We expect to stay until Tuesday (15 June) afternoon, then go to Manchester where I'll repeat the lecture — just about 60 years after we both came there.

Of course we expect to stay with Linde, but thanks very much for the offer to out us up. We very much want to see you of course. Sorry the time coincides with the Pugwash, and thank you for giving that up.

On the story of Bohr bringing Heisenberg's sketch to Los Alamos, this is quite wrong. Bohr gave it to Oppy and said: "the Germans talk about an atomic bomb, and this is what Heisenberg gave me." Bohr said nothing about slow neutrons (which is in accord with your remark that Bohr had pointed out that you could not get a powerful explosion with slow neutrons). Instead, Edward Teller and I, looking at the picture, said "this is obviously a reactor", and others concurred. We wondered what the Germans could possibly want to do with it? Drop it on London?

I was interested in your reaction to the Powers book. I also found it had many interesting bits of information that I did not know. But I did not find contradictions to my own knowledge — probably I did not read it as attentively as you. I did correct some (unimportant) scientific inaccuracies, which I wrote him about. When he quotes Robert Jungk, this is of course totally unreliable, but David Irving, I thought, was pretty good — even if in later years, D.I. became much too pro-German.

There will be lots to talk about. See you again in London and Oxford.

Yours,

Hans

[212] Hans Bethe to Rudolf Peierls

[Ithaca], 7.6.1994

Dear Rudi,
I hear from Linde that you are not at all well,[723] that you have trouble both with your heart and with your kidneys. I hope both of them will clear up soon and that you can go home and enjoy life.

Thank you for your statement about Fuchs coming to Los Alamos. It was helpful. I think the matter of that miserable book is fairly well under control; you may be interested in the enclosed Op.Ed. in the Washington Post. But I suppose the book has good sales nevertheless.

The book reminded me of the past, as did a story which (for some reason forgotten) I had to tell several times: Also we visited Chadwick in 1933–34, when he was disintegrating the deuteron together with Goldhaber. He bet that we could not make a theory of that process and on the way back to Manchester, on the long train ride, we did of course solve it. Do you remember?

During that year, 1933–1934, you introduced me to nuclear physics. Yes, I think you did — I was too much stuck in my old work, like order in alloys (Kellnerinnen, you remember),[724] while you were always open to new problems. So we worked together on the deuteron, neutron-proton scattering, and then we wondered about Wigner or Majorana forces. It was the basis of my work for several years at Cornell, including my three articles in Reviews of Modern Physics.

Altogether my year with you in Manchester was one of my most productive years, and maybe yours, too. Manchester was dirty and foggy, but I remember it as a happy time. I am amazed we could do so much work with Gaby just in her first year. She must have been an exemplary baby.

Later came Los Alamos, again a very productive time for both of us. And then 1948 when Rose and I visited you in Birmingham, and

[723]Linde Davidson, Rose Bethe's sister, also lived in Oxford, and had become a good friend of Rudolf Peierls'.

[724]See letters [33–34].

just at the time Dyson wrote me a letter that he had shown that QED converges in every order.

We had an interesting time in physics, and often together. Including just a year ago at the Royal Society, which you described so well. I could not hear the questions and you could not see the questioner.[725] With all our disabilities, I hope we'll see each other soon again.
Always yours,

Hans

[213] Rudolf Peierls to Hans Bethe

Oxford, 9.7.1994

Dear Hans,[726]
As you will understand I was not up to ... respond immediately to your wonderful letter, but by now I can try. No doubt we learnt much from each other. I am sure I learnt more from you than vice versa, but let us not argue. What I never learnt was ... I remember Genia's face when she found one morning that because of ... you had stopped some days' work and start again. You ask if I remember our bet with Chadwick. I certainly do, and it is written up in my memoirs. Do you remember Mr Shoenberg ..., also took offering ... him... first him to send first whenever you start to be employed.

Some recollections include, which are not related to physics. ... Think of your habit of numbering letters and pages of letter (what totals did you finally ...

Now after two weeks in hospital (heart, kidneys and lungs) and two weeks of convalescent homes I am starting a new life. It is clear that in the foreseeable future I shall not be able to look after myself. So the

[725]When Hans Bethe gave the Royal Society Bakerian Lecture in 1993, Rudolf Peierls chaired the proceedings. Bethe, who was hard of hearing, could not hear the questions, whereas Peierls, who was visually impaired, could not see the questioner.
[726]Rudolf Peierls wrote the following letter by hand. He was already frail and could not see well. Therefore much of the letter is illegible.

only options are: Get someone else to live with me in the flat or live in a retirement home. The flat's not really suitable for joint occupation, so I am trying the alternative for a few weeks. This is a beautiful estate in the middle of the country, roughly 5 miles from Oxford. I have a room with bathroom and meals, very great meals, are provided
Love to Rose,

Rudi

[214] Hans Bethe to Rudolf Peierls

[Ithaca], 8.9.1995

Dear Rudy,
I am very sorry you have so much trouble with various parts of our body. It must be very trying, and you have been very brave living with all this.

It reminds me how much our lives were connected, beginning with Sommerfeld's reading room where all the graduate students sat, and we giggled together so much. The Americans, Rabi and Condon, thought we were laughing at them.

Perhaps the most productive year in my life was the year I spent in your house in Didsbury. You introduced me to Nuclear Physics — I believe I had not realized that it was the coming subject. Our trip to Cambridge has become famous, when we saw the Chadwick-Goldhaber experiment. Chadwick challenged us: I bet you cannot find a theory for this photo-disintegration of the deuteron, and then we worked it out on the train back to Manchester. Physics was then ... easy.

Then there was our paper saying that the free neutrino would never be observed because the cross section was so small. Our cross section was correct but we never dreamed of powerful neutrino sources, reactors and supernovae.

And the work for Bragg on order in alloys. I did it laboriously by counting configurations — Genia called it "Kellnerinnen" later you did it professionally.

After the war was a most important period. We sent each other Ph.D's, good ones. In 1948 I may have visited you twice, early and late in the summer. I picked up Salpeter, the Austro-Australian, and he became a great astrophysicist. We were both intrigued by renormalisation theory in every order. You offered Dyson a job, when he came back to England.

Gerry Brown stayed with you for many years. He has now become my closest collaborator on supernovas.

It is good to remember all those days. We travelled together a great deal in physics. You did a lot in physics and you educated countless good physicists, mostly in Birmingham. I was greatly impressed at your 80th birthday when they all came to celebrate.

You had a full and good life, and I thank you for letting me participate in it.

Yours,

Hans

In Lieu of a Bibliography

The combined publication record of Hans Bethe and Rudolf Peierls amounts to well over 750 titles, including several books and substantial review articles and hundreds of mostly technical papers. Full bibliographical records of these publications can be found in:

R. H. Dalitz, 'Complete bibliography for Prof. Sir Rudolf Peierls', *Nuclear Physics* **A604**, 7–23 (1996)

S. Lee, 'Hans A. Bethe: Publications', *Nuclear Physics* **A762**, 13–49 (2005)

Full bibliographies are also included as electronic supplementary material to the Royal Society Memoirs of Rudolf Peierls and Hans Bethe:

Sabine Lee and Gerry Brown, 'Hans Albrecht Bethe. 2 July 1906–6 March 2005', *Bibliographical Memoirs of the Fellows of the Royal Society*, **53**, 1–20 (2007)

Sabine Lee, 'Rudolf Ernst Peierls. 5 June 1907–19 September 1995', *Bibliographical Memoirs of the Fellows of the Royal Society*, **53**, 265–284 (2007)

For further autobiographical and biographical information see the following publications:

Rudolf Peierls, *Bird of Passage: Recollections of a Physicist* (Princeton: Princeton University Press, 1985) [cited as Peierls, *Bird of Passage*]

Jeremy Bernstein, Hans Bethe, *Prophet of Energy* (New York: Basic Books, 1980) [cited as Bernstein, Bethe, *Prophet of Energy*]

Silvan S. Schweber, *In the Shadow of the Bomb: Oppenheimer, Bethe, and the Moral Responsibility of the Scientist* (Princeton: Princeton University Press, 2000)

Hans Bethe, 'My Life in Astrophysics', *Ann. Rev. Astron. Astrophysics* **41**, 1–14 (2003)

Gerald E. Brown and Chang-Hwan Lee (eds.), *Hans Bethe and His Physics* (Singapore: World Scientific, 1997) [cited as Brown and Lee, *Hans Bethe and His Physics*]

A selection of key publications can be found in:

R.H. Dalitz and Sir Rudolf Peierls (eds.), *Selected Scientific Papers of Sir Rudolf Peierls. With Commentary* (Singapore: World Scientific, 1997)

Rudolf E. Peierls, *Atomic Histories* (New York: Springer, 1997)

Hans A. Bethe, *Selected Works of Hans A. Bethe. With Commentary* (Singapore: World Scientific, 1997)

Other editions of interest and frequently cited in the references include:

Karl von Meyenn et al. (eds.), *Wolfgang Pauli; Briefwechsel mit Bohr, Einstein, Heisenberg u.a*, 4 vols, Berlin, Heidelberg, New York, Tokyo: Springer-Verlag, 1979 ff.) [cited as *Pauli Briefwechsel* and volume number]

Michael Eckert und Karl Märker (eds.), *Arnold Sommerfeld. Wissenschaftlicher Briefwechsel*, 2 vols. (Diepholz-München: GNT-Verlag, 2000 and 2004) [cited as *Sommerfeld Briefwechsel* and volume number]

Sabine Lee (ed.), Sir Rudolf Peierls: Selected Private and Scientific Correspondence, 2 vols. (Singapore: World Scientific, 2007 and 2008) [cited as Lee, *Selected Correspondence* and volume number]

Name Index

(Asterisk following the page number indicates a mention of the name in footnote of that page.)